Lecture Notes in Artificial Intelligence 6820

Edited by R. Goebel, J. Siekmann, and W. Wahlster

Subseries of Lecture Notes in Computer Science

Vicenç Torra Yasuo Narukawa
Jianping Yin Jun Long (Eds.)

Modeling Decisions for Artificial Intelligence

8th International Conference, MDAI 2011
Changsha, China, July 28-30, 2011
Proceedings

Springer

Volume Editors

Vicenç Torra
IIIA-CSIC, Campus Universitat Autònoma de Barcelona
08193 Bellaterra, Spain
E-mail: vtorra@iiia.csic.es

Yasuo Narukawa
Toho Gakuen, 3-1-10, Naka, Kunitachi, Tokyo, 186-0004, Japan
E-mail: narukawa@d4.dion.ne.jp

Jianping Yin
Jun Long
National University of Defense Technology
Yanwachi Street 137, Changsha, 410073, China
E-mail: jpyin,junlong@nudt.edu.cn

ISSN 0302-9743
e-ISSN 1611-3349
ISBN 978-3-642-22588-8
ISBN 978-3-642-22589-5 (eBook)
DOI 10.1007/978-3-642-22589-5
Springer Heidelberg Dordrecht London New York

Library of Congress Control Number: 2011931794

CR Subject Classification (1998): I.2, H.3, H.4, F.1, C.2, H.2.8

LNCS Sublibrary: SL 7 – Artificial Intelligence

Typesetting: Camera-ready by author, data conversion by Scientific Publishing Services, Chennai, India

Printed on acid-free paper

Springer is part of Springer Science+Business Media (www.springer.com)

Preface

This volume contains papers presented at the 8th International Conference on Modeling Decisions for Artificial Intelligence (MDAI 2011), held in Changsha, China, July 28-30. This conference followed MDAI 2004 (Barcelona, Catalonia, Spain), MDAI 2005 (Tsukuba, Japan), MDAI 2006 (Tarragona, Catalonia, Spain), MDAI 2007 (Kitakyushu, Japan), MDAI 2008 (Sabadell, Catalonia, Spain), MDAI 2009 (Awaji Island, Japan), and MDAI 2011 (Perpinyà, Catalonia, Spain, France) with proceedings also published in the LNAI series (Vols. 3131, 3558, 3885, 4617, 5285, 5861, and 6408).

The aim of this conference was to provide a forum for researchers to discuss theory and tools for modeling decision, as well as applications that encompass decision-making processes and information fusion-techniques.

The organizers received 51 papers from 10 different countries, from Europe, Asia, Australia and New Zealand, 19 of which are published in this volume. Each submission received at least two reviews from the Program Committee and a few external reviewers. We would like to express our gratitude to them for their work. The plenary talks presented at the conference are also included in this volume.

The conference was supported by the National University of Defense Technology, the China Computer Federation, the Catalan Association for Artificial Intelligence (ACIA), the European Society for Fuzzy Logic and Technology (EUSFLAT), the Japan Society for Fuzzy Theory and Intelligent Informatics (SOFT), the UNESCO Chair in Data Privacy, and the Spanish MEC (ARES - CONSOLIDER INGENIO 2010 CSD2007-00004).

May 2011

Vicenç Torra
Yasuo Narukawa
Jianping Yin
Jun Long

Organization

Modeling Decisions for Artificial Intelligence – MDAI 2011

General Chair

Jianping Yin	National University of Defense Technology, Changsha, China
Jun Long	National University of Defense Technology, Changsha, China

Program Chairs

Vicenç Torra	IIIA-CSIC, Bellaterra, Catalonia, Spain
Yasuo Narukawa	Toho Gakuen, Tokyo, Japan

Advisory Board

Bernadette Bouchon-Meunier
Didier Dubois
Lluis Godo
Kaoru Hirota
Janusz Kacprzyk
Sadaaki Miyamoto
Michio Sugeno
Ronald R. Yager

Program Committee

Gleb Beliakov
Ulrich Bodenhofer
Tomasa Calvo
Marc Daumas
Josep Domingo-Ferrer
Jozo Dujmovic
Michel Grabisch
Enrique Herrera-Viedma
Masahiro Inuiguchi
Hiroaki Kikuchi
Ivan Kojadinovic
Xinwang Liu

Xinjun Mao
Jean-Luc. Marichal
Rosa Meo
Radko Mesiar
Tetsuya Murai
Toshiaki Murofushi
Guillermo Navarro-Arribas
Michael Ng
Gabriella Pasi
Leszek Rutkowski
Susanne Saminger-Platz
Aida Valls
Zeshui Xu
Yuji Yoshida
Gexiang Zhang
En Zhu

Local Organizing Committee Chair

Danlin Yao
Wentao Zhao

Local Organizing Committee

Chunjiao Tan
Yong Li
Qiang Liu
Fayao Liu
Ming Zhu
Jiarun Lin

Additional Referees

Jordi Soria
Arnau Erola
Cristina Romero-Tris
Arnau Vives-Guasch
Jordi Pujol
Domenec Puig
Jordi Marés
Daniel Abril
David Nettleton
Sergi Martínez

Supporting Institutions

National University of Defense Technology
The China Computer Federation
The Catalan Association for Artificial Intelligence (ACIA)
The European Society for Fuzzy Logic and Technology (EUSFLAT)
The Japan Society for Fuzzy Theory and Intelligent Informatics (SOFT)
The UNESCO Chair in Data Privacy
The Spanish MEC (ARES - CONSOLIDER INGENIO 2010 CSD2007-00004)

Table of Contents

Invited Papers

Regular Papers

Aggregation Operators and Decision Making

Clustering

Computational Intelligence and Data Mining

Data Privacy

Online Social Honeynets: Trapping Web Crawlers in OSN

Jordi Herrera-Joancomartí and Cristina Pérez-Solà

Dept. d'Enginyeria de la Informació i les Comunicacions
Escola d'Enginyeria
Universitat Autònoma de Barcelona
08193 Bellaterra, Catalonia, Spain
{jherrera,cperez}deic.uab.cat

Abstract. Web crawlers are complex applications that explore the Web with different purposes. Web crawlers can be configured to crawl online social networks (OSN) to obtain relevant data about its global structure. Before a web crawler can be launched to explore the web, a large amount of settings have to be configured. This settings define the behavior of the crawler and have a big impact on the collected data. The amount of collected data and the quality of the information that it contains are affected by the crawler settings and, therefore, by properly configuring this web crawler settings we can target specific goals to achieve with our crawl. In this paper, we analyze how different scheduler algorithms affect to the collected data in terms of users' privacy. Furthermore, we introduce the concept of online social honeynet (OShN) to protect OSN from web crawlers and we provide an OShN proof-of-concept that achieve good results for protecting OSN from a specific web crawler.

Keywords: privacy, social networks, web crawling, graph mining, social honeynets.

1 Introduction

The increasingly popularity of online social networks (OSN) has lead them to become an important part of people's everyday communication. With millions of individuals who use OSN to share all kinds of contents, privacy concerns of how all this content is managed have arisen. Content shared in an OSN varies from trivial text messages to compromising photographies but, in either of those cases, users expect to control their shared data with their profile's visibility configuration. In addition to this personal data, users in OSN create relationships which can also be considered sensitive data from themselves. Moreover, the discovering of these relationships can also produce other data revelation, what makes link privacy an important issue to preserve in social networks.

OSN information can be obtained by crawling the profiles of users in the network. Web crawlers are complex applications that explore the Web with different purposes and they can be configured to crawl OSN to obtain both user and link

V. Torra et al. (Eds.): MDAI 2011, LNAI 6820, pp. 1–16, 2011.

information. When crawling online social networks, many choices have to be made in order to set up the crawler that is going to be used to obtain all the information from a social networking site. This configuration choices conform the crawler settings and, as we are going to see, they are the key to accomplish the desired crawling goal. Specifically, the choice of the next-node-to-crawl (determined by the scheduler algorithm) is a critical point, since it will determine largely which part of the network is going to be obtained and, therefore, which level of exposure will suffer the online social network users.

The contribution of this paper is twofold. On one hand, we detail the privacy implications that imply the election of different scheduler algorithms for web crawlers. On the other hand, we introduce the concept of Online Social Honeynet (OShN) to provide some level of protection against attacks performed by web crawlers. We provide a proof-of-concept of the feasibility to design appropriated OShN that can prevent specific web crawler configurations.

The rest of the paper is organized as follows. First, we present the state of the art and we describe the basic architecture of a web crawler with special emphasis in the scheduler module, detailing some of the scheduler algorithms that a crawler may implement. Later on, we discus the privacy threats that each of these scheduler algorithms suppose for the online social network users. After that, the concept of Online Social Honeynet is introduced in order to mitigate these privacy risks originated from the usage of web crawlers. Finally, we present the conclusions and provide some guidelines for further research.

2 State of the Art

The Web crawling problem has been widely studied in the scientific literature and in the practical arena. Architectures for web crawlers are proposed in [1] and [2]. These studies are centered on obtaining a fully scalable web crawler which can be used to crawl the entire Web. Detailed analysis on the bottlenecks of the crawling architectures can also be found in previous articles. Architectures for distributed web crawlers have been proposed too ([3] and [4]).

Web crawling scheduler algorithms have also been studied in depth, mostly for its use on Internet search engines. However, less attention has been paid, until now, to the specific scenario of crawling online social networks. At present time, studies in OSN web crawling deal with different related problems from algorithm performance to quality of collected data.

In [5], authors evaluate how different parameters of the crawler algorithms affect crawling efficiency (defined previously in [6]) as well as the quality of collected data. Biases produced by certain schedulers can be avoided by selecting the proper scheduler algorithm as is shown in [7], where a random sample of Facebook users is collected using a Metropolis-Hasting Random Walk (MHRW). They also demonstrate that metrics obtained with MHRW largely differ from those obtained with BFS, which remarks the importance of properly selecting the crawler scheduler algorithm based on the crawling purpose. Graphs retrieved by different collection techniques are also compared for Twitter network in [8].

Large-scale measurement of online social networks has been done in [9], where four of the most popular social networks were crawled in depth. Analyzing the obtained data the authors are able to confirm that online social networks satisfy the power-law, small-world and scale-free properties.

Although some of this crawling literature refers to users' privacy, comparisons made by different crawling algorithms are centered in their effect on classic graph metrics or on crawling efficiency, but, at our best knowledge, no work about how crawling algorithms affect user's privacy can be found in this crawling literature. On the other hand, a similar problem appears in privacy literature, where user's privacy is analyzed in detail but no references to crawler algorithms can be found.

Privacy implications of social networks have been a popular topic in recent years. Link privacy has been studied in [10] and [6]. In [10], Backstrom *et al* present several attacks on edge privacy. These attacks allow an adversary to reidentify a set of targeted users from a single anonymized copy of the network. In [6], edge privacy is studied from the point of view of the number of compromised accounts needed to expose as much nodes as possible depending on the lookahead of the network. Lookahead is defined as the distance from which a user can see his friends links.

Theoretical work centered on maintaining privacy when releasing network data sets has also been done. In [11], the authors quantify the privacy risks associated with different network release scenarios and propose an anonymization technique that leads to substantial reduction of the privacy threat. In [12], authors propose several strategies for preventing link re-identification in anonymized graphs. In [13], authors assume that an adversary knows the neighborhood of some target individuals and present an anonymization algorithm. In [14], other anonymization techniques are proposed, now considering that the adversary knows the degree of certain nodes a priori. Much effort has also been done in reidentification algorithms for anonymized social graphs in [15], where the authors present a deanonymization algorithm based on the usage of publicly available auxiliary information.

3 Web Crawling Architecture

Web crawlers are programs that automatically explore web pages in a methodical manner. Web crawlers start the search in one or more URLs, which are called seeds, and explore them in order to find new URLs to search for, until they reach a predefined termination condition. When used to crawl OSN, web crawlers start from an initial user, or list of users, and discover other users of the network by following their social relationships.

Although the architecture of a web crawler is not a fixed one and different solutions have been proposed to optimize the crawling process, the basic architecture of a web crawler can be explained by defining its five essential modules:

1. The **downloader** is the interface between the Web (or, in our case, the OSN that is being explored) and our crawler. Its job is to download a web page and pass it to the parser.

2. The **parser** is in charge of analyzing the page that has been downloaded and extract useful information and links to other pages.
3. The **storage device** keeps record of the crawled information, information about a user that can be found in his profile (*e.g.* name, location or birth date) together with links to other pages that are, in fact, links to other users' profiles which define user relationships inside the OSN.
4. The **queue** contains all the links to other users' profiles found when crawling every user that are awaiting to be explored.
5. The **scheduler** is responsible for selecting which user, from the ones in the queue, is going to be explored and communicating its decision to the downloader, completing the crawling cycle.

Based on the described crawling process of the OSN, users can be classified in three different categories: crawled, discovered or hidden. A **crawled user** is the one that all his profile's information and all his friends are known to the crawler (we denote by V_{crawl} the subset of crawled users). A **discovered user** is the one that his presence and at least one relationship is noticed by the crawler but is not a crawled user (V_{disc} denotes the subset of discovered users). Finally, a **hidden user** is the one that the crawler is not even aware of his existence (V_{hidd} denoting this subset). We also use $n_* = |V_*|$ to describe de cardinality of each set.

3.1 Scheduler Algorithms

The scheduler algorithm is the most critical part of a web crawler since its definition and configuration impacts in important aspects of a web crawler, like performance, efficiency or collected data. In this section, we describe different scheduler algorithms. The goal of this section is to provide a comprehensive description of the most frequently used scheduler algorithms in order to discuss, in next section, their implications on the collected data that, in fact, determines the users privacy. For that reason, no detailed measures on performance or efficiency are included. Interested readers can review [6] or [5] for an exhaustive study on these characteristics for different scheduler algorithms.

- **Breath-First Search (BFS)** algorithm acts as a simple queue, where the first nodes to be crawled are the first that have been discovered. Newly discovered nodes are appended to the end of the queue, thus previously discovered nodes are crawled sooner than the new ones.
- **Depth-First Search (DFS)** algorithm works as a traditional stack, where the first nodes to be crawled are the last ones that have been discovered. Newly discovered nodes are added at the top of the stack, thus they are going to be explored sooner than previously crawled nodes.
- **Greedy** algorithm selects as the next node to be crawled the one with the highest degree from all V_{disc} nodes. Depending on how this degree is computed, we can distinguish three different greedy algorithms:
  - **Real-degree greedy** takes its decisions based on the real degree of the nodes in the OSN. Notice that using the architecture described above,

information on the real degree of a node is unknown for discovered nodes so additional requests may have to be done to the OSN in order to use this scheduler. This real-degree greedy definition corresponds to the *hypothetical greedy* algorithm in [5] and would be called *highest-degree-crawler* in the [6] terms.

- **Explored-degree greedy** uses the actual known degree of the node in the explored subgraph G_{crawl} as the measure to select the next node to crawl. This definition of explored-degree greedy is the same that can be found in [5] under the mere *greedy* name.
- **Unseen-degree greedy** uses the unseen degree of a node, that is the real degree minus the explored one. Unseen degree corresponds to the number of friends of a node that the crawler is not aware of. This definition of unseen-degree is exactly the same of the *degree-greedy-crawler* used in [6].

– **Lottery** algorithm selects the next node to be crawled with a proportional probability with its degree. This gives more chance to high degree nodes to be selected while maintaining the possibility to select low degree ones. Lottery algorithm can be configured to use any of the previous degrees (real, explored or unseen) in order to make its decisions.

4 Privacy Threats Related to Crawling Activity

By crawling an OSN the corresponding social graph can be obtained. Such social graphs may provide very important information about the network and their users, since using appropriated graph mining techniques allows to discover important user characteristics.

When dealing with social graphs, two kinds of user's information can be extracted: node information and edge information. All data about a specific user is considered as node information. Node information includes all details provided in the user's profile on a specific OSN. Such data generally contains information like user name, age, nationality, current location, phone number, marital status, personal web site url and a thumbnail. Moreover, specific content OSN include other information in their users profile like photographies, music or books preferences.

The other kind of user information that can be obtained from the social graph is edge information. The mere existence of edges already offers information about users that are linked through them but, in some networks, this edges can be labeled, thus providing a more in depth information about the relations that they represent. A part from providing information of the relationships between different users, edges can also directly disclose node attributes. For instance, an edge representing a sentimental relationship between two individuals of the same sex would be revealing their sexual orientation.

Although both node attributes and edges may be considered sensitive information that the user wants to control, in this paper we focus on edge privacy since edges suppose an added risk to user's privacy in many different ways. In contrast to node attributes, which disclosure can be configured by the user, protecting edge information involve more than one user and, for that reason, it makes more difficult for the participating users to maintain control on the visibility of this relations [16]. On the other hand, relations between users can be used to detect communities. Communities are groups of nodes which are highly tight together within the network. Detecting and identifying these communities is a usual procedure in social network analysis, since communities facilitate the understanding of network data. Knowing to which communities does a user belong is an excellent way to gain information about the user: family, friends, college or work mates are subgroups that arise from a social network and can be detected from the graph structure itself. Since they do not need the explicit intervention of the user to be created, they entail a new risk for OSN users privacy. Moreover, it has been shown that users belonging to the same clique share common interests, believes or even food habits [17], which are in fact, node attributes. For this reason, node attributes can be induced from information known about other users in the same clique.

Furthermore, edge information has been proved to serve as auxiliary information for many deanonymization attacks ([10], [15]), which makes edges and its attributes an important information to care about. Relations that a user has with others describe that user in a quasi-unique form. Even when all labels have been removed from the graph, its structure is leaking information that can be used to reidentify the nodes. For instance, if an adversary knows how many friends does the victim have and which are the relations among them, the attacker may be able to find this subgraph inside an anonymized release of the whole graph and learn information about the victim and his friends.

4.1 Scheduler Implications on Privacy

It seems clear that the corresponding social graph of an OSN is a powerful tool to derive private information of the users. However, due to the actual size of OSN sites, crawling them entirely to obtain the corresponding social graph may not be an affordable option. Having to conform with the obtainment of a partial view, the concept of quality of the collected data of the crawler comes into play. The scheduler algorithm, together with the initial seed of the crawler, is the module of the crawler that determines the path to follow during the crawling process and then the exact data that will be finally retrieved from the OSN.

The quality of collected data is a difficult term to deal with since such quality depends on the objective of the crawling process.

In order to make a comprehensive analysis, we fixed three different and somehow opposite objectives for the crawler (from the attacker's point of view):

- **Objective A:** to determine all links and communities where a specific victim belongs to.
- **Objective B:** to discover general characteristics of the OSN, focused on identifying communities.
- **Objective C:** to discover the maximum number of nodes of the network.

Notice that while objective A is centered on attacking a single user, objectives B and C target the whole network but with different purposes in mind.

For each scheduler algorithm, we analyze the achievement of these objectives in terms of cohesive subgroups identification (A and B) or crawling efficiency (C). For cohesive subgroups identification, we focus on finding cliques and k-plexes [17] since this structure relaxes the strong familiarity conditions expressed in a clique but, at the same time, still provide the properties of reachability and robustness in the resulting cohesive group. For crawling efficiency, we will use the metric defined previously in [6], where efficiency is defined as the number of discovered nodes divided by the number of crawled nodes

Breath-First Search. Using a BFS algorithm with only one initial user as seed allows the crawler to explore the k-neighborhood of the seed, that is, to crawl all nodes at distance k (starting at $k = 1$) from the seed and, therefore, discover all nodes at distance $k+1$. Then, the collected data obtained using a BFS scheduler algorithm is of high quality regarding *Objective A*, since an accurate view of the OSN centered on the victims will be obtained.

However, BFS performs poorly with respect to *Objective B.* The sequentiality of the BFS with respect to the neighbor distance k does not allow to move the crawler to specific nodes belonging to interesting communities, and then the collected data cannot be taken as a representative of the OSN since it is focused on a particular part.

In BFS algorithm, no special attention is paid to higher degree nodes thus BFS does not offer advantages regarding *Objective C.*

Depth-First Search. As DFS scheduler algorithm tries to get as far as possible from the initial seed, neither the neighborhood of the seed nor subgroup structures will not be formed easily when the value n_{crawl} is low with respect n_V. In fact, cliques that are actually found by this crawling method will be small, usually with just 3 nodes. For that reason, collected data of a crawler with DFS scheduler algorithm does not provide quality information regarding neither *Objective A* nor *Objective B.*

DFS does not take into account node degrees neither, but crawling efficiency is slightly better for DFS than for BFS. The reason is that, as the crawler tries to get far away from the seed, crawled nodes tend to have a few friends in common, thus for the same n_{crawl} more n_{disc} are obtained. Thus DFS performs better than BFS with respect to *Objective C.*

Real-degree greedy. Real-degree greedy moves quickly through the largest degree node, and once reached, the algorithm provides large numbers of cliques and k-plexes since at each iteration a maximum number of edges are added to the crawled graph. For this reason, this algorithm provides a good data quality regarding *Objective B.* However, real-degree greedy is not suitable to reach *Objective A*, unless the victim is the highest degree node. In fact, higher degree nodes are very vulnerable against this scheduler algorithm since they are reached with few iterations independently of the used seed.

As first nodes selected to be crawled are the ones with higher degrees, graphs obtained with real-degree greedy always present a high mean degree, which is much more bigger than the real mean degree of the complete OSN. Selecting this high degree nodes leads to obtain high efficiency, thus this algorithm is adequate to reach *Objective C.*

Explored-degree greedy. In the explored-degree greedy, first nodes to be crawled are the ones that are more connected to already crawled ones. In contrast to the real-degree greedy, explored greedy also move towards the highest degree node but more slowly, finding the cliques and k-plexes that are in the path between the initial seed and the highest degree node. With these properties, explored-degree greedy algorithm is suitable to achieve *Objective B* although the speed at which cliques and k-plexes are discovered is much more lower that with real-degree greedy. Regarding *Objective A*, the explored-degree greedy does not provide a good strategy since it does not guarantee that the crawl is centered on the seed and then, the initial seed may not belong to the cohesive subgroups that are retrieved. However, in comparison with real-degree greedy, explored-degree greedy keeps the crawler closer to the seed and then, in terms of *Objective A*, explored-greedy performs better than real-greedy.

Unseen-degree greedy. The first users to be crawled with unseen-degree are the ones that have a high real degree and a small explored degree. In the first iterations of the crawler, unseen-degree and real-degree perform similar, moving quickly towards the highest degree node. At later stages of the crawler, the unseen-degree greedy achieves better efficiency since it discovers more new nodes than the real-degree. However, since the discovered nodes do not provide much information into the crawled graph until they are crawled, the numbers of cliques and k-plexes, and its sizes are equivalent to the ones obtained with real-degree. For that reason, performance of unseen-greedy with respect to *Objectives A* and *B* is equivalent to real-degree greedy.

Selecting the highest unseen degree node as the first node to crawl results in selecting the node that would lead the crawler to discover the maximum amount of new nodes when it is crawled. Then, unsee-degree greedy performs better than the above algorithms regarding *Objective C.*

Lottery. The random effect introduced in the lottery schedulers gives a chance to select low degree nodes as the next-node-to-crawl. As a consequence, for the same number of V_{crawl} nodes, lottery will discover more nodes than BFS, random list or DFS but less than greedy schedulers. So we can affirm that lottery performs better than BFS, random list and DFS regarding *Objective C* but worse than greedy. The same happens with found cliques and k-plexes when using the explored degree as a selection measure. In this case, lottery will find more cliques than DFS or random list but less than greedy algorithms. Much like the explored-degree greedy case, explored-degree lottery also presents the problem that the initial seed may not belong to the found cliques, which can suppose a problem when the pursued goal is *Objective A*.

Like in the greedy case, lottery tends to select as next node to crawl the ones with highest degrees (whatever the chosen degree is used), resulting on a higher mean degree in G_{crawl} than the actual graph G mean degree. However, this effect is less pronounced in the lottery case because its random component that gives a chance to low degree nodes to be selected. As a consequence, lottery performs worse than greedy algorithms regarding *Objective B*.

5 Online Social Honeynets

As we have seen, web crawling supposes a big risk for users privacy. OSN contain enormous amounts of personal data which is, in most cases, publicly available to anyone who is interested in it. Web crawlers can be used as a tool to collect all this data. For this reason, it is important to be able to defend an OSN from automated web crawlers which try to obtain information about its users.

The first trivial approach to avoid these risks is to deny the access to the network for web crawlers. In order to do so, it is needed to distinguish between web crawlers and other kinds of accesses (usually web browser requests) to the network. Although some web crawlers identify themselves as so via the User Agent field in the HTML protocol, it is easy to forge the requests in order to simulate that they are made by a common browser. Consequently, we can not rely on the HTML User Agent to tell the difference between web crawlers and non web crawlers.

It is also possible to try to forbid the access to web crawlers by banning the public access to the network. However, this is a difficult task to perform without affecting the usability of the network. It is possible to configure the network in such manner that only registered users are allowed to obtain information about other users. In addition, the information that a user can obtain of another user can be constrained depending on the distance between this users. For instance, a sample configuration may be to allow a user to obtain all the information that the network has of a direct friend, only the degree of a user which is a friend of a friend and none information at all about the rest of the users of the network.

However, even when the network is closed and the neighbors of a targeted user can only be obtained by users in the network at a fixed distance l of this targeted user, published studies [6] show different strategies to maximize the portion of the network discovered depending on the value of the lookahead l. All the presented attacks require that the attacker subverts some user accounts to obtain information of its friends. The authors show that for lookahead values higher than 2, the number of subverted accounts needed to discover the 80% of the nodes of the network is less than 100.

Furthermore, there are some OSN whose own properties or objectives make them impossible to be build under a closed paradigm network. This is the case, for example, of Twitter, whose slogan describes it as "the best way to discover what's happening on your world". How could be this accomplished by limiting the disclosure of all comments to just the users friends?

Another different approach to try to forbid the access for web crawlers is to try to limit the number of accesses to the network done by the same IP address.

Although this may seem a good strategy, it can be easily circumvented by using anonymizing techniques that mask the source IP address.

As we have seen, neither making the OSN a closed network nor limiting the number of accesses that can be done by the same IP address per unit of time are feasible solutions to our problem. For this reason, some other techniques have to be designed to limit the information that crawler may obtain from OSN. In traditional web crawling literature, web crawler traps are known to cause troubles to web crawlers [1]. Crawler traps are URLs that cause the crawler to crawl indefinitely. In the traditional Web, some crawler traps can be created unintentionally. For example, symbolic links within a file system can create cycles. Other crawler traps are produced intentionally. For instance, CGI programs that dynamically generate an infinite Web of documents. We propose a similar approach to protect OSN from web crawlers by introducing the idea of Online Social Honeynets.

Online Social Honeynets (OShN) are, much like traditional honeynets, a set of users in the network whose objective is to attract and defend the network from attackers that want to retrieve information from the network. Also like traditional honeynets, OShN consist of a set of users that appear to be part of the network with information of value to the attackers but they are actually isolated and monitored. OShN also extents the concept of Social Honeypot introduced in [18] where fake users are created in OSN to detect spam profiles and distinguish social spammers from legitimate users.

So given a social graph $G = (V, E)$ that represents an entire OSN, an OShN can be modeled as a social graph G_h, which consist on a fake set of users V_h, its relationships E_h, and a set of honeynet bridges E_b that will link the real graph G with our honeynet graph G_h (see Figure 1). Then, the disclosed network can be modeled as a social graph $G_d = (V_d, E_d)$ containing all nodes $V_d = V \cup V_h$ from both graphs and all edges $E_d = E \cup E_h \cup E_b$ from both graphs plus the honeynet bridges. Notice that we keep the edges defining the honeynet bridges outside G and G_h, since, as we describe later, such bridges play an important role for the objective of the OShN. Nodes in G_h incident to some edge in E_b are called exterior nodes while nodes in G_h without any connection to G will be called interior nodes.

Although it is obvious that the idea of OShN can be used for different purposes, our main goal is to design an OShN that may provide some protection from web crawlers, minimizing the useful information that the web crawler may obtain from the OSN.

In order to protect OSN from web crawlers, the OShN should be able, first of all, to attract web crawlers and, later on, to keep the crawler in the boundaries of the OShN, G_h. Notice that with this approach, OSN providers do not have to be concerned anymore about blocking the access to web crawlers since they will be attracted and trapped by the honeynet and, therefore, will not be able to obtain information of the real users of the network.

Let t_a be the time that our OShN needs to attract the crawler, that is the time needed to reach one of the honeynet nodes in V_h. Let t_t be the time our

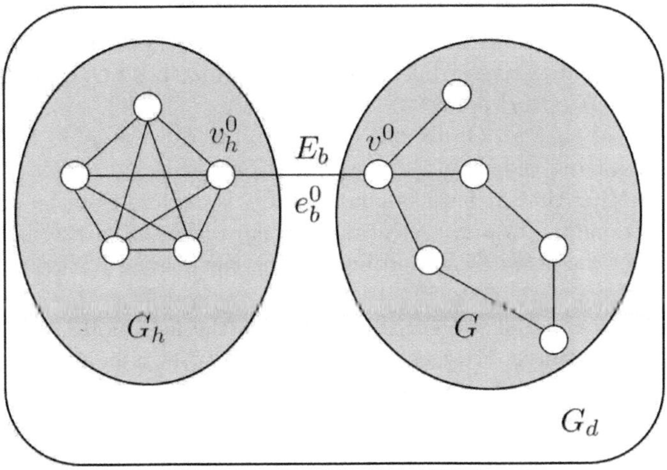

Fig. 1. Online Social Honeynet

OShN could trap the crawler. Then, t_a and t_t determine the amount of correct information the crawler may obtain from the OSN and our OShN design should focus on by minimizing t_a and maximizing t_t, since in this way we could achieve a good level of protection. The design of such a OShN that achieve such objectives is not an easy task an, obviously, it is not likely that a single OShN could provide such protection for all different web crawler configurations. In fact, as we discuss in the next subsection, the design of an effective OShN is related with the exact scheduler algorithm that the web crawler uses to crawl the OSN.

5.1 An Online Social Honeynet to Protect OSN from Greedy Schedulers

In this section, we present a proof-of-concept of an OShN in order to show the feasibility of the idea. We focus our OShN to be resistant against attacks of a web crawler configured using a real-degree greedy which represent a threat for OSN since it achieves a high efficiency rate as it has been proven in [5]. Furthermore, this algorithm is suitable to obtain a general view of the OSN, as it has been pointed out in the previous section 4 and provides an important number of cliques and k-plexes.

In order to define our OShN, we make the following assumptions. Firstly, our OShN is static in the sense that elements in V_h , E_h, and E_b remain unchanged during the crawling. Secondly, we assume that the OSN that we want to protect can be represented as a directed graph. Notice that such assumptions are very restrictive in the sense that a significant number of solutions can not be implemented with these constrains. However, we argue that our proof-of-concept become more reliable if the concept of OShN can be proven its effectiveness even under such constrained conditions.

Furthermore, a side objective of our OShN is to minimize the introduced noise. We want to design G_h and E_b in such manner that it allows us to accomplish the previously noted objectives while trying to minimize both the number of nodes $|V_h|$ and edges $|E_h \cup E_b|$ added to the network.

The first goal that an OShN has to accomplish is to be able to attract web crawlers minimizing the time t_a. The attraction of a web crawler to our OShN consists of getting that at least one node of G_h is explored by the crawler. This attraction is done by properly selecting the connections of our OShN to the rest of the nodes of the network E_b and by defining the degrees of the exterior nodes of G_h.

As we have seen, greedy algorithms select as the next node to crawl the one with the highest degree. For this reason, the node that is being explored at any time tends to have a bigger degree than the previously explored nodes.[1] So when a crawler is launched configured with a greedy algorithm, it will tend to explore the highest degree nodes of the network. Consequently, we would create the set of fake edges E_b between our OShN and the OSN so that they connect G_h with a number k of the highest degree nodes in G, ensuring that the crawler will discover those nodes when exploring the highest degree nodes of the network. This accomplish implicitly another goal that is minimizing the annoyances that the OShN cause to users. Since high degree users tend to have thousands of connections with other users, the impact of establishing a connection with G_h is minimum and, in fact, most of the users will not even be aware of this connection.[2] However, the time t_a to attract the crawler using this strategy, depends on the exact greedy algorithm. Note that, when defending the OSN from these particular scheduler algorithms, it is not needed to attach one node of G to more than one node in G_h since all the nodes of G_h connected with the same node in G would be discovered at the same time. However, it may be useful to connect the same node of G_h to many other nodes in G since that would let the crawler discover the node in G_h from different real nodes, reducing the time t_a.

Once the attraction has been done, and an exterior node of G_h has been discovered, we want to maximize the time t_t by forcing the crawler to discover more nodes from G_h and crawling all of them. While the crawler is inside G_h, no real nodes are crawled thus no node attributes of real nodes are ever disclosed. However, even when the crawler is inside G_h, some real nodes may be discovered, depending on the size of $|E_b|$. Since our OShN is designed towards protecting the OSN from real-degree and unseen-degree greedy algorithms, the best strategy to maximize t_t is to set the degrees of the exterior nodes of G_h to at least $max\{m_i + 1\}$ where m_i is the real degree of their neighbors in G. Using this strategy, the time t_t is exactly the time needed for the crawler to crawl all nodes in G_h. Then, we can defend the OSN from a crawler by assigning an arbitrary large number of nodes to G_h. Notice that trapping indefinitely the crawler in

[1] Note that this is true for the vast majority of starting nodes in the network. Some extreme cases, for instance, starting the crawl in the highest degree node, don't have this property.

[2] Deciding how to deal with uncooperative users is outside the scope of this paper.

G_h imply to assign an infinite number of nodes in G_h which is not feasible in our scenario since we have assumed that our OShN is not dynamic, in the sense that V_h , E_h, and E_b remain unchanged during the execution of the crawler.

There are many possible configurations that meet the above requirements. For instance, we can design G_h as a complete graph of d nodes where all nodes have degree $d-1$ except for $v_h^0 \in V_h$, which has degree d. The additional edge incident to this node is going to be our bridge edge $e_b^0 = (v^0, v_h^0) \in E_b$, which will link our honeynet G_h with the real graph G. As we want to ensure that the crawler is not able to escape from the honeynet until it has crawled all the nodes inside G_h, we will force that interior nodes of G_h have a higher degree than the node that has served as an entry point to the honeynet v^0. For this reason, we will set $d = max\{m_i + 2\}$, so the interior nodes of G_h will have one more link than the most connected node of G. Notice that doing so, the entry node v_h^0 has exactly the degree of v^0 plus 2. Even though a 1 point degree increment will be enough to force the crawler to crawl v_h^0 just after crawling v^0, incrementing it by 2 allows us to construct G_h in an easy manner, avoiding having to spend computational resources in the design of G_h. Figure 1 shows an example with $m_i = 3$.

5.2 Experimental Results

We have simulated the correctness of our OShN over the Flickr OSN, taking as a testbed the data collected by Mislove *et al.* in [9] which contains over 11 millions of users. This dataset is one of the most complete OSN data available and can be used as a testing set for OSN analysis. We have centered our experiments in the Flickr network, for which this dataset contains almost the 27% of nodes existing on the network at the time of the crawl (1,846,198 nodes) with its relations (22,613,981 links). Our experiments are done considering that our OSN graph G is exactly the Flickr graph that had been retrieved in [9]. The diameter of this graph G is 27, the radius is 13 and its mean degree is 12.24. The highest degree of a node in G is 26,185.

Since real-degree greedy is the scheduler algorithm used as a base point for the tests in [5], we have conducted our experiments with a crawler configured with this algorithm as a scheduler. Two termination conditions have been set for the crawler to stop his job: to reach 1,000 crawled nodes, $n_{crawl} = 1,000$, or to crawl the v_h^0 node, which would be the first node in G_h that has been crawled. Furthermore, another end crawling condition has been added when there are no V_{disc} nodes left to crawl, in case the initial seed belongs to an isolated component of the graph containing less than 1,000 nodes.

Assuming these settings, we have created our experimental OShN by generating a complete subgraph of $d = 26,187$ nodes, such that every node in the G_h has exactly degree 26,186 except for a node $v_h^0 \in V_h$, for which we set a degree of 26,187.

We have conducted 18,461 experiments (1% of the total number of nodes in the data testbed) in order to evaluate the attraction and trapping capacity of our OShN. For each experiment, we select a random node in the Flickr network and we launch a crawler using this node as initial seed and the configuration detailed above. In 12,283 of the conducted experiments, the 66.53%, the OShN was able to attract the web crawler and the crawler crawled the gateway node v_h^0. For these experiments, the crawler only need 5.09 hops (in mean) to reach v_h^0 from the initial seed. This value indicates that the time t_a for this proof-of-concept is really low. The leaked information obtained by the crawler is very low, since, in mean, only 5 nodes are crawled by the crawler and the mean number of discovered nodes is 12,645.60 nodes, that is less than the 0.7% of the entire network. Notice, however, that the implications of discovered nodes for link privacy are less strong since link information of discovered nodes is incomplete until they are not crawled.

A detailed analysis of the 6,178 experiments where the OShN could not attract the crawler shows that in all cases there is no path between the seed and v_h^0. The interesting point is that, for that seeds, the total number of nodes that the crawler is able to crawl is, in mean, 4.40 which imply that the isolated parts of the graph, where the crawler seed has been randomly chosen, are really small.

Obviously, regarding the design of the OShN, the trapping time t_t was maximum, in the sense that the ending condition was met before the crawler left the OShN.

6 Conclusions

In this paper, we have studied the effect that web crawlers may have on the information that can be retrieved from an OSN. We review some of the most relevant scheduler algorithms and describe their main properties. We discuss the private information that can be inferred from a social graph, focusing on the communities that can be identified in a graph by means of the connectivity of their members. Then, we analyze the impact of different schedulers algorithms regarding the information that the web crawler retrieve.

All this analysis shows the threat that web crawlers suppose for OSN information. For that reason, and assuming the difficulty to ban web crawlers from OSN, we introduce the concept of online social honeynet (OShN) as a mechanism to achieve some degree of protection against web crawlers. We provide a proof-of-concept of an OShN designed to protect the OSN from a web crawled with a real-degree greedy as a scheduler algorithm. Experimental data shows that the proposed protection is effective and that the amount of OSN data disclosed to the web crawler can be keep at lower levels. Although the proposed OShN only protects the OSN from a specific crawler configuration, it requires low $|E_b|$ values, which makes it easy to be implemented in real world environments.

We have provided some hints towards the construction of the honeynet graph and some of the conditions that force the the crawler to enter the honeynet once it has been discovered and that ensure that the crawler is not able to exit the honeynet once it is inside. However, a detailed analysis on the construction of the honeynet graph remains to be done. The exact construction of a graph that meets the requirements needed for the honeynet while minimizing the overhead introduced to the network is an interesting future work to proceed with.

Moreover, an interesting feature to require to our OShN is that it is not distinguishable from the rest of the network. In doing so we assure that web crawlers can not detect when they are inside the OShN.

On the other hand, in our discussions, we have assumed that our OShN is static in the sense that elements in V_h, E_h, and E_b remain unchanged during the execution of the crawler. However, a dynamic model can present some advantages to both the effectiveness of the protection as well as the resources used to hold the OShN. Designing such OShN is further work to be done.

Acknowledgements. This work was partially supported by the Spanish MCYT and the FEDER funds under grants TSI2007-65406-C03-03 "E-AEGIS", TIN2010-15764 "N-KHRONOUS", and CONSOLIDER CSD2007-00004 "ARES".

References

1. Heydon, A., Najork, M.: Mercator: A scalable, extensible web crawler. World Wide Web 2, 219–229 (1999)
2. Brin, S., Page, L.: The anatomy of a large-scale hypertextual web search engine. Computer Networks and ISDN Systems 30, 107–117 (1998)
3. Shkapenyuk, V., Suel, T.: Design and implementation of a high-performance distributed web crawler. In: Proc. of the Int. Conf. on Data Engineering, pp. 357–368 (2002)
4. Boldi, P., Codenotti, B., Santini, M., Vigna, S.: Ubicrawler: a scalable fully distributed web crawler. Softw. Pract. Exper. 34, 711–726 (2004)
5. Ye, S., Lang, J., Wu, F.: Crawling online social graphs. In: Proceedings of the 2010 12th International Asia-Pacific Web Conference, APWEB 2010, pp. 236–242. IEEE Computer Society, Washington, DC, USA (2010)
6. Korolova, A., Motwani, R., Nabar, S.U., Xu, Y.: Link privacy in social networks. In: CIKM 2008: Proceeding of the 17th ACM Conference on Information and Knowledge Management, pp. 289–298. ACM, New York (2008)
7. Gjoka, M., Kurant, M., Butts, C.T., Markopoulou, A.: A walk in facebook: Uniform sampling of users in online social networks (2009)
8. Krishnamurthy, B., Gill, P., Arlitt, M.: A few chirps about twitter. In: WOSP 2008: Proceedings of the First Workshop on Online Social Networks, pp. 19–24. ACM, New York (2008)
9. Mislove, A., Marcon, M., Gummadi, K.P., Druschel, P., Bhattacharjee, B.: Measurement and analysis of online social networks. In: IMC 2007: Proceedings of the 7th ACM SIGCOMM Conference on Internet Measurement, pp. 29–42. ACM, New York (2007)
10. Backstrom, L., Dwork, C., Kleinberg, J.: Wherefore art thou r3579x?: anonymized social networks, hidden patterns, and structural steganography. In: WWW 2007: Proceedings of the 16th International Conference on World Wide Web, pp. 181–190. ACM, New York (2007)
11. Hay, M., Miklau, G., Jensen, D., Weis, P., Srivastava, S.: Anonymizing social networks. Technical report (2007)
12. Zheleva, E., Getoor, L.: Preserving the privacy of sensitive relationships in graph data. In: Bonchi, F., Ferrari, E., Malin, B., Saygın, Y. (eds.) PInKDD 2007. LNCS, vol. 4890, pp. 153–171. Springer, Heidelberg (2008)

13. Zhou, B., Pei, J.: Preserving privacy in social networks against neighborhood attacks. In: 2008 IEEE 24th International Conference on Data Engineering, pp. 506–515. IEEE, Los Alamitos (2008)
14. Liu, K., Terzi, E.: Towards identity anonymization on graphs. In: SIGMOD 2008: Proceedings of the 2008 ACM SIGMOD International Conference on Management of Data, pp. 93–106. ACM, New York (2008)
15. Narayanan, A., Shmatikov, V.: De-anonymizing social networks. In: SP 2009: Proceedings of the 2009 30th IEEE Symposium on Security and Privacy, pp. 173–187. IEEE Computer Society, Washington DC, USA (2009)
16. Pérez-Solà, C., Herrera-Joancomartí, J.: OSN: When multiple autonomous users disclose another individual's information. In: International Conference on P2P, Parallel, Grid, Cloud, and Internet Computing, pp. 471–476. IEEE Computer Society, Fukuoka (2010)
17. Wasserman, S., Faust, K.: Social network analysis: methods and applications. In: Structural Analysis in the Social Sciences, vol. 8. Cambridge University Press, Cambridge (1994)
18. Lee, K., Caverlee, J., Webb, S.: The social honeypot project: protecting online communities from spammers. In: Proceedings of the 19th International Conference on World wide web, WWW 2010, pp. 1139–1140. ACM, New York (2010)

Cost-Sensitive Learning

Zhi-Hua Zhou

National Key Laboratory for Novel Software Technology
Nanjing University, Nanjing 210093, China
zhouzh@lamda.nju.edu.cn

In conventional classification settings, the classifiers generally try to maximize the *accuracy* or minimize the *error rate*, both are equivalent to minimizing the number of mistakes in classifying new instances. Such a setting is valid when the costs of different types of mistakes are equal. In real-world applications, however, the costs of different types of mistakes are often unequal. For example, in intrusion detection, the cost of mistakenly classifying an intrusion as a normal access is usually far larger than that of mistakenly classifying a normal access as an intrusion, because the former type of mistakes will result in much more serious losses.

In cost-sensitive learning, rather than simply minimizing the number of mistakes, the goal is to minimize the *total cost*. Roughly speaking, there are two types of misclassification costs, i.e., class-dependent or example-dependent costs. The former assumes that the costs are associated with classes, that is, every class has its own misclassification cost; the latter assumes that the costs are associated with examples, that is, every example has its own misclassification cost. In most real tasks it is feasible to get the cost of misclassifying one class to another class, e.g., by querying domain experts, while only in some special tasks it is easy to get the cost for every training example. In this talk we will focus on the class-dependent misclassification costs.

The most fundamental and popular approach to cost-sensitive learning is `Rescaling`, or called `Rebalance`. This approach tries to rebalance the classes such that the influences of different classes are in proportion to their costs. For example, the `Rescaling` approach can be realized by *resampling*, where the lower-cost class examples can be under-sampled such that the number of examples of the lower-cost and higher cost classes are in proportion to their misclassification costs, respectively. In addition to *resampling*, the `Rescaling` approach can also be realized in other forms, such as *reweighting* the training examples or *threshold-moving* of the decision boundaries. Notice that `Rescaling` is an essential procedure for handling unequal costs; indeed, most cost-sensitive learning approaches can be regarded as different realizations of `Rescaling` with different base learners.

Though `Rescaling` works very well in two-class classification problems, it was found that it often fails in multi-class problems. In this talk, we will analyze why this phenomenon occurs, and introduce an updated `Rescaling` approach. Then, we will discuss on how to handle inexact cost information; this is an important and challenging problem since it is usually difficult to get exact cost information

V. Torra et al. (Eds.): MDAI 2011, LNAI 6820, pp. 17–18, 2011.

in real-world tasks, yet previous cost-sensitive learning studies assumed that exact costs of different types of misclassifications are known. We will also briefly introduce cost-sensitive face recognition, and a task involving other types of unequal costs such as feature extraction cost.

References

1. Zhou, Z.H., Liu, X.Y.: On multi-class cost-sensitive learning. Computational Intelligence 26, 232–257 (2010)
2. Zhou, Z.H., Liu, X.Y.: Training cost-sensitive neural networks with methods addressing the class imbalance problem. IEEE Transactions on Knowledge and Data Engineering 18, 63–77 (2006)
3. Liu, X.Y., Zhou, Z.H.: Learning with cost intervals. In: Proceedings of the 16th ACM SIGKDD Conference on Knowledge Discovery and Data Mining, Washington, DC, pp. 403–412 (2010)
4. Zhang, Y., Zhou, Z.H.: Cost-sensitive face recognition. IEEE Transactions on Pattern Analysis and Machine Intelligence 32, 1758–1769 (2010)
5. Liu, L.P., Yu, Y., Jiang, Y., Zhou, Z.H.: TEFE: A time-efficient approach to feature extraction. In: Proceedings of the 8th IEEE International Conference on Data Mining, Pisa, Italy, pp. 423–432 (2008)

Evolving Graph Structures for Drug Discovery

Keith C.C. Chan

Department of Computing
The Hong Kong Polytechnic University
Hung Hom, Kowloon, Hong Kong
cskcchan@comp.polyu.edu.hk

Abstract. Computer-Aided Drug Discovery (CADD) is concerned with the use of computational techniques to determine drug structures with certain desirable properties. Evolutionary algorithms (EAs) have been proposed to evolve drug molecules by mimicking chemical reactions that cause the exchange of chemical bonds and components between molecules. For these EAs to perform their tasks, known molecular components, which can serve as building blocks for the drugs to be designed, and known chemical rules, which govern chemical combination between different components, have to be introduced before an evolutionary process can take place. To automate drug molecular design without such prior knowledge and constraints, we need a special EA that can evolve molecular graphs with minimal background knowledge. In this talk, we present one such EA that can evolves graph structures used to represent drug molecules. We show how *molecular fingerprints* can be used to evaluate the "fitness" of an evolved drug structure obtained at each generation during the evolutionary process. We also show how the discovering of *privileged structures* in many drug molecules and the use of *ligand docking* and *binding affinity* can be used as alternatives for fitness evaluating in an EA for drug design. We show how the results obtained using the proposed EA may lead to a promising approach for CADD.

V. Torra et al. (Eds.): MDAI 2011, LNAI 6820, p. 19, 2011.

Fuzzy Measures and Comonotonicity on Multisets

Yasuo Narukawa[1,2], Klara Stokes[3], and Vicenç Torra[4]

[1] Toho Gakuen,
3-1-10, Naka, Kunitachi, Tokyo, 186-0004, Japan
[2] Department of Computational Intelligence and Systems Science
Tokyo Institute of Technology
4259 Nagatuta, Midori-ku, Yokohama 226-8502, Japan
[3] Universitat Rovira i Virgili
Dept. of Computer Engineering and Maths,
UNESCO Chair in Data Privacy
Av. Països Catalans 26,
43007 Tarragona, Catalonia, Spain
[4] IIIA, Institut d'Investigació en Intel·ligència Artificial
CSIC, Consejo Superior de Investigaciones Científicas
Campus UAB s/n, 08193 Bellaterra, Catalonia, Spain
narukawa@d4.dion.ne.jp, klara.stokes@urv.cat,
vtorra@iiia.csic.es

Abstract. Fuzzy measures on multisets are studied. We show that a class of multisets can be represented as a subset of positive integers. Comonotonicity for multisets are defined. We show that a fuzzy measure on multisets with some comonotonicity condition can be represented by generalized fuzzy integral.

Keywords: Fuzzy measure, multiset, Choquet integral, Sugeno integral, Generalized fuzzy integral.

1 Introduction

Fuzzy measures [15] are set functions $\mu : \wp(X) \to [0,1]$ that permits us to represent interactions between the elements of the set. Fuzzy measures are often used in conjunction with fuzzy integrals to aggregate information from several sources [17].

Several extensions and variations for fuzzy measures exist. E.g. [4,10] deal with discrete fuzzy measures (i.e., $\mu : \wp(X) \to L$ where L is an ordinal scale), [5] considered set-valued measures.

In this paper we discuss another type of extension. Here, fuzzy measures are defined for multisets. Multisets (or bags) [6,11,19] are a generalization of sets in which multiple appearances of an element is permitted.

The structure of the paper is as follows. In Section 2 we review some previous results needed in this paper. In Section 3 we discuss about the representation of fuzzy measures on multisets. In Section 4 we introduce comonotonicity on multisets and give some results. In Section 5 we have a proposition that a fuzzy measure on multisets can be represented by a generalized fuzzy integral when some comonotonicity conditions hold.

V. Torra et al. (Eds.): MDAI 2011, LNAI 6820, pp. 20–30, 2011.

2 Preliminaries

In this section we review the concepts of fuzzy measures and two fuzzy integrals: the Choquet and Sugeno integrals.

2.1 Choquet Integral and Sugeno Integral

We present here the definition of fuzzy measures and the ones of Choquet and Sugeno integrals. In this paper we will use $\vee$ and $\wedge$ to denote, respectively, the maximum and the minimum.

Definition 1. *Let X be a universal set and $\mathscr{X}$ be a subset of 2^X with $\emptyset \in \mathscr{X}$ and $X \in \mathscr{X}$. Then, $(X, \mathscr{X})$ is called a fuzzy measurable space. We say that a function $f : X \to \mathbb{R}^+$ is $\mathscr{X}$-measurable if $\{x|f(x) \geq a\} \in \mathscr{X}$ for all a.*

Definition 2. *[3] Let f and g be $\mathscr{X}$-measurable functions on X; then, we say that f and g are comonotonic if*

$$f(x) < f(y) \Rightarrow g(x) \leq g(y)$$

for $x, y \in X$.

Definition 3. *[15] Let $(X, \mathscr{X})$ be a fuzzy measurable space; then, a fuzzy measure μ on $(X, \mathscr{X})$ is a real valued set function, $\mu : \mathscr{X} \longrightarrow \mathbb{R}^+$ with the following properties.*

(i) $\mu(\emptyset) = 0$, $\mu(X) = k$ *where* $k \in (0, \infty)$.
(ii) $\mu(A) \leq \mu(B)$ *whenever* $A \subseteq B$, $A, B \in \mathscr{X}$.

A triplet $(X, \mathscr{X}, \mu)$ is said to be a fuzzy measure space.

Definition 4. *Let μ be a fuzzy measure on $(X, \mathscr{X})$, we say that:*

(i) *μ is a one to one fuzzy measure, if $A \neq B$ implies $\mu(A) \neq \mu(B)$;*
(ii) *μ is a distorted additive measure if there exist a strictly monotone function f and an additive measure m such that $\mu = f \circ m$.*

Definition 5. *[2,12] Let $(X, \mathscr{X}, \mu)$ be a fuzzy measure space and let f be a $\mathscr{X}$-measurable function; then, the Choquet integral of f with respect to μ is defined by*

$$(C)\int f d\mu := \int_0^\infty \mu_f(r) dr,$$

where $\mu_f(r) = \mu(\{x|f(x) \geq r\})$.

Definition 6. *[1] For any $r > 0$ and $A \in \mathscr{X}$, the basic simple function $b(r, A)$ is defined by $b(r, A)(x) = r$ if $x \in A$ and $b(r, A)(x) = 0$ if $x \notin A$.*

Then, we say that a function f is a simple function if it can be expressed as

$$f := \sum_{i=1}^{n} b(a_i, A_i) \text{ for } a_i > 0 \tag{1}$$

where $A_1 \supsetneq A_2 \supsetneq \cdots \supsetneq A_n$, $A_i \in \mathscr{X}$.

Expression 1 is called a comonotonic additive representation of f. f can also be expressed as $f := \vee_{i=1}^{n} b(a'_i, A_i)$ for $a'_1 > \cdots > a'_n > 0$, where $A_1 \supsetneq A_2 \supsetneq \cdots \supsetneq A_n$, $A_i \in \mathscr{X}$. This expression is called a comonotonic maxitive representation of f.

Then, when a $\mathscr{X}$-measurable function f is a simple function with a comonotonic additive representation, we have that the following equation holds:

$$(C)\int f d\mu = \sum_{i=1}^{n} a_i \mu(A_i).$$

Definition 7. *[14,15] Let $(X, \mathscr{X}, \mu)$ be a fuzzy measure space and let $f : X \to [0,\infty)$ be a $\mathscr{X}$-measurable function; then, the Sugeno integral of f with respect to μ is defined by*

$$(S)\int f d\mu := \sup_{r\in[0,\infty)} [r \wedge \mu_f(r)].$$

When f is a simple function with a comonotonic maxitive representation, the Sugeno integral can be written as

$$(S)\int f d\mu = \bigvee_{i=1}^{n} (a_i' \wedge \mu(A_i)).$$

2.2 Generalized Fuzzy Integral

In this section, we define a generalized fuzzy integral in terms of a pseudo-addition $\oplus$ and a pseudo-multiplication $\boxdot$. Formally, $\oplus$ and $\boxdot$ are binary operators that generalize addition and multiplication, and also max and min. We want to recall that generalized fuzzy integrals have been investigated by Benvenuti et al. in [1].

Note that we will use $k \in (0,\infty)$ in the rest of this paper.

Definition 8. *A pseudo-addition $\oplus$ is a binary operation on $[0,k]$ or $[0,\infty)$ fulfilling the following conditions:*

(A1) $x \oplus 0 = 0 \oplus x = x$.
(A2) $x \oplus y \leq u \oplus v$ *whenever* $x \leq u$ *and* $y \leq v$.
(A3) $x \oplus y = y \oplus x$.
(A4) $(x \oplus y) \oplus z = x \oplus (y \oplus z)$.
(A5) $x_n \to x, y_n \to y$ *implies* $x_n \oplus y_n \to x \oplus y$.

A pseudo-addition $\oplus$ is said to be strict if and only if $x \oplus y < x \oplus z$ whenever $x > 0$ and $y < z$, for $x,y,z \in (0,k)$; and it is said to be Archimedean if and only if $x \oplus x > x$ for all $x \in (0,k)$.

Definition 9. *A pseudo-multiplication $\boxdot$ is a binary operation on $[0,k]$ or $[0,\infty)$ fulfilling the conditions:*

(M1) *There exists a unit element* $e \in (0,k]$ *such that* $x \boxdot e = e \boxdot x = x$.
(M2) $x \boxdot y \leq u \boxdot v$ *whenever* $x \leq u$ *and* $y \leq v$.
(M3) $x \boxdot y = y \boxdot x$.
(M4) $(x \boxdot y) \boxdot z = x \boxdot (y \boxdot z)$.
(M5) $x_n \to x, y_n \to y$ *implies* $x_n \boxdot y_n \to x \boxdot y$.

Example 1

(i) The maximum operator $x \vee y$ is a non Archimedean pseudo-addition on $[0,k]$.
(ii) The sum $x+y$ is an Archimedean pseudo-addition on $[0,\infty)$.
(iii) The Sugeno operator $x +_\lambda y := 1 \wedge (x+y+\lambda xy)$ $(-1 < \lambda < \infty)$ is an Archimedean pseudo-addition on $[0,1]$.

Proposition 1. *[9] If a pseudo-addition $\oplus$ is Archimedean, then there exists a continuous and strictly increasing function $g : [0,k] \to [0,\infty]$ such that $x \oplus y = g^{(-1)}(g(x) + g(y))$, where $g^{(-1)}$ is the pseudo-inverse of g defined by*

$$g^{(-1)}(u) := \begin{cases} g^{(-1)}(u) & \text{if } u \leq g(k) \\ k & \text{if } u > g(k). \end{cases}$$

The function g is called an additive generator of $\oplus$.

Definition 10. *Let μ be a fuzzy measure on a fuzzy measurable space $(X, \mathscr{X})$; then, we say that μ is a $\oplus$-measure or a $\oplus$-decomposable fuzzy measure if $\mu(A \cup B) = \mu(A) \oplus \mu(B)$ whenever $A \cap B = \emptyset$ for $A, B \in \mathscr{X}$.*

A $\oplus$-measure μ is called normal when either $\oplus = \vee$, or $\oplus$ is Archimedean and $g \circ \mu$ is an additive measure. Here, g corresponds to an additive generator of $\oplus$.

Definition 11. *Let $k \in (0,\infty)$, let $\oplus$ be a pseudo-addition on $[0,k]$ or $[0,\infty)$ and let $\boxdot$ be a pseudo-multiplication on $[0,k]$ or $[0,\infty)$; then, we say that $\boxdot$ is $\oplus$-fitting if*

(F1) $a \boxdot x = 0$ *implies* $a = 0$ *or* $x = 0$,
(F2) $a \boxdot (x \oplus y) = (a \boxdot x) \oplus (a \boxdot y)$.

Under these conditions, we say that $(\oplus, \boxdot)$ is a pseudo-fitting system.

Let $\oplus$ be a pseudo-addition; then, we define its pseudo-inverse $-_\oplus$ as

$$a -_\oplus b := \inf\{c | b \oplus c \geq a\}$$

for all $(a,b) \in [0,k]^2$.

Definition 12. *[16] Let μ be a fuzzy measure on a fuzzy measurable space $(X, \mathscr{X})$, and let $(\oplus, \boxdot)$ be a pseudo-fitting system. Then, when μ is a normal $\oplus$-measure, we define the pseudo-decomposable integral of a measurable simple function f on X such that $f = \oplus 1_{i=1}^n b(r_i, D_i)$ where $D_i \cap D_j \neq \emptyset$ for $i \neq j$, as follows:*

$$(D) \int f d\mu := \oplus_{i=1}^n r_i \boxdot \mu(D_i).$$

Since μ is an $\oplus$-measure, it is obvious that the integral is well defined.

Definition 13. *Let μ be a fuzzy measure on a measurable space $(X, \mathscr{X})$, and let $(\oplus, \boxdot)$ be a pseudo-fitting system. Then, the generalized fuzzy integral (GF-integral) of a measurable simple function $f := \oplus_{i=1}^n b(a_i, A_i)$, with $a_i > 0$ and $A_1 \supsetneq A_2 \supsetneq \ldots A_n$, $A_i \in \mathscr{X}$, is defined as follows:*

$$(GF) \int f d\mu := \oplus_{i=1}^n a_i \boxdot \mu(A_i).$$

The GF-integral of a simple function is well defined [1].

The next proposition follows from the definition of the pseudo-inverse $-_{\oplus}$, the generalized t-conorm integral (Definition 13), and the t-conorm integral (Definition 12).

Proposition 2. *Let μ be a fuzzy measure on a fuzzy measurable space $(X, \mathscr{X})$, and let $(\oplus, \boxdot)$ be a pseudo-fitting system. Then, if μ is a normal $\oplus$-measure, the generalized fuzzy integral coincides with the pseudo-decomposable integral.*

Example 2

(i) When $\oplus = +$ and $\boxdot = \cdot$, the generalized fuzzy integral is a Choquet integral.
(ii) When $\oplus = \vee$ and $\boxdot = \wedge$, the generalized fuzzy integral is a Sugeno integral.

Let f, g be comonotonic measurable functions. Then, since for all $a, b > 0$ either $\{x|f(x) \geq a\} \subseteq \{x|g(x) \geq b\}$ or $\{x|f(x) \geq a\} \supseteq \{x|g(x) \geq b\}$, the following theorem can be proved.

Theorem 1. *[13] Let $(X, \mathscr{X}, \mu)$ be a fuzzy measure space and let $(\oplus, \boxdot)$ be a pseudo-fitting system. Then, for comonotonic measurable functions f, and g, we have*

$$(GF)\int (f \oplus g) d\mu = (GF)\int f d\mu \oplus (GF)\int g d\mu.$$

We call this property the comonotonic $\oplus$-additivity of a generalized fuzzy integral.

2.3 Multisets

Let X be a universal set. Then, a multiset M over X is characterized by the count function $C_M : X \to N := \{0, 1, 2, \ldots\}$, where C_M corresponds to the number of occurrences of the object $x \in X$.

We denote by $\mathscr{M}(X)$ the class of multisets of X.

Example 3. *Let $X := \{a, b, c\}$ and $M := \{a, a, a, b, b\}$. Then $C_M(a) = 3$, $C_M(b) = 2$, $C_M(c) = 0$.*

The multiset M in Example 3 can also be represented as $M = \{3/a, 2/b\}$ or $M = \{(a, 3), (b, 2)\}$.

Definition 14. *Let $M, N \in \mathscr{M}(X)$. Then, we define:*

- *the inclusion of multisets by*

$$M \subseteq N \Leftrightarrow C_M(x) \leq C_N(x)$$

  for all $x \in X$;
- *the equality of multisets $M = N$ by*

$$C_M(x) = C_N(x)$$

  for all $x \in X$.

Let $M \in \mathscr{M}(X)$. Then $\mathscr{P}(M)$ denotes the class of subsets of the multiset M, that is,

$$\mathscr{P}(M) := \{N | N \subseteq M, N \in \mathscr{M}(X)\}.$$

Proposition 3. *Let $|X| = n$ and $M \in \mathscr{M}(X)$. If $M = \{(a_i, k_i) | i = 1, 2, \ldots n\}$, then*

$$|\mathscr{P}(M)| = \prod_{i=1}^{n} (k_i + 1).$$

Example 4. *Let $M = \{a, a, a, b, b\}$. Then*

$$\begin{aligned} \mathscr{P}(M) = \{ \ & M_1 = \emptyset, \quad M_2 = \{a,a\}, \quad M_3 = \{a,a,a\}, \quad M_4 = \{a,a,a,b\}, \\ & M_5 = \{a\}, \quad M_6 = \{a,b\}, \quad M_7 = \{a,a,b\}, \quad M_8 = \{a,a,b,b\}, \\ & M_9 = \{b\}, M_{10} = \{b,b\}, M_{11} = \{a,b,b\}, M_{12} = \{a,a,a,b,b\}\}. \end{aligned}$$

Definition 15. *Let $A, B \in \mathscr{M}(X)$. We define some binary operations on $\mathscr{M}(X)$. Definitions include union, intersection and addition of two multisets.*

(i) $C_{A \cup B}(x) = C_A(x) \vee C_B(x)$
(ii) $C_{A \cap B}(x) = C_A(x) \wedge C_B(x)$
(iii) $C_{A+B}(x) = C_A(x) + C_B(x)$
(iv) $C_{A \oplus B}(x) = C_A(x) \oplus C_B(x)$
(v) $C_{A \boxdot B}(x) = C_A(x) \boxdot C_B(x)$

where $x \in X$ and C_A is a count function of A.

Proposition 4. *Let $A, B \in \mathscr{M}(X)$. We have*

$$A \cap B \subseteq A \cup B \subseteq A + B$$

Example 5. *Let $X := \{a, b, c\}$ and $A := \{a, a, b\}$, $B := \{a, b, b, c\}$. Then we have*

(i) $A \cup B = \{a, a, b, b, c\}$
(ii) $A \cap B = \{a, b\}$
(iii) $A + B = \{a, a, a, b, b, b, c\}$
(iv) $A \oplus B = \{a, a, b, b, c\}$ *when* $\oplus = \vee$
(v) $A \boxdot B = \{a, b\}$ *when* $\oplus = \wedge$

3 Representation of Fuzzy Measures for Finite Multisets

Let X be a finite universal set, $|X| = n$, and $\mathbb{P}$ be the set of prime numbers, that is, $\mathbb{P} := \{2, 3, 5, 7, \ldots\}$. Since X is a finite set, there exists a one to one mapping φ_X from X to a subset of $\mathbb{P}$. That is,

$$\varphi_X : X \to \{p_1, p_2, \ldots, p_n\}.$$

Let $M \in \mathscr{M}(X)$, then, we have an induced one to one mapping Φ_X from $\mathscr{M}(X)$ to a subset S of natural numbers by

$$\Phi_X(M) := \prod_{i=1}^{n} \varphi_X(x_i)^{C_M(x_i)}.$$

We say that $\Phi_X(M)$ is a natural number representation of the multiset M.

Example 6. *Let $X := \{a,b,c\}$ and $\varphi_X(a) = 2, \varphi_X(b) = 3, \varphi_X(c) = 5$. Then,*

$$\Phi_X(A) := 2^{C_A(a)}3^{C_A(b)}5^{C_A(c)}$$

for $A \in \mathscr{M}(X)$. For example, if $A := \{a,a,b,c\}$, then

$$\Phi_X(A) = 2^2 \cdot 3 \cdot 5 = 60.$$

Let $M \in \mathscr{M}(X)$. We have

$$\Phi_X(\mathscr{P}(M)) := \left\{ \prod_{i=1}^{n} \varphi_X(x_i)^{C_A(x_i)} | \ A \in \mathscr{P}(M) \right\}.$$

Proposition 5. *Let $M \in \mathscr{M}(X)$; then, $\Phi_X(\mathscr{P}(M))$ is the set of divisors of $\Phi_X(M)$.*

Example 7. *Let $X := \{a,b,c\}$, let $\varphi_X(a) = 2, \varphi_X(b) = 3, \varphi_X(c) = 5$, and let the multiset M be defined by $M := \{a,a,a,b,b,c\}$. Then, $\Phi_X(M) := 2^3 3^2 5^1 = 120$, and*

$$\Phi_X(\mathscr{P}(M)) := \{1,2,3,4,5,6,8,10,12,15,24,30,40,60,120\}.$$

Definition 16. *[18] Let X be a reference set, let M be a multiset on X such that $M \neq \emptyset$; then, a function μ from $(M, \mathscr{P}(M))$ to $[0,1]$ is a fuzzy measure if the following holds:*

- *$\mu(\emptyset) = 0$ and $\mu(M) = k$ for $k \in (0,\infty)$,*
- *$\mu(A) \leq \mu(B)$ when $A \subseteq B$ and $B \subseteq M$.*

We have the next proposition.

Proposition 6. *Let Φ_X be a natural number representation of multisets. Then, for any non-decreasing function such that $\rho(1) = 0$, and $\rho(\Phi_X(M)) > 0$, $\rho \circ \Phi_X$ is a fuzzy measure.*

Example 8. *Let Φ_X be a natural number representation of multisets. Then we have that $\mu := \rho \circ \Phi_X$ with $\rho(x) = \log(x)$ is a fuzzy measure.*

Conversely, we can represent any fuzzy measure in terms of a natural number representation Φ_X and a distortion function.

Proposition 7. *Let μ be a one to one fuzzy measure. There exists a function $f_\mu : \mathbb{R}^+ \to \mathbb{R}$ such that $\mu = f_\mu \circ \Phi_X$.*

In this case, we say that f_μ is a representation function of a one to one fuzzy measure.

Proposition 8. *Let μ be a one to one fuzzy measure. If a representation function f_μ is monotone, then μ is a distorted additive measure.*

The representations of fuzzy measures in terms of mappings on the set of prime numbers are further studied in [18].

4 Comonotonicity of Multisets

We start this section with a definition of the concept of comonotonicity of multisets.

Definition 17. *Let M,N be multisets on X. Then, we say that M and N are comonotonic if $C_M(x_1) < C_M(x_2)$ implies $C_N(x_1) \leq C_N(x_2)$ for all pairs $x_1,x_2 \in X$.*

Let M,N be two universal sets. Then, since $C_M(x) = 0$ or 1 and $C_N(x) = 0$ or 1, M and N are comonotonic if and only if $M \subseteq N$ or $M \supseteq N$. However, in general there exist comonotonic multisets $M,N \in \mathcal{M}(X)$ such that $M \not\subseteq N$ and $M \not\supseteq N$. Indeed, a multiset M which satisfies $C_M(x_i) = C_M(x_j)$ for all $x_i,x_j \in X$ is comonotonic with any multiset $N \in \mathcal{M}(X)$. In particular, the emptyset is comonotonic with any set $N \in \mathcal{M}(X)$. Finally, $M \subseteq N$ does not imply the comonotonicity of M and N.

Proposition 9. *Comonotonicity of multisets is a reflexive and symmetric binary relation. However, it is not an equivalence relation, since it is not transitive.*

Example 9. *Let $X := \{a,b,c\}$ and consider the natural number representation of multisets Φ_X induced by $\varphi_X(X) = \{2,3,5\}$. Let*

$$\begin{aligned}
M_1 &:= \{a,a,a,b,b,c\}, & \Phi_X(M_1) &= 360;\\
M_2 &:= \{a,a,b,c\}, & \Phi_X(M_2) &= 60;\\
M_3 &:= \{a,b,c\}, & \Phi_X(M_3) &= 30;\\
M_4 &:= \{c,c,c\}, & \Phi_X(M_4) &= 125,\\
M_5 &:= \{b,b,b,b,c,c,c\}, & \Phi_X(M_5) &= 10125.
\end{aligned}$$

Then M_1 and M_2 are comonotonic. We have that $M_3 \not\subseteq M_4$ and $M_3 \not\supseteq M_4$, but M_3 and M_4 are comonotonic anyway. If comonotonicity had been an equivalence relation, then the fact that M_2 and M_3 are comonotonic and the transitivity of the equivalence relation would have implied the comonotonicity of M_2 and M_4. However M_2 and M_4 are not comonotonic, hence illustrating the abscence of transitivity. Finally $M_4 \subseteq M_5$ but M_4 and M_5 are not comonotonic.

Let M be a multiset and $\mathcal{P}(M)$ be the class of submultisets of M. Because of the symmetry of the comonotonicity, we can decompose $\mathcal{P}(M)$ in blocks of pairwise comonotonic multisets, as will be described in the following.

Example 10. *Let $X := \{a,b\}$ and $M := \{a,a,b\}$. We can decompose $\mathcal{P}(M)$ in blocks of pairwise comonotonic multisets as*

$$\mathcal{P}(M) = \mathbf{M}_1 \cup \mathbf{M}_2,$$

with

- $\mathbf{M}_1 := \{\emptyset,\{a\},\{a,a\},\{a,b\},\{a,a,b\}\}$ *and*
- $\mathbf{M}_2 := \{\emptyset,\{b\},\{a,b\}\}$.

Then the members of each M_i (i=1,2) are pairwise comonotonic. Observe that $M_1 \cap M_2 \neq \emptyset$.

Consider the natural number representation of multisets Φ_X induced by the function $\varphi_X(X) := \{2,3\}$. Then $\Phi_X(M) = 2^2 3 = 12$ and we have

$$\Phi_X(\mathscr{P}(M)) = \Phi_X(M_1) \cup \Phi_X(M_2),$$

with

- $M_1 := \{1,2,4,6,12\}$ *and*
- $M_2 := \{1,3,6\}$.

Lemma 1. *Let $\mathscr{F} \subseteq \mathscr{M}(X)$ be a family of multisets over X. Then, there exist sets (which we call blocks) M_i $i = 1,2,\ldots,k$ of elements of $\mathscr{F}$, such that $\mathscr{F} = \cup_{1 \leq i \leq k} M_i$ and such that the members of a block M_i are pairwise comonotonic. Also, these blocks can always be chosen to be maximal with respect to inclusion.*

We say that a decomposition of $\mathscr{F}$ in comonotonic blocks, which is maximal with respect to inclusion, is a maximal comonotonic block decomposition of $\mathscr{F}$ and we call each M_i a comonotonic block of $\mathscr{F}$. If $\mathscr{F} = \mathscr{P}(M)$ for some multiset M over X, then we abuse notation and talk about the block decomposition of M and the comonotonic blocks of M.

Let M_1 and M_2 be two comonotonic blocks of $\mathscr{M}(X)$. We say that M_1 and M_2 are different if $M_1 \triangle M_2 \neq \emptyset$ where $M_1 \triangle M_2 := (M_1 \cap M_2^c) \cup (M_2 \cap M_1^c)$ and where M^c is the standard complement.

Definition 18. *Let $\mathscr{F} \subseteq \mathscr{M}(X)$ be a family of multisets over X and let $\mathscr{F} = \cup_{1 \leq i \leq l} M_i$ be a maximal comonotonic block decomposition of $\mathscr{F}$. The positive integer l is is said to be the variety of $\mathscr{F}$.*

Suppose that $X = \{a_1, a_2, \ldots, a_n\}$. It is not hard to see that the variety of $\mathscr{M}(X)$ coincides with the number of permutations of $C_M(a_1), C_M(a_2), \ldots, C_M(a_n)$, proving the next proposition.

Proposition 10. *Suppose that $|X| = n$. The variety of $\mathscr{M}(X)$ is $n!$.*

Example 11. *Let $X := \{a,b,c\}$. Proposition 10 says that the variety of $\mathscr{M}(X)$ is 6. Indeed the maximal comonotonic block decomposition of $\mathscr{M}(X)$ is*

$$\mathscr{M}(X) = \bigcup_{i=1}^{6} M_i$$

with

$$M_1 := \{M | C_M(a) \leq C_M(b) \leq C_M(c)\}$$
$$M_2 := \{M | C_M(a) \leq C_M(c) \leq C_M(b)\}$$
$$M_3 := \{M | C_M(b) \leq C_M(a) \leq C_M(c)\}$$
$$M_4 := \{M | C_M(b) \leq C_M(c) \leq C_M(a)\}$$

$$M_5 := \{M | C_M(c) \leq C_M(a) \leq C_M(b)\}$$
$$M_6 := \{M | C_M(c) \leq C_M(b) \leq C_M(a)\}.$$

Let $M = \{a,a,b,c\}$, then, following Section 3 we use the natural number representation of M induced by the function $\varphi_X(\{a,b,c\}) = \{2,3,5\}$, and we get

$$M_1 \cap \mathscr{P}(M) := \{\emptyset, \{c\}, \{b,c\}, \{a,b,c\}\} = \{1,5,15,30\}$$
$$M_2 \cap \mathscr{P}(M) := \{\emptyset, \{b\}, \{b,c\}, \{a,b,c\}\} = \{1,3,15,30\}$$
$$M_3 \cap \mathscr{P}(M) := \{\emptyset, \{c\}, \{a,c\}, \{a,b,c\}\} = \{1,5,10,30\}$$
$$M_4 \cap \mathscr{P}(M) := \{\emptyset, \{a\}, \{a,a\}, \{a,c\}, \{a,a,c\}, \{a,b,c\}, \{a,a,b,c\}\} = \{1,2,4,10,20,30,60\}$$
$$M_5 \cap \mathscr{P}(M) := \{\emptyset, \{b\}, \{a,b\}, \{a,b,c\}\} = \{1,3,6,30\}$$
$$M_6 \cap \mathscr{P}(M) := \{\emptyset, \{a\}, \{a,a\}, \{a,b\}, \{a,a,b\}, \{a,b,c\}, \{a,a,b,c\}\} = \{1,2,4,6,12,30,60\}.$$

5 Extension of Fuzzy Measure to Multisets

Even if X is a finite set, $\mathscr{M}(X)$ is infinite. The problem is how to define a fuzzy measure on $\mathscr{M}(X)$. We give a partial solution to this problem.

If X is a finite set, then 2^X is also a finite set. We can define a fuzzy measure μ on 2^X.

Definition 19. *Let M be a multiset on X, then we can represent a count function C_M by $C_M := \oplus_{i=1}^{n} b(a_i, A_i)$, with $a_i > 0$ and $A_1 \supsetneq A_2 \supsetneq \ldots A_n$, $A_i \in \mathscr{X}$. Then we can define an extension $\bar{\mu}$ of the fuzzy measure μ to a multiset M by*

$$\bar{\mu}(M) := \oplus_{i=1}^{n} a_i \boxdot \mu(A_i) = (GF) \int C_M d\mu$$

with $a_i > 0$ and $A_1 \supsetneq A_2 \supsetneq \ldots A_n$, $A_i \in \mathscr{X}$.
We say that $\bar{\mu}$ is a comonotonic $(\oplus, \boxdot)$-extension of μ.

As a corollary of Theorem 1 we get the following proposition.

Proposition 11. *Let $M, N \in \mathscr{M}(X)$, and let μ be a fuzzy measure on 2^X. If M and N are comonotonic, then $\bar{\mu}(M \oplus N) = \bar{\mu}(M) \oplus \bar{\mu}(N)$.*

Given a fuzzy measure ν on $\mathscr{M}(X)$ we have the next theorem, which can be seen as the converse of Definition 19.

Theorem 2. *If ν on $\mathscr{M}(X)$ is comonotonic $\oplus$-additive, then there exists a fuzzy measure μ on 2^X such that*

$$\nu(M) = (GF) \int C_M d\mu$$

for $M \in \mathscr{M}(X)$.

6 Conclusion

In this paper we have extended fuzzy measures on multisets. We have also introduced comonotonicity of multisets and given some examples.

Acknowledgements. Partial support by the Spanish MEC (projects ARES – CONSOLIDER INGENIO 2010 CSD2007-00004 –, eAEGIS – TSI2007-65406-C03-02 –, and RIPUP – TIN2009-11689) is acknowledged. One author is partially supported by the FPU grant (BOEs 17/11/2009 and 11/10/2010) and by the Government of Catalonia under grant 2009 SGR 1135. The authors are with the UNESCO Chair in Data Privacy, but their views do not necessarily reflect those of UNESCO nor commit that organization.

References

1. Benvenuti, P., Mesiar, R., Vivona, D.: Monotone set functions-based integrals. In: Pap, E. (ed.) Handbook of Measure Theory, pp. 1329–1379. Elsevier, Amsterdam (2002)
2. Choquet, G.: Theory of capacities. Ann. Inst. Fourier 5, 131–295 (1953-1954)
3. Dellacherie, C.: Quelques commentaires sur les prolongements de capacités, Séminaire de Probabilités 1969/1970. Lecture Notes in Mathematics, Strasbourg, vol. 191, pp. 77–81 (1971)
4. Grabisch, M.: *k*-order additive discrete fuzzy measures and their representation. Fuzzy Sets and Systems 92(2), 167–189 (1997)
5. Guo, C., Zhang, D.: On set-valued fuzzy measures. Information Sciences 160, 13–25 (2004)
6. Hickman, J.L.: A note on the concept of multiset. Bulletin of the Australian Mathematical Society 22, 211–217 (1980)
7. Klement, E.P., Mesiar, R., Pap, E.: Triangular Norms. Kluwer Academic Publishers, Dordrecht (2000)
8. Klement, E.P., Mesiar, R., Pap, E.: Integration with respect to decomposable measures, based on a conditionally distributive semiring on the unit interval. Int. J. of Unc., Fuzziness and Knowledge Based Systems 8(6), 701–717 (2000)
9. Ling, C.H.: Representation of associative functions. Publ. Math. Debrecen 12, 189–212 (1965)
10. Marichal, J.-L., Roubens, M.: Entropy of discrete fuzzy measures. Int. J. of Unc., Fuzz. and Knowledge Based Systems 8(6), 625–640 (2000)
11. Miyamoto, S.: Generalizations of multisets and rough approximations. Int. J. of Intel. Syst. 19, 639–652 (2004)
12. Murofushi, T., Sugeno, M.: An interpretation of fuzzy measures and the Choquet integral as an integral with respect to a fuzzy measure. Fuzzy Sets and Systems 29, 201–227 (1989)
13. Narukawa, Y., Torra, V.: Multidimensional generalized fuzzy integral. Fuzzy Sets and Systems 160, 802–815 (2009)
14. Ralescu, D., Adams, G.: The fuzzy integral. J. Math. Anal. Appl. 75, 562–570 (1980)
15. Sugeno, M.: Theory of fuzzy integrals and its applications, Ph. D. Dissertation, Tokyo Institute of Technology (1974)
16. Sugeno, M., Murofushi, T.: Pseudo-additive measures and integrals. J. Math. Anal. Appl. 122, 197–222 (1987)
17. Torra, V., Narukawa, Y.: Modeling decisions: information fusion and aggregation operators. Springer, Heidelberg (2007)
18. Torra, V., Stokes, K., Narukawa, Y.: Fuzzy Measures on Multisets (submitted)
19. Yager, R.R.: On the theory of bags. Int. J. of General Systems 13, 23–37 (1986)

A Parallel Fusion Method for Heterogeneous Multi-sensor Transportation Data

Yingjie Xia[1,2], Chengkun Wu[3], Qingjie Kong[2], Zhenyu Shan[1], and Li Kuang[1]

[1] Hangzhou Institute of Service Engineering, Hangzhou Normal University, 310012 Hangzhou, P.R. China
[2] Department of Automation, School of Electronic, Information, and Electrical Engineering, Shanghai Jiao Tong University, 200240 Shanghai, P.R. China
[3] Manchester Interdisciplinary Biocentre, University of Manchester, M1 7DN Manchester, United Kingdom
{Xiayingjie,shanzhenyu,kuangli}@zju.edu.cn, chengkun.wu@postgrad.manchester.ac.uk, qjkong@sjtu.edu.cn

Abstract. Information fusion technology has been introduced for data analysis in intelligent transportation systems (ITS) in order to generate a more accurate evaluation of the traffic state. The data collected from multiple heterogeneous traffic sensors are converted into common traffic state features, such as mean speed and volume. Afterwards, we design a hierarchical evidential fusion model (HEFM) based on D-S Evidence Theory to implement the feature-level fusion. When the data quantity reaches a large amount, HEFM can be parallelized in data-centric mode, which mainly consists of region-based data decomposition by quadtree and fusion task scheduling. The experiments are conducted to testify the scalability of this parallel fusion model on accuracy and efficiency as the numbers of decomposed sub-regions and cyberinfrastructure computing nodes increase. The results show that significant speedups can be achieved without loss in accuracy.

Keywords: Information Fusion, Intelligent Transportation Systems, Cyber-infrastructure, Parallelization.

1 Introduction

The advanced traveler information system (ATIS) [1] is one of the most important traveling guide systems, which reaches the application of information technology in intelligent transportation systems (ITS). In ATIS, a large amount of transportation data is collected, processed and transmitted to agencies and travelers aiming at instant demonstration of the traffic state, automatic traffic control and guidance, etc. A primary goal of ATIS is to provide a real-time and accurate road network traffic state in large-scale urban regions.

Recently, more and more transportation data come from different types of traffic sensors, e.g., loop detectors [2, 3], probe vehicles [4], cameras [5] and cell phones [6]. Among those sensors, loop detectors and probe vehicles are widely used in urban settings, however both of them still have inherent drawbacks [7]. Two principle

V. Torra et al. (Eds.): MDAI 2011, LNAI 6820, pp. 31–42, 2011.

shortcomings for loop detector are high failure ratio and inaccurate conversion to traffic state features, and two primary disadvantages for probe vehicles are poor statistical representation and errors in the map-matching process. The heterogeneity of sensors and their respective drawbacks require a fusion on collected data to evaluate the traffic state more accurately.

Information fusion technology, which stems from military applications, has been introduced into ITS recently [8]. Its aim is to obtain a more accurate and comprehensive traffic state evaluation through combining data from multiple types of traffic sensors. As utilizing different sorts of input data, the information fusion can work at pixel-level, feature-level or decision-level [9]. In this paper, we propose a hierarchical evidential fusion model (HEFM) based on D-S Evidence Theory [11]. This model takes into account overcoming the deficiencies of Evidence Theory in case of conflict evidences, and its implementation can also be readily parallelized. The HEFM is designed to be a kind of feature-level fusion, and our first step is a conversion from the collected raw data to traffic state features, such as mean speed and volume.

Another grand challenge is brought up by the temporal complexity of fusing transportation data generated by a large number of sensors. Parallel computing technology is of great importance in order to reach the goal of accelerating computation which is required by our application context. The implementation is parallelized based on the division of region-aware transportation data. The whole computing task can be divided into a workload-balanced set of sub-tasks to feed the high performance computing infrastructure. The ultimate goal is to achieve a real-time fusion of data from heterogeneous multi-sensors in the traffic monitoring systems of urban road networks.

This paper is organized as follows: some related work on information fusion and the utilization of parallel computing in ITS is reviewed in Section 2; Section 3 proposes the hierarchical evidential fusion model and Section 4 presents its parallelized implementation; Experimental results on accuracy and efficiency are demonstrated in Section 5; Finally, a conclusion with remarks on future work is given in Section 6.

2 Related Work

Information fusion technology was firstly used in ITS in Sunmer's work [12], which shows its great significance. Since then, various methods have been presented. Cheu et al. [13] implemented a neural-network-based model for fusion and achieved good performance in traffic simulation. However, this model requires a large training set of real values, which would be infeasible in practice. Choi and Chung [14] designed a fusion algorithm based on fuzzy regression for estimating link travel time. The algorithm is over specialized to fit all links of road network. Automatic incident detection (AID) [15] and advanced driver assistance systems (ADAS) [16] are two other applications using information fusion technology for traffic management.

Parallel computing is a form of computation in which many calculations are carried out simultaneously. Different types of parallel computing, such as cluster computing, grid computing, and general purpose GPU computing, have been utilized in ITS. Most often related work includes parallel implementation of analysis and modeling of traffic flows [17], parallelized information retrieval on transportation data [18], and parallel implementation of a transportation network model [19], etc. To the best of our knowledge, there is little research work on the parallel fusion of transportation data. Whereas, such an effort becomes critical when we need to process a massive amount of data collected from heterogeneous multi-source traffic sensors. Therefore, to achieve accurate and real-time evaluation of traffic state, we design a fusion model specific for transportation data and parallelizing its implementation.

3 Hierarchical Evidential Fusion Model

Since our hierarchical evidential fusion model is based on D-S Evidence Theory, this section mainly consists of three parts: a brief introduction of D-S Evidence Theory, the architecture of HEFM and its embodied algorithm.

3.1 A Brief Introduction of D-S Evidence Theory

D-S Evidence Theory is a mathematical theory that allows to combine evidence from different sources and to arrive at a degree of belief and plausibility represented by a belief function and a plausibility function respectively. Let $\Omega=\{\omega_1,\omega_2,...,\omega_N\}$ be a set of discernment, in which all elements are assumed to be mutually exclusive and exhaustive. A Basic Probability Assignment (BPA) is a function m that is defined on the power set of Ω, $2^\Omega=\{A \mid A\subseteq\Omega\}$, and maps to values in $[0,1]$, such that $m(\Phi)=0$ where Φ denotes the empty set and $\sum_{A\subseteq\Omega} m(A)=1$.

The belief function (bel) and the plausibility function (pl) are then defined as the following functions on 2^Ω:

$$\begin{cases} bel(A)=\sum_{B\subseteq A} m(B) \\ pl(A)=\sum_{B\cap A\neq\Phi} m(B) \end{cases} \quad \forall A,B\subseteq\Omega \tag{1}$$

in which $bel(A)$ represents the sum of masses in all subsets of A, and $pl(A)$ corresponds to the sum of masses committed to those subsets, which do not discredit A.

Multiple evidences can be fused by using Dempster's combination rules as follows:

$$m(C)=\begin{cases} 0 & A\cap B=\Phi \\ \dfrac{1}{1-K}\sum_{A\cap B=C,\forall A,B\subseteq\Omega} m_i(A)\cdot m_j(B) & A\cap B\neq\Phi \end{cases} \tag{2}$$

where

$$K = \sum_{A \cap B = \Phi, \forall A, B \subseteq \Omega} m_i(A) \cdot m_j(B) \tag{3}$$

is the conflict factor between two evidences, i and j.

3.2 Architecture of HEFM

The architecture of HEFM is displayed in Fig. 1. It mainly consists of four parts: data collection, feature conversion, hierarchical evidential fusion and output. The source data are collected from multiple heterogeneous sensors, and converted into two traffic state features, mean speed and volume. Then the features are fused in hierarchy by D-S Evidence Theory for adapting to parallelization. The fusion part can be further divided into two levels, sub-fusion and main-fusion, which make use of the same algorithm on different input. Finally, the fusion results are used to evaluate the traffic states which are regarded as the output of HEFM. The architecture places its main difficulties on feature conversion and hierarchical evidential fusion, whose details will be specified in the following section.

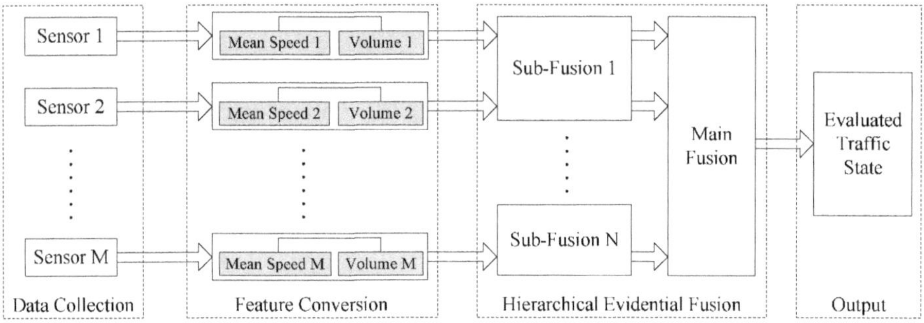

Fig. 1. Architecture of HEFM

3.3 Algorithm of HEFM

Feature Conversion Algorithm. The source transportation data are collected from two kinds of widely-used traffic sensors, loop detector and GPS. Their raw data can be converted into two main traffic state features, mean speed and volume. Therefore, the analysis of the feature conversion algorithm will be briefly presented on two features for two kinds of sensors.

(a) Mean Speed Estimation by Loop Detector
A method based on traffic wave theory [10] is used to estimate mean speed on loop detector data. The structure of the source data includes detector ID, phase, cycle, flow, saturation, and time occupancy. The length of the detector is 1.5 meters, and its collected data are uploaded once per cycle of red and green lights. The mean speed $\overline{v}$ along a link of its length L and vehicle volume per cycle q can be calculated as follows:

The time for the i th vehicle passing the whole link is

$$t_i = (L - i \cdot l_v) / u_1 + (i \cdot l_v + l_c) / u_2 + t_r \tag{4}$$

where l_v and l_c are the mean length of the vehicle and the loop detector respectively, u_1 and u_2 represent the respective mean speed of the vehicle before entering the queue and driving away from the queue, t_r is the time length of a red light period. All the variables can be derived or calculated from source data. Therefore, the mean speed can be calculated by

$$\overline{v} = q \cdot L / \sum_{i=1}^{q} t_i \tag{5}$$

(b) Volume Estimation by Loop Detector

A single loop detector can estimate the volume m, the number of vehicles that pass the detector during a fixed sample period, as follows:

$$m(i, j) = \frac{n(i, j)}{T} \tag{6}$$

where i indexes lane, j indexes the sample time period, $n(i, j)$ represents the number of vehicles that pass over the detector in lane i during time period j, and T is the sampling period. Therefore, a road with multiple lanes can calculate its volume by summing up volumes of all contained lanes.

(c) Mean Speed Estimation by GPS

The source data collected from GPS-equipped probe vehicles can also be used to estimate mean speed. The structure of the data includes vehicle ID, position coordinates, time, velocity, moving direction, etc. We take three processing steps on source data: coordinates transforming, map matching and curve approximating. The main algorithm lies in the step of curve approximating. The mean speed $\overline{v}_0$ along the link of its length L at time t_0 can be calculated by the following equation:

$$\overline{v}_0(t_0) = (\int_0^L v(l,t)|_{t=t_0} \, dl) / L \tag{7}$$

where $v(l,t)$ represents the speed at space l and time t which can be derived from GPS. Therefore, the spatial-temporal mean speed along the whole link during a time period from $(k-1)T$ to kT can be calculated by

$$\overline{v} = (\int_{(k-1)T}^{kT} \overline{v}_0(t) dt) / L \tag{8}$$

(d) Volume Estimation by GPS

The traffic volume collected by GPS is just a sampling of real volume. Through classifying the links based on their attributes of region, width, etc. and assuming that links in the same class have similar scale between real volume and GPS-collected volume, we can estimate the real volume m of link i_c as follows:

$$m(i_c) = \frac{m_r(j_c)}{g(j_c)} \cdot g(i_c) \tag{9}$$

where $g(i_c)$ and $g(j_c)$ are sampling volumes from GPS along the links i_c and j_c both of which belong to the same class c, and $m_r(j_c)$ is the real volume of one selected link j_c which can be easily estimated by some other methods, such as camera or manual work.

Hierarchical Evidential Fusion Algorithm. Since heterogeneous sensors can estimate the traffic state features with different accuracy and reliability, we use fusion to overcome the conflicts of evidences and compensate for the deficiencies of sensors mutually. Both hierarchies use the same algorithm of D-S Evidence Theory with different scales of input data. The algorithm is as follows:

$$m(C_t) = m_1(A_{1,t}) \oplus m_2(A_{2,t}) \cdots \oplus m_X(A_{X,t}) = \frac{\sum_{\hbar_{i=1}^{X} A_{i,t} = C_t} \left(\prod_{i=1}^{X} m_i(A_{i,t}) \right)}{1 - \sum_{\hbar_{i=1}^{X} A_{i,t} = \Phi} \left(\prod_{i=1}^{X} m_i(A_{i,t}) \right)} \tag{10}$$

where $m(C_t)$ denotes the integrated result of the fusion system at time t, and $m_i(A_{i,t})$, $i = 1, 2, ..., X$ represents the BPA extracted from the data collected by *ith* sensor at time t.

The sub-fusion system deals with the sensor data in region-based divisions, and main fusion system integrates the results from the sub-fuison systems, especially much care is taken about links in boundary regions. Finally, we can evaluate the traffic state following a certain decision rule, such as maximum belief or maximum plausibility.

4 Parallelized Implementation

The region-based division for hierarchical evidential fusion inherently enables an efficient parallelized implementation of HEFM. This is a data-centric parallelization problem, which can be efficiently solved by data decomposition. Specifically, a Morton ordered quadtree [20] is constructed to decompose regions full of links and produce adjustable, scalable fusion tasks that are sensitive to underlying data distributions. The tasks are scheduled among cyberinfrastructure resources for workload balancing. The scheduling strategy takes into account comprehensive consideration of the computing capacity of each resource and the computation intensity of each task. Supposing that the resources can be viewed as with the same capacity, the region-based divisions should be decomposed following the rule of equal computation intensity.

4.1 Quadtree

A quadtree is a representation of a regular partition of space where regions are split recursively until achieving a tradeoff between computation intensity and the number of divisions. Each quadtree division, also referred to as block or cell, always covers a portion of space that forms a quad. Various quadtrees have been defined, differing in the rules that govern data decomposition, the type of data being indexed, and other details [21]. Quadtree algorithms have also been implemented in parallel [22] though not in the specific cyberinfrastructure.

A basic quadtree in two-dimensional space is a *4-way* branching tree that represents a recursive decomposition of space wherein at each level a square subspace is divided into four equal-size squares. By traversing all the leaf nodes of the quadtree, the decomposed divisions can be linked as an ordered linear list. Actually, only leaf nodes are stored in the list because they contain all the required information to support the flexible region-based decomposition of traffic feature data.

4.2 Region-Based Decomposition of Traffic Feature Data

The traffic feature data, mean speed and volume, are associated with geographical information which represents the coordinates of corresponding links. Therefore, a quadtree- and region-based decomposition can be applied to traffic feature data, and it can produce scalable geographical workloads which can be allocated to available computational resource, like a cluster. The algorithm, *Quad_Decompose*, addresses the decomposition challenges which focus on finding efficient data partitions that are assigned to each cluster node as the running task. In *Quad_Decompose*, the number of links at each quadtree node is used to determine the level of recursive division. When the algorithm executes, those decomposed regions with higher densities of traffic feature data are recursively decomposed until the density reaches a specified threshold. The threshold is determined by evaluating the specific computation intensity and computing capacity. Fig. 2 shows an example of decomposing a region by quadtree based on the intensity of fusing traffic feature data in that region. The threshold is set as 40 links, and each sub-region is tagged with *A-B* where *A* represents the level number of recursive division and *B* stands for the number of links in that sub-region.

Although the region is easily decomposed, a link can not be broken off between two neighbored sub-regions when evaluating its traffic state. In our approach, we duplicate the data of that kind of links, and conduct their fusion redundantly. This is because we plan to continue our research on coordinative traffic state affection of neighboring links which can be used to filter the exceptional data. The duplication

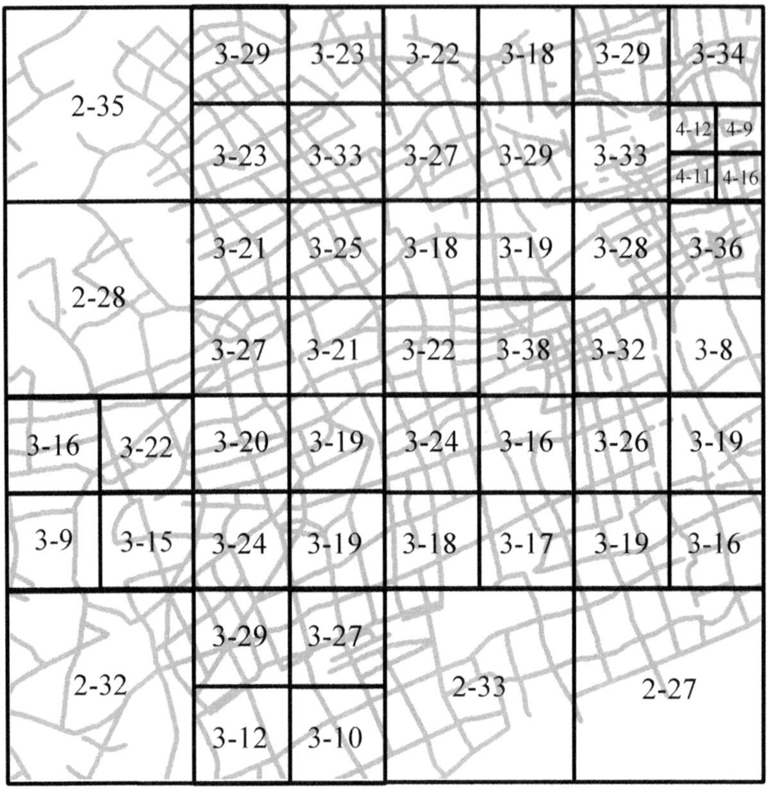

Fig. 2. Fusion-Intensity-Based Region Decomposition by Quadtree

makes the data in sub-regions more complete and well prepared for this future research work.

4.3 Fusion Task Scheduling

Dynamic fusion task scheduling is infeasible because of the unpredictable nature of network traffic and job queues on the cluster nodes. Consequently, a static strategy has been adopted to schedule fusion tasks according to region-based decomposition. In our research, the cluster has a homogeneous architecture which means that each cluster node holds the same computing capacity, so the scheduling can be easily done by allocating each node to fuse the traffic feature data in each sub-region.

In HEFM, there are two levels of fusion, sub-fusion and main-fusion. Multiple sub-fusion systems, which take charge of fusing data in sub-regions, are distributed to slave nodes of cluster. The sub-fusion results are fused by main-fusion system, which runs on master node of cluster. Since we duplicate boundary data when decomposing regions, the main-fusion system does not need to consider the inconsistency issue of boundary-thru links. The scheduling can be implemented by Globus Resource Allocation Management (GRAM) [23] package deployed on our cluster, and achieve workload-balance among all nodes.

5 Experimental Results and Analysis

5.1 Experiments Setup

The experiments are carried out on the cluster of Zhejiang University Campus Grid (ZJU-CG), which combines high performance computing resources in the university, and provides a platform for solving scientific computation problems. The fusion tasks on the data divisions are simultaneously executed on 24 computing nodes of Dawning TC4000L, each with Intel Xeon Dual Core 2.4GHz, 2G memory and Redhat Linux 9.0 OS. As a test case, the region of Shanghai downtown with 393 roads is decomposed into three different granularities, 16, 64 and 256 sub-regions, which are implemented as different depths of the quadtree. The transportation data of GPS and Scats loop detectors on those roads are collected every 20 seconds from 8:00 to 17:00, and transformed into mean speed and volume. The tests are run on 1 node, 4 nodes, 16 nodes and 24 nodes respectively. Both accuracy and efficiency for parallel fusion of mean speed and volume on heterogeneous multi-sensor transportation data are evaluated in the experiments.

5.2 Results and Analysis

Accuracy. The parallel fusion on a large amount of ITS data is implemented as the data-centric parallelization, which is arguably the easiest parallel strategy to adopt for migrating from serial to parallel programming. Since data-centric parallelization focuses on applying the algorithm to multiple divisions concurrently, the accuracy of the fusion results of both traffic state features will not be lost, especially supported by duplicating boundary data.

Efficiency. The efficiency experiments are conducted under 16, 64 and 256 decompositions of the region by different numbers of computing nodes. We set the measurement T as the reciprocal of computing time t,

$$T = 30\,\text{min}/t, \tag{11}$$

and evaluate its speedups through fusion parallelization for mean speed and volume respectively. The experimental results are shown in Fig. 3.

In (a), (b) and (c) of Figure 3, T increases as the number of computing nodes increases, to represent their efficiency speedups, while by using 16 and 24 nodes for fusing traffic data of 16 sub-regions their efficiency is similar. This is because the amount of nodes greater than 16 is large enough allocated for all sub-regions whose scalability has reached its ceiling limit.

According to the definition of T, the average slope of the curve stands for the speedup rate of computing efficiency. In experiments among 16, 64 and 256 sub-regions, as the amount of sub-regions increases, the curve becomes flat which means that the speedup rate decreases. This is caused by the incremental overhead for boundary duplication of decompositions and their main fusion.

Since all source data of volume are of the integer type while most source data of mean speed are of float type, and the computation complexity for a float variable is

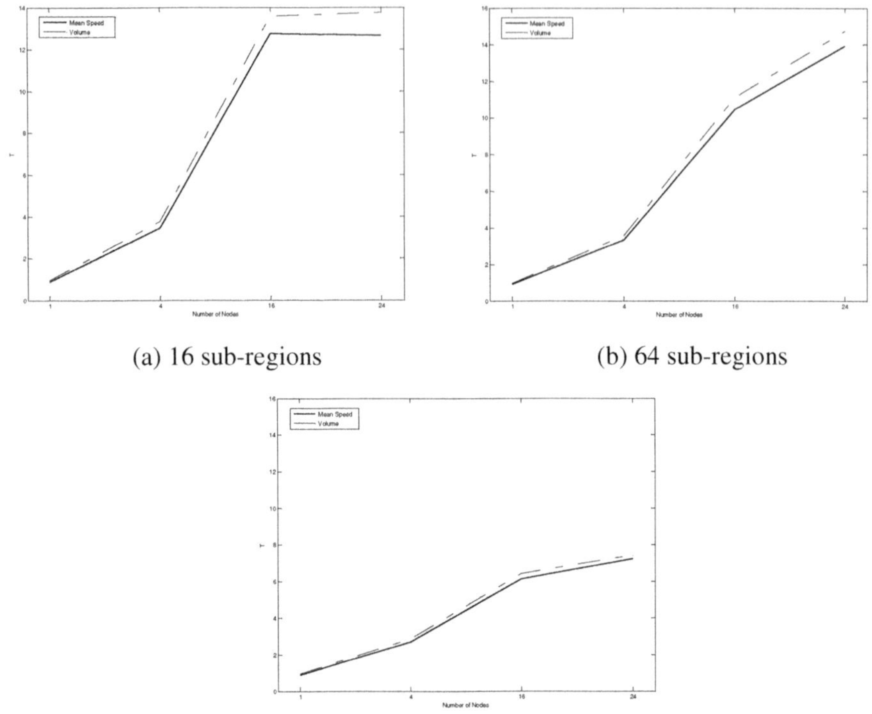

(a) 16 sub-regions (b) 64 sub-regions

(c) 256 sub-regions

Fig. 3. Efficiency Experiments under Multiple Setups

slightly bigger than for an integer variable, therefore T values of volume fusion are always a little bit greater than those of mean speed fusion in all (a), (b) and (c) of Figure 3.

6 Conclusions and Future Work

The general goal of the research reported in this paper was to investigate the performance of a parallelized hierarchical evidential fusion model based on D-S Evidence Theory for ITS data. The data collected by multiple heterogeneous sensors, such as Scats loop detectors and GPS probe vehicles, were converted to two common traffic state features: mean speed and volume. A large amount of traffic data were decomposed based on regions through the quadtree algorithm, and a static task scheduling strategy was developed and evaluated when used to implement data-centric parallel fusion on cyberinfrastructure computing resources. A cluster with its scheduling package GRAM, was used to conduct the experiments on the region of Shanghai downtown with 393 roads.

The results showed that for a dataset with region-based decomposition and data-centric parallelization, the fusion accuracy can be maintained as the number of decomposed sub-regions changes. Moreover, in terms of computing efficiency, its speedup scaled well when more computing nodes are available. As expected, the same amount of nodes allocated to more sub-regions reduced the speedup rate for more overhead of duplication and main fusion. Consequently, significant speedups were achieved for the implementation of our parallel fusion.

Future research will investigate not only data-centric parallelization, but also algorithm-centric parallelization for ITS data fusion. We will also examine some other traffic state features, such as length of vehicle queue and pedestrian amount. Finally, further research and experiments will be conducted to evaluate a logical quad-tree using adjustable geographical bounder because it is anticipated to overcome the drawback of duplicated calculations on the roads across the boundaries of decomposed quads.

Acknowledgments. This research work is supported by National Natural Science Foundation of China under grant number 61002009, Science and Technology Planning Project of Zhejiang Province under grant number 2010C31018, and Scientific Research Fund of Hangzhou Normal University under grant number HSKQ0042. The authors also do appreciate the helpful assistance from Center for Service Engineering in Hangzhou Normal University.

References

1. Kumar, P., Singh, V., Reddy, D.: Advanced Traveler Information System for Hyderabad City. IEEE Transactions on Intelligent Transportation Systems 6(1), 26–37 (2005)
2. Dailey, D.: A Statistical Algorithm for Estimating Speed from Single Loop Volume and Occupancy Measurements. Transportation Research Part B: Methodological 33(5), 313–322 (1999)
3. Coifman, B.: Improved Velocity Estimation using Single Loop Detectors. Transportation Research Part A: Policy and Practice 35(10), 863–880 (2001)
4. Quiroga, C.A., Bullock, D.: Travel Time Studies with Global Positioning and Geographic Information Systems: An Integrated Methodology. Transportation Research Part C: Emerging Technologies 6(1/2), 101–127 (1998)
5. Cho, Y., Rice, J.: Estimating Velocity Fields on a Freeway from Low-resolution Videos. IEEE Transactions on Intelligent Transportation Systems 7(4), 463–469 (2006)
6. Sohn, K., Hwang, K.: Space-Based Passing Time Estimation on a Free-Way using Cell Phones as Traffic Probes. IEEE Transactions on Intelligent Transportation Systems 9(3), 559–568 (2008)
7. El Faouzi, N.E., Lefevre, E.: Classifiers and Distance-Based Evidential Fusion for Road Travel Time Estimation. In: Dasarathy, B.V. (ed.) Multisensor, Multisource Information Fusion: Architectures, Algorithms, and Applications 2006. SPIE, vol. 6242, pp. 1–16. SPIE, Bellingham (2006)
8. El Faouzi, N.E.: Data Fusion in Road Traffic Engineering: An Overview. In: Dasarathy, B.V. (ed.) Multisensor, Multisource Information Fusion: Architectures, Algorithms, and Applications 2004. SPIE, vol. 5434, pp. 360–371. SPIE, Bellingham (2004)
9. Steinberg, A.N., Bowman, C.L., White, C.E.: Revisions to the JDL Data Fusion Model. In: Dasarathy, B.V. (ed.) Multisensor, Multisource Information Fusion: Architectures, Algorithms, and Applications 1999. SPIE, vol. 3719, pp. 430–441. SPIE, Bellingham (1999)

10. Gartner, N.H., Messer, C., Rathi, A.K.: Monograph on Traffic Flow Theory. Fed. Highway Admin. (1996)
11. Murphy, R.R.: Dempster-Shafer Theory for Sensor Fusion in Autonomous Mobile Robots. IEEE Transactions on Robotics and Automation 14(2), 197–206 (1998)
12. Sumner, R.: Data Fusion in PathFinder and TravTek. In: 2nd IEEE Vehicle Navigation and Information Systems Conference, pp. 71–75. IEEE Press, New York (1991)
13. Cheu, R.L., Lee, D.H., Xie, C.: An Arterial Speed Estimation Model Fusing Data from Stationary and Mobile Sensors. In: 4th IEEE International Conference on Intelligent Transportation Systems, pp. 573–578. IEEE Press, New York (2001)
14. Choi, K., Chung, Y.: A Data Fusion Algorithm for Estimating Link Travel Time. Journal of Intelligent Transportation Systems: Technology, Planning, and Operations 7(3/4), 235–260 (2002)
15. Klein, L.: Dempster-Shafer Data Fusion at the Traffic Management Center. In: 79th Annual Transportation Research Board Meeting, Washington, Paper No. 00–1211 (2000)
16. Hellinga, B.R., Fu, L.P.: Reducing Bias in Probe-Based Arterial Link Travel Time Estimates. Transportation Research Part C: Emerging Technologies 10(4), 257–273 (2002)
17. Nagel, K., Rickert, M.: Parallel Implementation of the TRANSIMS Micro-Simulation. Parallel Computing 27(12), 1611–1639 (2001)
18. Krishnan, R., Hodge, V., Austin, J., Polak, J.W.: A Computationally Efficient Method for Online Identification of Traffic Control Intervention Measures. In: 42nd Annual Meeting of the Universities Transport Study Group, pp. 1–11. Plymouth (2010)
19. O'Cearbhaill, E.A., O'Mahony, M.: Parallel Implementation of a Transportation Network Model. Journal of Parallel and Distributed Computing 65(1), 1–14 (2005)
20. Samet, H.: Applications of Spatial Data Structures. Addison Wesley, MA (1990)
21. Samet, H.: The Quadtree and Related Hierarchical Data Structures. ACM Computing Surveys 20(4), 187–260 (1984)
22. Hoel, E.G., Samet, H.: Data-Parallel Primitives for Spatial Operations. In: 24th International Conference on Parallel Processing, Oconomowoc, pp. 184–191 (1995)
23. Nabrzyski, J., Schopf, J.M., Weglarz, J.: Grid Resource Management. Kluwer Publishing, Netherlands (2003)

A Dynamic Value-at-Risk Portfolio Model

Yuji Yoshida

Faculty of Economics and Business Administration, University of Kitakyushu
4-2-1 Kitagata, Kokuraminami, Kitakyushu 802-8577, Japan
yoshida@kitakyu-u.ac.jp

Abstract. A mathematical dynamic portfolio allocation model with uncertainty is discussed. Introducing a value-at-risk under a condition, this paper formulates value-at-risks in a dynamic stochastic environment. By dynamic programming approach, an optimality condition of the optimal portfolio for dynamic value-at-risks is derived. It is shown that the optimal time-average value-at-risk is a solution of the optimality equation under a reasonable assumption, and an optimal trading strategy is obtained from the equation. A numerical example is given to illustrate our idea.

1 Introduction

From the financial crisis in October 2008, we have learned the importance of the estimation regarding the drastic decline of asset prices in the market. The criterion and stable portfolio technique for dynamical drastic declines are major topics in financial application. In this paper, we present a risk optimal allocation portfolio model, and then we need to discuss the aggregation of risk criteria over time.

Portfolio is very useful for hedging the risk in asset management finance and it is used to make asset management stable. As a classical portfolio theory, *Markowitz's mean-variance model* is studied by many researchers and fruitful results have been achieved, and the variance-minimizing is also important to minimize the risk in portfolio ([9,12,14,15]). Recently, *value-at-risk (VaR)* is used widely in finance to estimate the risk of worst-scenarios. VaR is a risk-sensitive criterion based on percentiles, and it is one of the standard criteria in asset management ([11]). VaR is a kind of risk values of the asset prices at a specified risk-level probability and it is for selecting portfolios to get rid of bad scenarios in investment. VaR is also strongly related to the bankruptcy and the falling in the asset prices ([7]). Many researchers and financial traders usually use VaR by numerical approximations since it is not easy to analyze the VaR portfolios mathematically ([11]). The difficulty of the analysis comes from the properties of the criterion. Because Markowitz's mean-variance criterion and variance-minimizing criterion are represented by quadratic programming, but VaR criterion in portfolio is neither linear nor quadratic. In this paper, a *dynamic VaR portfolio selection problem model* is proposed in order to optimize both of VaR and the expected rates of return. In the proposed portfolio model,

V. Torra et al. (Eds.): MDAI 2011, LNAI 6820, pp. 43–54, 2011.

owing to VaR we can maximize the expected rate of return after due consideration of the worst-scenarios. This paper derives analytical solutions for the VaR portfolio problem under uncertainty. The risk criterion is composed by the sum of unexpected short-term risks which occur suddenly in each period. Introducing value-at-risk on a condition, we derive the optimality equation and the optimal trading strategies for the dynamic model by dynamic programming.

In the next section, we introduce a dynamic portfolio model. In Section 3, this paper formulates value-at-risks in a dynamic stochastic environment and we introduce a value-at-risk under a condition for the optimization problem. In Section 4, we discuss a portfolio optimization for dynamic value-at-risks and its computation. Finally, in the last section, a numerical example is given to illustrate our idea.

2 A Dynamic Portfolio Model

In this section, we explain a portfolio model with n stocks, where n is a positive integer. Let $\mathbb{T} := \{0, 1, 2, \cdots, T\}$ be the time space with an expiration date T, and $\mathbb{R}$ denotes the set of all real numbers. Let (Ω, P) be a probability space, where P is a non-atomic probability on a sample space Ω. For an asset $i = 1, 2, \cdots, n$, a *stock price process* $\{S_t^i\}_{t=0}^T$ is given by *rates of return* R_t^i as follows. Let

$$S_t^i := S_{t-1}^i (1 + R_t^i) \tag{1}$$

for $t = 1, 2, \cdots, T$, where $\{R_t^i\}_{t=1}^T$ is assumed to be an integrable sequence of independent real random variables. Hence $w_t = (w_t^1, w_t^2, \cdots, w_t^n)$ is called a *portfolio weight vector* if it satisfies $w_t^1 + w_t^2 + \cdots + w_t^n = 1$, and further a portfolio $(w_t^1, w_t^2, \cdots, w_t^n)$ is said to *allow for short selling* if $w_t^i \geq 0$ for all $i = 1, 2, \cdots, n$. Then the rate of return with a portfolio $(w_t^1, w_t^2, \cdots, w_t^n)$ is given by

$$R_t := w_t^1 R_t^1 + w_t^2 R_t^2 + \cdots + w_t^n R_t^n. \tag{2}$$

Therefore, the reward at time $t = 1, 2, \cdots, T$ follows

$$S_t := S_{t-1} \sum_{i=1}^n w_t^i (1 + R_t^i) = S_{t-1}(1 + R_t). \tag{3}$$

In this paper, we present a dynamic portfolio model for stock price processes $\{S_t^i\}_{t=1}^T$. The falling of asset prices is one of the most important risks in stock markets. In this section, we discuss a portfolio model where the risk is estimated by the rate of falling. Regarding the asset (3) with the portfolio w_t, the theoretical *bankruptcy* at time t occurs on scenarios ω satisfying $S_t(\omega) \leq 0$, i.e. it follows $1 + R_t(\omega) \leq 0$ from (3). Similarly, for a constant δ satisfying $0 \leq \delta \leq 1$, a set of sample paths

$$\{\omega \in \Omega \mid 1 + R_t(\omega) \leq 1 - \delta\} = \{\omega \in \Omega \mid R_t(\omega) \leq -\delta\} \tag{4}$$

is the event of scenarios where the asset price S_t will fall from the current price S_{t-1} to a lower level than $100(1-\delta)\,\%$ of the current price S_{t-1}, i.e. the rate of falling is $100\,\delta\,\%$. The parameter δ is called *the rate of falling*. Then the probability of falling is also given by

$$p_\delta := P(R_t \le -\delta). \tag{5}$$

For example, p_δ denotes the probability of the falling below par value if '$\delta = 0$' and it indicates the probability of the bankruptcy if '$\delta = 1$'. In this paper, we discuss dynamic portfolios regarding the rate of falling δ.

For a positive probability p, a value-at-risk (VaR) regarding the rate of return R_t at the probability p is given by a real number v satisfying

$$P(R_t \le v) = p \tag{6}$$

since P is non-atomic. The value-at-risk v is the upper bound of the rate of return R_t at the worst scenarios under a given risk probability p, and then the value-at-risk v in (6) is denoted by $\mathrm{VaR}_p(R_t)$. From (5) and (6), for a risk probability $p = p_\delta$, the rate of falling is

$$\delta = -\mathrm{VaR}_p(R_t). \tag{7}$$

To minimize the rate of falling derived from (7) under a random environment, in next section we discuss the fundamental properties of value-at-risks. In this paper, we deal with a portfolio model where the value-at-risk v in (6) has the following representation.

$$(\mathrm{VaR}\ v) = (\text{the mean}) - (\text{a positive constant } \kappa) \times (\text{the standard deviation}), \tag{8}$$

where the positive constant κ is given corresponding to the probability p (Fig.1). One of the most popular sufficient condition for (8) is what the distribution of the rate of return R_t is Gaussian ([3,11]).

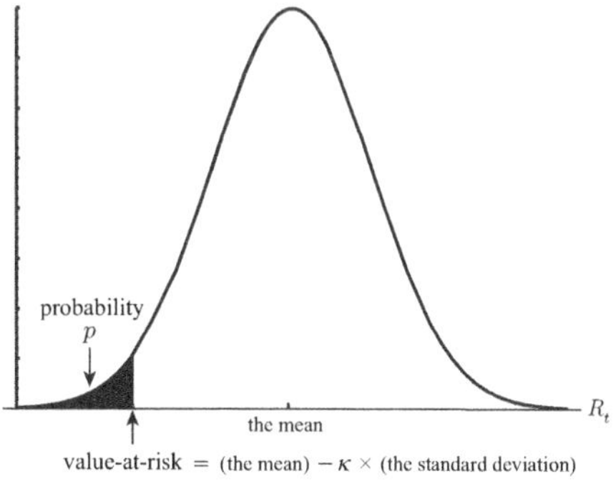

Fig. 1. Value-at-risk v at a probability p

3 Value-at-Risks in a Dynamic Stochastic Environment

First we introduce mathematical notations of value-at-risk for real random variables to apply it to the rates of return (2). Let $\mathcal{X}$ be the set of all integrable real random variables X on Ω with a continuous distribution function $x \mapsto F_X(x) := P(X < x)$ for which there exists a non-empty open interval I such that $F_X(\cdot) : I \mapsto (0,1)$ is a strictly increasing and onto. Then there exists a strictly increasing and continuous inverse function $F_X^{-1} : (0,1) \mapsto I$. We note that $F_X(\cdot) : I \mapsto (0,1)$ and $F_X^{-1} : (0,1) \mapsto I$ are one-to-one and onto, and we put $F_X(\inf I) := \lim_{x \downarrow \inf I} F_X(x) = 0$ and $F_X(\sup I) := \lim_{x \uparrow \sup I} F_X(x) = 1$. Then, the *value-at-risk (VaR)* at a risk probability p is given by the $100p$-percentile of the distribution function F_X.

$$\mathrm{VaR}_p(X) := \begin{cases} \inf I & \text{if } p = 0 \\ \sup\{x \in I \mid F_X(x) \leq p\} & \text{if } 0 < p < 1 \\ \sup I & \text{if } p = 1. \end{cases} \tag{9}$$

Then we have $F_X(\mathrm{VaR}_p(X)) = p$ and $\mathrm{VaR}_p(X) = F_X^{-1}(p)$ for $0 < p < 1$. The following preliminary results are important when we apply the value-at-risk to the rates of return (2).

Lemma 1. ([20]). *Let $X, Y \in \mathcal{X}$ and let p be a positive probability. Then the value-at-risk* VaR_p *defined by (9) has the following properties.*

(i) *If $X \leq Y$, then* $\mathrm{VaR}_p(X) \leq \mathrm{VaR}_p(Y)$.
(ii) $\mathrm{VaR}_p(\zeta X) = \zeta\, \mathrm{VaR}_p(X)$ *for $\zeta > 0$.*
(iii) $\mathrm{VaR}_p(X + \theta) = \mathrm{VaR}_p(X) + \theta$ *for $\theta \in \mathbb{R}$.*

From Eq. (3), the value-at-risk for the reward S_t at time t is given by

$$\mathrm{VaR}_p(S_t) = \mathrm{VaR}_p\left(S_{t-1} \sum_{i=1}^{n} w_t^i (1 + R_t^i)\right) = \mathrm{VaR}_p\left(S_{t-1}(1 + R_t)\right). \tag{10}$$

To discuss (10), we introduce a value-at-risk based on conditional expectations. Let $\mathcal{G}$ be a sub-σ-field of $\mathcal{M}$. Define a map $x \mapsto F_X(x \mid \mathcal{G}) := P(X < x \mid \mathcal{G}) = E(1_{\{X<x\}} \mid \mathcal{G})$. We define a *value-at-risk of $X (\in \mathcal{X})$ under a condition $\mathcal{G}$* at a risk probability p by

$$\mathrm{VaR}_p(X \mid \mathcal{G}) := \begin{cases} \inf I & \text{if } p = 0 \\ \sup\{x \in I \mid F_X(x \mid \mathcal{G}) \leq p\} & \text{if } 0 < p < 1 \\ \sup I & \text{if } p = 1. \end{cases} \tag{11}$$

Then we note that $\mathrm{VaR}_p(X \mid \mathcal{G})$ is a random variable adapted to the σ-field $\mathcal{G}$ since $\{\omega \mid \mathrm{VaR}_p(X \mid \mathcal{G})(\omega) > y\} = \bigcup_{x \in \mathbb{Q}: x > y} \{\omega \mid F_X(x \mid \mathcal{G})(\omega) \leq p\} \in \mathcal{G}$ for all $y \in \mathbb{R}$ from the continuity of the map $x \mapsto F_X(x \mid \mathcal{G})$. We also have $F_X(\mathrm{VaR}_p(X \mid \mathcal{G}) \mid \mathcal{G}) \leq p$ for $0 < p < 1$. The value-at-risk under the condition $\mathcal{G}$ has the same properties as Lemma 1, which are listed in the following lemma.

Lemma 2. *Let $\mathcal{G}$ be a sub-σ-field of $\mathcal{M}$. Let p be a probability satisfying $0 < p < 1$ and let $X, Y \in \mathcal{X}$. Then,* $\mathrm{VaR}_p(\cdot \mid \mathcal{G})$ *defined by (11) has the following properties:*

(i) *If $X \leq Y$, then* $\mathrm{VaR}_p(X \mid \mathcal{G}) \leq \mathrm{VaR}_p(Y \mid \mathcal{G})$.
(ii) $\mathrm{VaR}_p(\zeta X \mid \mathcal{G}) = \zeta\, \mathrm{VaR}_p(X \mid \mathcal{G})$ *for* $\zeta \geq 0$.
(iii) $\mathrm{VaR}_p(X + \theta \mid \mathcal{G}) = \mathrm{VaR}_p(X \mid \mathcal{G}) + \theta$ *for* $\theta \in \mathbb{R}$.

Next the following lemma shows particular properties for the value-at-risk $\mathrm{VaR}_p(\cdot \mid \mathcal{G})$ under the condition $\mathcal{G}$.

Lemma 3. *Let $X, Y, Z \in \mathcal{X}$ be random variables such that Y and Z are independent. Let $\mathcal{G}$ be a sub-σ-field of $\mathcal{M}$ such that $\mathcal{G} := \sigma(Z)$, where $\sigma(Z)$ is the σ-field generated by the random variable Z. Let p be a probability satisfying $0 < p < 1$. Then,* $\mathrm{VaR}_p(\cdot \mid \mathcal{G})$ *defined by (11) has the following properties:*

(i) $\mathrm{VaR}_p(Y \mid \mathcal{G}) = \mathrm{VaR}_p(Y)$.
(ii) $\mathrm{VaR}_p(Z \mid \mathcal{G}) = Z$.
(iii) $\mathrm{VaR}_p(ZX \mid \mathcal{G}) = Z\mathrm{VaR}_p(X \mid \mathcal{G})$ *if* $Z \geq 0$.
(iv) $\mathrm{VaR}_p(X + Z \mid \mathcal{G}) = \mathrm{VaR}_p(X \mid \mathcal{G}) + Z$.

In Eq. (2), portfolio weights $w_t = (w_t^1, w_t^2, \cdots, w_t^n)$ are decided sequentially and predictably. We note that the risk of S_t is related to the information $\mathcal{M}_{t-1}$ up to time $t-1$. Then, the value-at-risk of S_t under information $\mathcal{M}_{t-1}$ at a probability level p is

$$\begin{aligned} \mathrm{VaR}_p(S_t \mid \mathcal{M}_{t-1}) &= \mathrm{VaR}_p\left(S_{t-1} \sum_{i=1}^{n} w_t^i (1 + R_t^i) \mid \mathcal{M}_{t-1} \right) \\ &= \mathrm{VaR}_p\left(S_{t-1}(1 + R_t) \mid \mathcal{M}_{t-1} \right). \end{aligned} \tag{12}$$

The term (12) means the risk of worst scenarios which occur on the transition from time $t-1$ to time t. Therefore, taking the sum of the risks which occur at each time, this paper deals with the following dynamic portfolio problem regarding the total of value-at-risks (12) under information $\{\mathcal{M}_{t-1}\}_{t=1}^{T}$. Let β be a constant satisfy $0 < \beta \leq 1$.

Dynamic Portfolio Problem 1 (D1): Maximize the total value-at-risk

$$E\left(\sum_{t=1}^{T} \beta^{t-1} \mathrm{VaR}_p(S_t \mid \mathcal{M}_{t-1}) \right) \tag{13}$$

with portfolio weights $w_t = (w_t^1, w_t^2, \cdots, w_t^n)$ satisfying $w_t^1 + w_t^2 + \cdots + w_t^n = 1$ and $w_t^i \geq 0$ $(i = 1, 2, \cdots, n; t = 1, 2, \cdots, T)$.

In financial systems, one of the most important topics is how to aggregate the risks at each time. We note that the criterion (13) is different from

$$\sum_{t=1}^{T} \beta^{t-1} \mathrm{VaR}_p(S_t). \tag{14}$$

The risk criterion (13) is composed by the sum of unexpected short-term risks which occur suddenly in each period from time $t-1$ to time t. On the other hand, the risk criterion (14) is composed by the sum of the long-term risks in the period from the beginning up to time t. Then, since $\mathrm{VaR}_p(S_t)$ may contain not only the risk at time t but also the potential risks up to time $t-1$, the sum $\sum_{t=1}^{T} \beta^{t-1} \mathrm{VaR}_p(S_t)$ in (14) estimates the potential risks multiple times. Therefore in this paper we use the criterion (13). Now we represent the *total value-at-risk* (13) by

$$A(X_1, X_2, \cdots, X_T) := E\left(\sum_{t=1}^{T} \beta^{t-1} \mathrm{VaR}_p(X_t \mid \mathcal{M}_{t-1}) \right) \tag{15}$$

for $(X_1, X_2, \cdots, X_T) \in \mathcal{X}^T$. The total value-at-risk (13) can be seen as an aggregation of the value-at-risks $\mathrm{VaR}_p(X_t \mid \mathcal{M}_{t-1})$ of X_t.

Proposition 1. *The total value-at-risk $A(X_1, X_2, \cdots, X_T)$ is defined by (15) has the following properties: For $(X_1, X_2, \cdots, X_T), (Y_1, Y_2, \cdots, Y_T) \in \mathcal{X}^T$,*

(i) *If $X_t \le Y_t$ for all $t = 1, 2, \cdots .T$, then $A(X_1, X_2, \cdots, X_T) \le A(Y_1, Y_2, \cdots, Y_T)$.*
(ii) *$A(\zeta X_1, \zeta X_2, \cdots, \zeta X_T) = \zeta\, A(X_1, X_2, \cdots, X_T)$ for $\zeta \ge 0$.*
(iii) *$A(X_1 + \theta, X_2 + \theta, \cdots, X_T + \theta) = A(X_1, X_2, \cdots, X_T) + \theta(1-\beta^T)/(1-\beta)$ for $\theta \in \mathbb{R}$.*

For simplicity we take $S_0 = 1$, and then, by Lemmas 2 and 3, Dynamic Portfolio Problem 1 (D1) is reduced to the following problem.

Dynamic Portfolio Problem 2 (D2): Maximize the total value-at-risk

$$\begin{aligned} &\sum_{t=1}^{T} \beta^{t-1} \prod_{s=1}^{t-1} (1 + E(R_s)) \cdot (1 + \mathrm{VaR}_p(R_t)) \\ &= \sum_{t=1}^{T} \beta^{t-1} \prod_{s=1}^{t-1} \left(1 + E\left(\sum_{i=1}^{n} w_s^i R_s^i \right) \right) \cdot \left(1 + \mathrm{VaR}_p \left(\sum_{i=1}^{n} w_t^i R_t^i \right) \right) \end{aligned} \tag{16}$$

with portfolios $w_t = (w_t^1, w_t^2, \cdots, w_t^n)$ satisfying $w_t^1 + w_t^2 + \cdots + w_t^n = 1$ and $w_t^i \ge 0$ $(i = 1, 2, \cdots, n; t = 1, 2, \cdots, T)$.

Define the set of portfolios by $\mathcal{W} := \{w_t = (w^1, w^2, \cdots, w^n) \in \mathbb{R}^n \mid w^1 + w^2 + \cdots + w^n = 1 \text{ and } w^i \ge 0\, (i = 1, 2, \cdots, n)\}$.

Theorem 1. *The optimal VaR for (16) in Dynamic Portfolio Problem 2 (D2) is given by v_1 which is defined inductively by the sequence $\{v_t\}$ of sub-total-sum value-at-risks after time $t-1$ satisfying the following backward optimality equations:*

$$v_{t-1} = \max_{(w^1,w^2,\cdots,w^n)\in\mathcal{W}} \left\{ 1 + \mathrm{VaR}_p\left(\sum_{i=1}^n w^i R_{t-1}^i\right) + \beta\left(1 + \sum_{i=1}^n w^i E(R_{t-1}^i)\right) v_t \right\} \tag{17}$$

for $t = 2, 3, \cdots, T$, and

$$v_T := \max_{(w^1,w^2,\cdots,w^n)\in\mathcal{W}} \left\{ 1 + \mathrm{VaR}_p\left(\sum_{i=1}^n w^i R_T^i\right) \right\}. \tag{18}$$

From Theorem 1, in the next section we focus on (17) and (18) regarding the value-at-risk portfolio at each time.

4 A Portfolio Optimization for Value-at-Risks

First we estimate the rate of return for a portfolio ([20]). Let $t = 1, 2, \cdots, T$. Let the mean, the variance and the covariance of the rate of return R_t^i, which is given in (2), respectively by

$$\begin{aligned} \mu_t^i &:= E(R_t^i), \\ (\sigma_t^i)^2 &:= E((R_t^i - \mu_t^i)^2), \\ \sigma_t^{ij} &:= E((R_t^i - \mu_t^i)(R_t^j - \mu_t^j)) \end{aligned}$$

for $i, j = 1, 2, \cdots, n$. Hence we assume that the determinant of the variance-covariance matrix $\Sigma_t := [\sigma_t^{ij}]$ is not zero and there exists its inverse matrix Σ_t^{-1}. This assumption is natural and it can be realized easily by taking care of the combinations of assets. For a portfolio $w = (w^1, w^2, \cdots, w^n)$ satisfying $w^1 + w^2 + \cdots + w^n = 1$ and $w^i \geq 0$ $(i = 1, 2, \cdots, n)$, we calculate the expectation and the variance regarding $R_t = w^1 R_t^1 + w^2 R_t^2 + \cdots + w^n R_t^n$. The expectation μ_t of the rate of return R_t with the portfolio w is

$$\mu_t := E(R_t) = \sum_{i=1}^n w^i E(R_t^i) = \sum_{i=1}^n w^i \mu_t^i. \tag{19}$$

On the other hand, the variance $(\sigma_t)^2$ of the rate of return R_t with the portfolio w is

$$(\sigma_t)^2 := E((R_t - \mu_t)^2) = \sum_{i=1}^n \sum_{j=1}^n w^i w^j \sigma_t^{ij}, \tag{20}$$

where $(\sigma_t^i)^2 = \sigma_t^{ii}$ for $i = 1, 2, \cdots, n$. Therefore, for a given positive probability p, the value-at-risk $\mathrm{VaR}_p(R_t)$ of the rate of return R_t is evaluated as

$$\mathrm{VaR}_p(R_t) = \sum_{i=1}^n w^i \mu_t^i - \kappa \sqrt{\sum_{i=1}^n \sum_{j=1}^n w^i w^j \sigma_t^{ij}} \tag{21}$$

with a positive constant κ in (8). Let

$$\mu_t := \begin{bmatrix} \mu_t^1 \\ \mu_t^2 \\ \vdots \\ \mu_t^n \end{bmatrix}, \quad \Sigma_t := \begin{bmatrix} \sigma_t^{11} & \sigma_t^{12} & \cdots & \sigma_t^{1n} \\ \sigma_t^{21} & \sigma_t^{22} & \cdots & \sigma_t^{2n} \\ \vdots & \vdots & \ddots & \vdots \\ \sigma_t^{n1} & \sigma_t^{n2} & \cdots & \sigma_t^{nn} \end{bmatrix}, \quad \mathbf{1} := \begin{bmatrix} 1 \\ 1 \\ \vdots \\ 1 \end{bmatrix},$$

$$A_t := \mathbf{1}^{\mathrm{T}} \Sigma_t^{-1} \mathbf{1}, \qquad B_t := \mathbf{1}^{\mathrm{T}} \Sigma_t^{-1} \mu_t, \qquad C_t := \mu_t^{\mathrm{T}} \Sigma_t^{-1} \mu_t, \qquad \Delta_t := A_t C_t - B_t^2,$$

where T denotes the transpose of a vector. Now we discuss the following VaR portfolio problem without allowance for short selling. The following form (22) comes from the value-at-risk $\mathrm{VaR}_p(R_t)$ given in (21).

VaR-portfolio problem (VP): Let $t = 1, 2, \cdots, T$. Maximize the value-at-risk

$$\mathrm{VaR}_p(R_t) = \sum_{i=1}^{n} w^i \mu_t^i - \kappa \sqrt{\sum_{i=1}^{n} \sum_{j=1}^{n} w^i w^j \sigma_t^{ij}} \tag{22}$$

with respect to portfolios $w = (w^1, w^2, \cdots, w^n)$ satisfying $w^1 + w^2 + \cdots + w^n = 1$ and $w^i \geq 0$ for $i = 1, 2, \cdots, n$.

We have the following analytical solutions regarding VaR-portfolio problem (VP).

Lemma 4 ([20]). *Let $t = 1, 2, \cdots, T$. Let A_t and Δ_t be positive. Let the constant κ satisfy $\kappa^2 > C_t$. Then the following (i) and (ii) hold.*

(i) *The solution of VaR-portfolio problem (VP) is given by*

$$w^* := \xi \Sigma_t^{-1} \mathbf{1} + \eta \Sigma_t^{-1} \mu_t, \tag{23}$$

and then the corresponding VaR is

$$\mathrm{VaR}_p\left(\sum_{i=1}^{n} w^{*i} R_t^i \right) = \frac{B_t - \sqrt{A_t \kappa^2 - \Delta_t}}{A_t}, \tag{24}$$

where $w^* = (w^{*1}, w^{*2}, \cdots, w^{*n})$, $\gamma := \frac{B_t}{A_t} + \frac{\Delta_t}{A_t \sqrt{A_t \kappa^2 - \Delta_t}}$, $\xi := \frac{C_t - B_t \gamma}{\Delta_t}$ and $\eta := \frac{A_t \gamma - B_t}{\Delta_t}$.

(ii) *Further, if $\Sigma_t^{-1} \mathbf{1} \geq \mathbf{0}$ and $\Sigma_t^{-1} \mu_t \geq \mathbf{0}$, then the portfolio (23) satisfies $w^* \geq \mathbf{0}$, i.e. w^* is a portfolio without allowance for short selling. Here, $\mathbf{0}$ denotes the zero vector.*

Assume that the rates of return R_t^i $(i = 1, 2, \cdots, n)$ have normal distributions. Then, for a risk probability p, we put a constant κ in (8) by

$$\kappa := -\Phi^{-1}(p), \tag{25}$$

where Φ^{-1} is the inverse function of the cumulative normal distribution function

$$\Phi(z) := \frac{1}{\sqrt{2\pi}} \int_{-\infty}^{z} e^{-\frac{t^2}{2}}\, \mathrm{d}t, \qquad z \in \mathbb{R}. \tag{26}$$

Condition (V). It holds that $1 + \beta v_t > 0$ for all $t = 1, 2, \cdots, T$.

If Condition (V) is not satisfied at some time t, it means that the the portfolio is bankrupt at the time t. Under Condition (V), Theorem 1 is written as following theorem by (19) – (21).

Theorem 2. *Suppose Condition (V) is satisfied. The optimal VaR v_1 in Theorem 1 is given by the sequence $\{v_t\}$ of sub-total-sum value-at-risks after time $t-1$ satisfying the following backward optimality equations:*

$$v_{t-1} := \max_{(w^1,w^2,\cdots,w^n)\in\mathcal{W}} (1+\beta v_t) \left(1 + \sum_{i=1}^{n} w^i \mu_{t-1}^i - \frac{\kappa}{1+\beta v_t} \sqrt{\sum_{i=1}^{n}\sum_{j=1}^{n} w^i w^j \sigma_{t-1}^{ij}} \right) \tag{27}$$

for $t = 2, 3, \cdots, T$, and

$$v_T := \max_{(w^1,w^2,\cdots,w^n)\in\mathcal{W}} \left(1 + \sum_{i=1}^{n} w^i \mu_T^i - \kappa \sqrt{\sum_{i=1}^{n}\sum_{j=1}^{n} w^i w^j \sigma_T^{ij}} \right). \tag{28}$$

Applying Lemma 4 to Theorem 2, we obtain the following results.

Theorem 3. *Suppose Condition (V) is satisfied. Assume R_t^i ($t = 1, 2, \cdots, T; i = 1, 2, \cdots, n$) have normal distributions, and let $\kappa := -\Phi^{-1}(p)$ in (25). Then the optimal VaR v_1 in Theorem 1 is calculated by the sequence $\{v_t\}$ of sub-total-sum value-at-risks after time $t-1$ satisfying the following backward optimality equations:*

$$v_{t-1} = \frac{(A_{t-1} + B_{t-1})(1 + \beta v_t) - \sqrt{A_{t-1}\kappa^2 - \Delta_{t-1}(1 + \beta v_t)^2}}{A_{t-1}} \tag{29}$$

for $t = 2, 3, \cdots, T$ and

$$v_T = \frac{A_T + B_T - \sqrt{A_T \kappa^2 - \Delta_T}}{A_T}. \tag{30}$$

Corollary 1. *Suppose Condition (V) is satisfied. Let A_t and Δ_t be positive for $t = 1, 2, \cdots, T$. Put $\kappa_{t-1} := \frac{\kappa}{1+\beta v_t}$ ($t = 2, 3, \cdots, T$) and $\kappa_T := \kappa = -\Phi^{-1}(p)$. Assume κ_{t-1} satisfies $\kappa_{t-1}^2 > C_{t-1}$ ($t = 2, 3, \cdots, T$). Then the following (i) and (ii) hold.*

(i) *The optimal portfolios of (17) and (18) in Theorem 1 are given by*

$$w_t := \xi \Sigma_t^{-1}\mathbf{1} + \eta \Sigma_t^{-1}\mu_t, \quad t = 1, 2, \cdots, T, \tag{31}$$

where $\gamma_t := \frac{B_t}{A_t} + \frac{\Delta_t}{A_t\sqrt{A_t\kappa_t^2 - \Delta_t}}$, $\xi_t := \frac{C_t - B_t\gamma_t}{\Delta_t}$ *and* $\eta_t := \frac{A_t\gamma_t - B_t}{\Delta_t}$ *for* $t = 1, 2, \cdots, T$.

(ii) *Further, if* $\Sigma_t^{-1}\mathbf{1} \geq \mathbf{0}$ *and* $\Sigma_t^{-1}\mu_t \geq \mathbf{0}$ *for* $t = 1, 2, \cdots, T$*, then the portfolio (31) satisfies* $w_t \geq \mathbf{0}$*, i.e.* w_t *is a portfolio without allowance for short selling.*

5 A Numerical Example

In this session, a simple numerical example is shown to explain the significance of the results obtained in previous sections. We consider a model with 4 assets, i.e. we put $n = 4$. Give the vector of the expected rate of return $\mu_t = [\mu_t^i]$ and the variance-covariance matrix $\Sigma_t = [\sigma_t^{ij}]$ in Table 1. Then we can easily calculate the constants A_t, B_t, C_t and Δ_t as $A_t = \mathbf{1}^{\mathrm{T}}\Sigma_t^{-1}\mathbf{1} = 14.7016$, $B_t = \mathbf{1}^{\mathrm{T}}\Sigma_t^{-1}\mu_t = 0.784825$, $C_t = \mu_t^{\mathrm{T}}\Sigma_t^{-1}\mu_t = 0.0456314$ and $\Delta_t = A_tC_t - B_t^2 = 0.054904$. In actual financial management, these data should be estimated from up-to-the current asset prices in stock market. As for (8), we assume that the distributions of the rate of return R_t^i is Gaussian. First, we discuss a risk probability 1% in the lower part of the Gaussian distribution, and then the corresponding constant is $\kappa = 2.326$, which is given in (25). Then, the conditions in Theorems 2 and 3 are satisfied. By Eq. (31) in Corollary 1, we easily obtain the optimal portfolio $w^* = (0.229604, 0.215551, 0.25200, 0.302845)$, which is optimal for the VaR-portfolio (VP).

Table 1. Expected rates of return and a variance-covariance matrix

Asset	μ_t^i
$i=1$	0.05
$i=2$	0.07
$i=3$	0.06
$i=4$	0.04

σ_t^{ij}	$j=1$	$j=2$	$j=3$	$j=4$
$i=1$	0.35	0.03	0.02	−0.08
$i=2$	0.03	0.25	−0.06	0.08
$i=3$	0.02	−0.06	0.33	−0.02
$i=4$	−0.08	0.08	−0.02	0.24

Since v_t is the total-sum of discounted value of falling from time t to the terminal time T, we put the discounted time-average value of falling V_t^T from time t to the terminal time T as follows.

$$V_t^T := \frac{v_t}{\sum_{s=1}^{T-t}\beta^s}$$

for $t = 1, 2, \cdots, T$. Let an initial amount of investment $S_0 = 1$, the terminal time $T = 20$ and with a time-discount weight $\beta = 0.95$. Then we obtain the discounted time-average value of falling for the portfolio process: $V_1^{20} = 0.711737$.

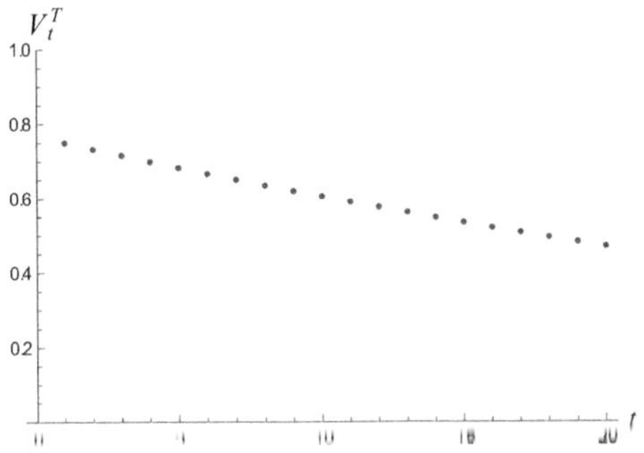

Fig. 2. The discounted time-average values of falling V_t^T after time t ($T = 20, \beta = 0.95$)

References

1. Artzner, P., Delbaen, F., Eber, J.-M., Heath, D.: Coherent measures of risk. Mathematical Finance 9, 203–228 (1999)
2. Boot, J.C.G.: Quadratic Programming. North-Holland, Amsterdam (1964)
3. El Chaoui, L., Oks, M., Oustry, F.: Worst-case value at risk and robust portfolio optimization: A conic programming approach. Operations Research 51, 543–556 (2003)
4. Gaivoronski, A., Pflug, G.C.: Value-at-risk in portfolio optimization: Properties and computational approach. Journal of Risk 7(2), 1–31 (2005)
5. Jorion, P.: Value at Risk: The New Benchmark for Managing Financial Risk, 3rd edn. McGraw-Hill, New York (2007)
6. Korn, R., Korn, E.: Options Pricing and Portfolio Optimization Modern Models of Financial Mathematics. Amer. Math. Soc. (2001)
7. Kumar, P.R., Ravi, V.: Bankruptcy prediction in banks and firms via statistical and intelligent techniques - A review. European J. Oper. Res. 180, 1–28 (2007)
8. Kusuoka, S.: On law invariant coherent risk measures. Advances in Mathematical Economics 3, 83–95 (2001)
9. Markowitz, H.: Mean-Variance Analysis in Portfolio Choice and Capital Markets. Blackwell, Oxford (1990)
10. Merton, R.C.: Lifetime portfolio selection under uncertainty: The continuous case. Reviews of Economical Statistics 51, 247–257 (1969)
11. Meucci, A.: Risk and Asset Allocation. Springer, Heidelberg (2005)
12. Pliska, S.R.: Introduction to Mathematical Finance: Discrete-Time Models. Blackwell Publ., New York (1997)
13. Rockafellar, R.T., Uryasev, S.P.: Optimization of conditional value-at-risk. Journal of Risk 2, 21–42 (2000)
14. Ross, S.M.: An Introduction to Mathematical Finance. Cambridge Univ. Press, Cambridge (1999)
15. Steinbach, M.C.: Markowitz revisited: Mean-variance model in financial portfolio analysis. SIAM Review 43, 31–85 (2001)
16. Tasche, D.: Expected shortfall and beyond. Journal of Banking Finance 26, 1519–1533 (2002)

17. Yoshida, Y.: The valuation of European options in uncertain environment. European J. Oper. Res. 145, 221–229 (2003)
18. Yoshida, Y.: A discrete-time model of American put option in an uncertain environment. European J. Oper. Res. 151, 153–166 (2003)
19. Yoshida, Y., Yasuda, M., Nakagami, J., Kurano, M.: A discrete-time portfolio selection with uncertainty of stock prices. In: De Baets, B., Kaynak, O., Bilgiç, T. (eds.) IFSA 2003. LNCS (LNAI), vol. 2715, pp. 245–252. Springer, Heidelberg (2003)
20. Yoshida, Y.: An estimation model of value-at-risk portfolio under uncertainty. Fuzzy Sets and Systems 160, 3250–3262 (2009)
21. Yoshida, Y.: A perception-based portfolio under uncertainty: Minimization of average rates of falling. In: Torra, V., Narukawa, Y., Inuiguchi, M. (eds.) MDAI 2009. LNCS (LNAI), vol. 5861, pp. 149–160. Springer, Heidelberg (2009)
22. Yoshida, Y.: An average value-at-risk portfolio model under uncertainty: A perception-based approach by fuzzy random variables. Journal of Advanced Computational Intelligence and Intelligent Informatics 15, 56–62 (2011)
23. Zmeškal, Z.: Value at risk methodology of international index portfolio under soft conditions (fuzzy-stochastic approach). International Review of Financial Analysis 14, 263–275 (2005)

Modelling Heterogeneity among Experts in Multi-criteria Group Decision Making Problems

Ignacio J. Pérez[1], Sergio Alonso[2], Francisco J. Cabrerizo[3],
Jie Lu[4], and Enrique Herrera-Viedma[1]

[1] Dept. of Computer Science and Artificial Intelligence, University of Granada, Spain
ijperez@decsai.ugr.es, viedma@decsai.ugr.es
[2] Dept. of Software Engineering, University of Granada, Spain
zerjioi@ugr.es
[3] Dept. of Software Engineering and Computer Systems, Distance Learning University of Spain (UNED), Madrid, Spain
cabrerizo@issi.uned.es
[4] Faculty of Information Technology, University of Technology, Sydney, Australia
jielu@it.uts.edu.au

Abstract. Heterogeneity in group decision making problems has been recently studied in the literature. Some instances of these studies include the use of heterogeneous preference representation structures, heterogeneous preference representation domains and heterogeneous importance degrees. On this last heterogeneity level, the importance degrees are associated to the experts regardless of what is being assessed by them, and these degrees are fixed through the problem. However, there are some situations in which the experts' importance degrees do not depend only on the expert. Sometimes we can find sets of heterogeneously specialized experts, that is, experts whose knowledge level is higher on some alternatives and criteria than it is on any others. Consequently, their importance degree should be established in accordance with what is being assessed. Thus, there is still a gap on heterogeneous group decision making frameworks to be studied. We propose a new fuzzy linguistic multi-criteria group decision making model which considers different importance degrees for each expert depending not only on the alternatives but also on the criterion which is taken into account to evaluate them.

keywords: Group decision making, multi-criteria decision making, heterogeneous decision frameworks, linguistic approach.

1 Introduction

Decision making is as very common activity all over the world. Usually, it is performed by people who have to consider some criteria in order to derive the best option from a feasible set. But sometimes, alternatives and criteria are imprecise, contradictory or belong to a wide range. In this case an expert can not make a decision on his own and it is necessary that a group of experts, with a high collective knowledge level on these particular criteria, take part in the decision process. Thus, we interpret the decision process in the framework of group decision making (GDM).

V. Torra et al. (Eds.): MDAI 2011, LNAI 6820, pp. 55–66, 2011.

GDM models are used to obtain the best solution(s) for a problem according to the information provided by some decision makers. Usually, each decision maker (expert) may approach the decision process from a different angle, but they have a common interest in reaching an agreement on taking the best decision. Concretely, in a GDM problem we have a set of different alternatives to solve the problem and a set of experts which are usually required to provide their preferences about the alternatives [1,2,3,4,5].

Furthermore, there are GDM problems in which, to evaluate the alternatives, the experts have to take into account the value of some criteria that define the features of each alternative. Multi-criteria decision making (MCDM) refers to making a decision (e.g., evaluation, prioritization, and selection) over the available alternatives that are characterized by multiple, usually conflicting, criteria [6]. In such decision situation the aim is to find a set of alternatives that, considering all the criteria, solves the problem in the best way. Multi-criteria group decision making (MCGDM), which combines MCDM and GDM methods, has been proved to be a very effective technique to increase the level of overall satisfaction for the final decision across the group and particularly in evaluation decision-making such as evaluating products, developing policies, selecting employees, and arranging various resources [7,8,9,10,11,12,13,14].

Due to the wide range of different problems that can be solved with GDM models, in recent years, these models have been studied and improved in order to deal with non-homogeneous frameworks. In particular, we can find in the literature some heterogeneous GDM models at three different levels: i) heterogeneity at the preference representation structure level (orders, utility functions or preference relations) [15,16], ii) heterogeneity at the preference representation domain level (numeric, linguistic, multi-granular, interval numbers) [17,18,19,20] and iii) heterogeneity at the importance degree of experts and criteria level [21,22].

On this third studied heterogeneity level, the importance degrees associated to the experts are fixed through the problem. However, there are some situations in which experts have an heterogeneous knowledge of the problem environment. Thus, their importance degrees can not be associated regardless what is being evaluated and it should be different for each criterion. Therefore, it is still necessary to study and improve the existing heterogeneous GDM models.

Accordingly, we propose to tackle heterogeneous MCGDM problem based on non-homogeneous frameworks with heterogeneously specialized experts' preferences. To do so, we assume that experts give their assessments about the alternatives in natural language, using preference relations [15] as the preferences representation structure on a fuzzy linguistic domain. For this reason, we use a *fuzzy linguistic modelling* [4] to represent the experts' opinions. This kind of modelling is an approximate technique which represents qualitative aspects as linguistic values by means of *linguistic variables*[23], that is, variables whose values are not numbers but words or sentences in a natural or artificial language. To compute the quality assessments we use computing with words tools based on linguistic aggregation operators.

The aim of this paper is to present a new model of MCGDM selection process based on heterogeneously specialized experts' preferences, where the set of experts is established depending on the different criteria of the problem. Moreover, an expert's opinion will have different importance level according to the criterion which is taken into account and the assessed alternatives.

In order to do this, the paper is set out as follows. Preliminaries are presented in Section 2. Section 3 defines the new fuzzy linguistic MCGDM selection process based on heterogeneously specialized experts' preferences and finally, Section 4 draws our conclusions.

2 Preliminaries

In this section we present some considerations about MCGDM problems, heterogeneity in group decision making and the basis of a fuzzy linguistic approach.

2.1 MCGDM Problems

In a GDM problem we have a finite set of feasible alternatives. $X = \{x_1, x_2, \ldots, x_n\}$, $(n \geq 2)$ and the best alternatives from X have to be identified using the information given by a set of experts, $E = \{e_1, e_2, \ldots, e_m\}$, $(m \geq 2)$, according to a set of criteria $C = \{c_1, c_2, \ldots, c_p\}$, $(p \geq 2)$.

Resolution methods for GDM problems are usually composed by two different processes [4] (see Figure 1):

1. *Consensus process:* Clearly, in any decision process, it is preferable that the experts reach a high degree of consensus on the solution set of alternatives. Thus, this process refers to how to obtain the maximum degree of consensus or agreement among the experts on the solution alternatives.
2. *Selection process:* This process consists in how to obtain the solution set of alternatives from the opinions on the alternatives given by the experts. Furthermore, the selection process is composed of two different phases:
   (a) Aggregation phase: This phase uses an aggregation operator in order to transform the individual preferences on the alternatives into a collective preference.
   (b) Exploitation phase: This phase uses choice functions [24] in order to transform the collective preference into a partial ranking of alternatives that helps to make the final decision.

In this paper, we center our attention only in the selection process, where the experts will provide their preferences about the set of alternatives on each criteria by using words in natural language by means of the fuzzy linguistic approach and a ranking of alternatives is obtained according to experts' preferences.

2.2 Heterogeneity in Group Decision Making

Recently, several authors have studied and approached MCGDM problems from different angles, showing that this kind of problems are not always homogeneous. We can classify them into three different heterogeneity levels.

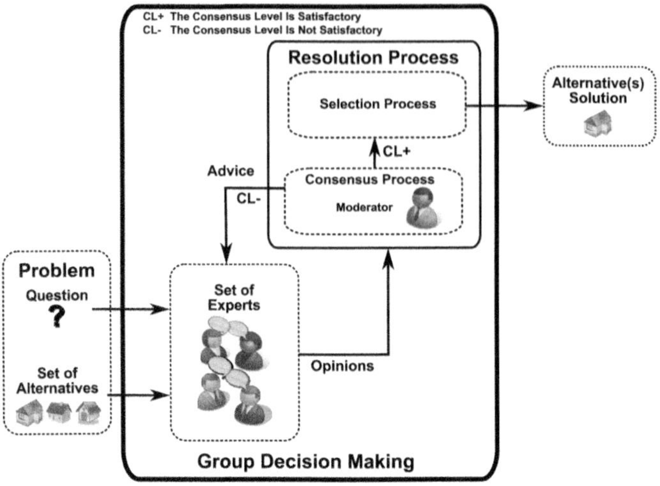

Fig. 1. *Resolution process of a GDM problem*

1. The first heterogeneity level studied in the literature [15,16,25,26], is focused on the preference representation structures. Usually, each expert e_h provides his/her preferences on the alternatives $X = \{x_1, x_2, ..., x_n\}$, separately according to each criteria $C = \{c_1, c_2, ..., c_p\}$ by means of different preference's representation format, the most commonly used are:
   - *Preference orderings of alternatives:* $O^k = \{o^k(1), ..., o^k(n)\}$, where $o^k(\cdot)$ is a permutation function over the index set, $\{1, ..., n\}$, for the expert, e_k, defining an ordered vector of alternatives, from best to worst.
   - *Utility functions:* $U^k = \{u_1^k, ..., u_n^k\}$, $u_i^k \in [0, 1]$, where u_i^k represents the utility evaluation given by the expert e_k to x_i.
   - *Fuzzy preference relations:* $P^k \subset X\text{x}X$, with a membership function, $\mu_{P^k}: X\text{x}X \rightarrow [0, 1]$, where $\mu_{P^k}(x_i, x_j) = p_{ij}^k$ denotes the preference degree of x_i over x_j.
   - *Multiplicative preference relations:* $A^k \subset X\text{x}X$, where the intensity of preference, a_{ij}^k, is measured using a ratio scale, particularly the 1/9 to 9 scale.

   Fuzzy preference relations are widely used in this kind of problems because they are more informative than preference orderings or utility functions [15], allowing the comparison of the alternatives in a pair by pair basis. Thus, users have much more freedom at giving their preferences and they can gain expressivity against other preference representations. When cardinality of X is small, the preference relation may be conveniently represented by an $n \times n$ matrix $P^{hs} = (p_{ij}^{hs})$.
2. The second heterogeneity level is focused on the preference representation domain (numeric, linguistic, multi-granular, interval numbers) [17,18,19,20,26,27,28]. We propose the use of a fuzzy linguistic approach that is presented in Section 2.3.
3. Finally, the third heterogeneity level [21,22], deals with some classical heterogeneous decision scenarios, where every expert has an associated weight value in order to model their different importance levels or knowledge degrees. Furthermore,

in some multi-criteria decision scenarios, we can find criteria with different weight values [12]. In these situations, the experts' or criteria's weight values are always established a priori, regardless what is being evaluated.
However, there are heterogeneous situations in which the knowledge degree of each expert is different depending on the criterion. In such a way, the opinion of an expert specialized on a certain criterion should be more important than one of another expert not familiarized with this same criterion. So, an expert could have different relevance according to what is being evaluated at every moment. Consequently, in this paper, we propose a new approach to model an experts' weights establishment mechanism in accordance with the criterion which is taken into account and the assessed alternatives.

2.3 Fuzzy Linguistic Approach

Several authors have provided interesting results on GDM with the help of fuzzy theory [1,5,8,17,18], but there are situations in which the information cannot be assessed precisely in a quantitative form but may be in a qualitative one. For example, when attempting to qualify phenomena related to human perception, we are often led to use words in natural language instead of numerical values. In other cases, precise quantitative information cannot be stated because either it is unavailable or the cost for its computation is too high and an "approximate value" can be applicable, eg. when evaluating the speed of a car, linguistic terms like *fast*, *very fast* or *slow* can be used instead of numeric values [17,29]. The use of Fuzzy Sets Theory has given very good results for modelling qualitative information [23].

Fuzzy linguistic modelling is a tool based on the concept of linguistic variable [23] to deal with qualitative assessments. It has proven its usefulness in many problems, e.g., in quality evaluation, information retrieval models, decision making, and so on [30,31,32]. Ordinal fuzzy linguistic modelling [4] is a very useful kind of fuzzy linguistic approach proposed as an alternative tool to the traditional fuzzy linguistic modelling which simplifies the computing with words process as well as linguistic aspects of problems. It is defined by considering a finite and totally ordered label set $S = \{s_i\}, i \in \{0, ..., g\}$ in the usual sense, i.e., $s_i \geq s_j$ if $i \geq j$, and with odd cardinality (usually 7 or 9 labels). The mid term represents an assessment of "approximately 0.5", and the rest of the terms are placed symmetrically around it. The semantics of the label set is established from the ordered structure of the label set by considering that each label for the pair (s_i, s_{g-i}) is equally informative [17]. For example, we can use the following set of seven labels to represent the linguistic information:

$S = \{$ N=*Null*, VL=*Very Low*, L=*Low*, M=*Medium*, H=*High*, VH=*Very High*, P=*Perfect*$\}$.

In any linguistic model we also need some management operators to deal with linguistic information. An advantage of the ordinal fuzzy linguistic modeling is the simplicity and speed of its computational model. It is based on the symbolic computational model [4] and acts by direct computation on labels by taking into account the order of such linguistic assessments in the ordered structure of labels. Usually, the ordinal fuzzy linguistic model for computing with words is defined by establishing i) a negation operator, ii) comparison operators based on the ordered structure of linguistic terms, and

iii) adequate aggregation operators of ordinal fuzzy linguistic information. In most ordinal fuzzy linguistic approaches the negation operator is defined from the semantics associated to the linguistic terms as

$$NEG(s_i) = s_j \mid j = (g - i)$$

and there are defined two comparison operators of linguistic terms:

1. *Maximization operator:* $MAX(s_i, s_j) = s_i$ if $s_i \geq s_j$; and
2. *Minimization operator:* $MIN(s_i, s_j) = s_i$ if $s_i \leq s_j$.

Using these operators it is possible to define automatic and symbolic aggregation operators of linguistic information, as for example the Linguistic Ordered Weighted Averaging (LOWA) operator [33]. Sometimes, the different items that we need to aggregate have as well an associated weight. It has to be taken into account on the aggregation operator selection. So, in these situations, we can use the Linguistic Induced Ordered Weighted Averaging (L-IOWA) operator that is a linguistic version of the IOWA operator [34,35,36].

Definition 1. *L-IOWA operator is defined as follows [37]:*

$$\Phi_W(\langle u_1, s_{\alpha_1}\rangle, \ldots, \langle u_n, s_{\alpha_n}\rangle) = w_1 \cdot s_{\gamma_1} \oplus w_2 \cdot s_{\gamma_2} \oplus \ldots \oplus w_n \cdot s_{\gamma_n} = s_{round(\bar{\gamma})}$$

where $\bar{\gamma} = \sum_{j=1}^{n} w_j \cdot \gamma_j$, $w = (w_1, w_2, \ldots, w_n)$ *is a weighting vector, such that* $w_j \in [0,1]$, $\sum_{j=1}^{n} w_j = 1$, s_{γ_j} *is the* s_{α_i} *value of the pair* $\langle u_i, s_{\alpha_i}\rangle$ *having the* jth *largest* u_i, *and* u_i *in* $\langle u_i, s_{\alpha_i}\rangle$ *is referred to as the order inducing variable and* s_i *as the linguistic argument variable.*

A natural question in the definition of this operator is how to obtain the associated weighting vector. In [38], an expression to obtain W that allows to represent the concept of fuzzy majority [39] by means of a fuzzy linguistic non-decreasing quantifier Q [40] was defined:
$w_i = Q(i/n) - Q((i-1)/n), i = 1, ..., n.$

3 A New Fuzzy Linguistic MCGDM Model Based on Heterogeneous Experts' Opinions

In this section we present a new approach to deal with MCGDM problems in which the main characteristic is the heterogeneous knowledge level of each expert among the multiple criteria. Usually, experts become to reach an excessive specialization on some specific aspects of their own field (heterogeneously specialized experts). So, each of them has a different knowledge level that depends on the criterion that is being taken into account to assess the alternatives. This characteristic is important for the problem management and has to be taken into account not only on the choice of experts, but also on the combination of their individual preferences in order to obtain a more realistic and appropriate collective preference on each criterion.

Consequently, we propose a new computation model composed of two different steps:

1. To obtain an appropriate set of heterogeneously specialized experts and their preferences.
2. To compute the ranking of alternatives through a fuzzy linguistic MCGDM selection process based on heterogeneously specialized experts' preferences.

3.1 Obtaining the Appropriate Set of Heterogeneously Specialized Experts and Their Preferences

When the field of the decision is large and non homogeneous, there are multiple criteria and different kinds of experts together in the problem framework. Therefore, the choice of experts becomes an important task. To do so in an appropriate way, we need to know the experts' typology or the kind of specialization of each expert before starting the decision making process. This requirement is necessary in order to select experts to cover every criterion with knowledge enough. For example, in library evaluation, to fully evaluate the quality of an university library, it is necessary to collect not only students' opinions who just use study resources but also researchers' preferences who usually are focused on research resources or staff's opinions who knows much better the quality of the space resources.

In these situations, it is necessary an initial approach to the problem in order to get the different alternatives and criteria. Once we know this information and the different specialization of each expert we can start the choice of experts step. In this way, we propose to select a suitable set of experts with heterogeneous knowledge enough, covering all the decision criteria, from any expert database. To do so, it is quite important the figure of the moderator, who is a person that has a deep knowledge about the problem (alternatives and criteria), and he is able to select a suitable and balanced set of experts.

For example, in order to select the best university library from a set of them for a particular use (to study for an exam, to research on a new topic...) it is clear that the alternatives are the different libraries of the university and the criteria to evaluate them are the resources offered by each one (space resources, electronic resources, paper resources or human resources).

At this point, is the moderator who select m experts of an experts' database that previously agree to take part in this kind of studies. To have a suitable group of experts, the set has to be balanced, that is, it seems reasonable to have the same number of experts of each kind. Thus, in our example, the set will be composed of z students, z researchers and z staff members in order to have a thorough collective knowledge of every criterion.

In addition, to be fair and correctly manage the heterogeneous collective knowledge, we propose that the opinion of an expert specialized on a certain criterion is more important than one of another expert not familiarized with this same criterion. In such a way, the weight values not only depend on the experts, but also on the criterion followed by the expert to assess the alternatives. Particularly, in academic libraries, this situation is frequently presented because students, researchers and staff members use the library in a different way, according to their own purposes. For instance, a student knows the main drawbacks and advantages of the space resources much better than a researcher because he is using it everyday while researchers usually work in their own offices.

On the other hand, a researcher has a more deep insight on the quality level of research resources like international journals or database access than a standard student.

To model this situation it is necessary to properly define the above mentioned experts' typologies and with them to define the importance of the experts' opinions on the alternatives over each criterion. The moderator is responsible for the weights refinement in each particular case by assigning an expert's weight value to each kind of experts for each single criterion ecw.

Finally, the last moderator's task is to assign the relevance of each criterion for the particular problem by mean of some importance values. In this way, if the library selection is performed with the aim of studying for an exam, the space resources is the most important criterion but if the use of the library will be to research on new technologies the library side lose relevance being more important the electronic resources. These importance values will be treated later as criteria's weight values cw by the aggregation operator.

The ordinal fuzzy linguistic modeling approach lets use a set of linguistic labels as weight values instead of numbers. So, these weights ecw and cw can be expressed using any set of linguistic labels. A feasible set of l labels could be the next: $S = \{s_1 = VeryLow, s_2 = Low, s_3 = Medium, s_4 = High, s_5 = VeryHigh\}$.

Once we have selected the most suitable sets of alternatives, criteria and experts, the decision making process starts with the collection of every expert's opinion. Thus, each expert must give his own assessments on every alternative for each criterion.

We assume that each expert e_h provides his/her preferences $\{P^{h1}, P^{h2}, ..., P^{hp}\}$ by means of p fuzzy linguistic preference relations (FLPR) characterized by a membership function [4]:

$$\mu_P : X \times X \longrightarrow S$$

where S is a set of linguistic labels and p is the number of criteria.

For instance, by using the set of seven labels introduced in Section 2, an expert e_h could provide the following FLPR on a set of four alternatives according to the criterion c_s.

$$P^{hs} = \begin{pmatrix} - & N & H & M \\ P & - & L & M \\ L & H & - & VL \\ M & M & VH & - \end{pmatrix}$$

According to $p^{hs}_{24} = M$ and $p^{hs}_{21} = P$, e_h considers that on the criterion c_s, alternatives x_2 and x_4 are at the same level but x_2 is better than x_1 respectively.

3.2 Fuzzy Linguistic MCGDM Selection Process with Heterogeneously Specialized Experts

In order to obtain a collective assessment from the whole group of experts, the individual opinions have to be computed using an aggregation operator.

When each expert has provided all his preferences (FLPRs) on the alternatives for each criterion, we can obtain a ranking of them by applying a selection process [1,4]. The selection process consists of two different phases:

1. **Aggregation of individual heterogeneously specialized experts' FLPRs on multiple criteria:**
The aggregation phase defines a collective preference relation, $P^c = (p_{ij}^c)$, obtained by means of the aggregation of all individual linguistic preference relations $\{P^{11}, P^{12}, \ldots, P^{1p}, P^{21}, P^{22}, \ldots, P^{2p}, \ldots, P^{mp}\}$. It indicates the global preference among every pair of alternatives according to all the experts' opinions taking into account the whole set of criteria.
To To deal with this situations, we propose a fuzzy linguistic MCGDM aggregation process with two different phases:i) aggregation of individual FLPRs on each criterion and ii) aggregation of collective FLPRs on each criterion.
    (a) Aggregation of individual FLPRs on each criterion:
    At this point, in order to aggregate the individual preferences taking into account every criteria and the heterogeneous knowledge degrees, the first step is to obtain a collective preference relation over each criterion c_s, $P^{cs} = (p_{ij}^{cs})$, obtained by means of the aggregation of all individual linguistic preference relations $\{P^{1s}, P^{2s}, \ldots, P^{ms}\}$. It indicates the global preference among every pair of alternatives according to the criterion c_s.
    Thus, to compute each collective fuzzy linguistic preference degree P^{cs} according to the knowledge level of each expert, we propose to use the L-IOWA operator with the linguistic experts' weight values ecw as the values of the order inducing variable, i.e.,
    $$p_{ij}^{cs} = \Phi_W(\langle ecw^{1s}, p_{ij}^{1s}\rangle, \ldots, \langle ecw^{ms}, p_{ij}^{ms}\rangle)$$
    (b) Aggregation of collective FPLRs on each criterion:
    Once all the individual FLPRs P^{hs} have been aggregated obtaining a collective FLPR P^{cs} for each criterion, this second aggregation step defines a collective preference relation, $P^c = (p_{ij}^c)$, computed by means of the aggregation of all collective FLPRs obtained in the previous step $\{P^{c1}, P^{c2}, \ldots, P^{cp}\}$. It indicates the global preference among every pair of alternatives according to all of different criteria.
    The aggregation operator of this step depends on the importance of the criteria, therefore, we propose to use again the L-IOWA operator with the linguistic criteria's weight values cw as the values of the order inducing variable, i.e.,
    $$p_{ij}^c = \Phi_W(\langle cw^1, p_{ij}^{c1}\rangle, \ldots, \langle cw^p, p_{ij}^{cp}\rangle)$$
2. **Exploitation of the collective FLPR:**
This phase transforms the global information about the alternatives, P^c, into a global ranking of them. In such a way, the set of solution alternatives is obtained. The global ranking is obtained applying these two choice degrees of alternatives on the collective preference relation QGDD and QGNDD. These degrees can be studies in more detail in [24]:

4 Conclusions

In this paper, we have presented a new approach, to deal with MCGDM problems, in which the main contribution is the possibility of join with more accuracy heterogeneously specialized experts' opinions. It has sense when each expert has different knowledge level among the different aspects of the discussion field. To do so, we propose a balanced selection of experts, the use of FLPRs as format of preferences representation and the use of a proper aggregation operator to model the heterogeneity among experts. Using this model, the heterogeneous knowledge of the different kinds of expert is managed with more accuracy over each particular situation instead of doing it over the whole problem. In such a way, better results and decisions can be obtained.

In the future, we will use incomplete information models and mobile technologies in order to present a dynamic decision making process in which the different elements of the problem (alternatives, experts, weights and so on) can be changed through the time.

Acknowledgements. This paper has been developed with the financing of FEDER funds in FUZZYLING Project TIN200761079, FUZZYLING-II Project TIN201017876, PETRI Project PET20070460, Andalusian Excellence Project TIC-05299, and project of Ministry of Public Works 90/07.

References

1. Herrera, F., Herrera-Viedma, E., Verdegay, J.: A sequential selection process in group decision making with a linguistic assessment approach. Information Sciences 85(4), 223–239 (1995)
2. Herrera, F., Herrera-Viedma, E., Verdegay, J.: A model of consensus in group decision making under linguistic assessments. Fuzzy Sets and Systems 78(1), 73–87 (1996)
3. Herrera, F., Herrera-Viedma, E., Verdegay, J.: Linguistic measures based on fuzzy coincidence for reaching consensus in group decision making. International Journal of Approximate Reasoning 16, 309–334 (1997)
4. Herrera, F., Herrera-Viedma, E.: Linguistic decision analisis: steps for solving decision problems under linguistic information. Fuzzy Set and Systems 115, 67–82 (2000)
5. Kacprzyk, J., Fedrizzi, M.: Multiperson decision making models using fuzzy sets and possibility theory. Kluwer Academic Publishers, Dordrecht (1990)
6. Triantaphyllou, E.: Multi-criteria decision making methods: a comparative study. Kluwer Academic Publishers, Dordrecht (2000)
7. Choi, D.H., Ahn, B., Kim, S.: Multicriteria group decision making under incomplete preference judgments: using fuzzy logic with a linguistic quantifier. International Journal of Intelligent Systems 22(6), 641–660 (2007)
8. Fodors, J., Roubens, M.: Fuzzy preference modelling and multicriteria decision support. Kluwer Academic Publishers, Dordrecht (1994)
9. Fu, G.: A fuzzy optimization method for multicriteria decision making: an application to reservoir flood control operation. Expert Systems with Applications 31(1), 145–149 (2008)
10. Gheorghe, R.A., Bufardi, A., Xirouchakis, P.: Fuzzy multicriteria decision aid method for conceptual design. Cirp Annals-Manufacturing Technology 54(1), 152–154 (2005)
11. Lu, J., Zhang, G., Ruan, D.: Intelligent multi-criteria fuzzy group decision making for situation assessments. Soft. Computing 12(3), 289–299 (2008)
12. Lu, J., Zhu, Y., Zeng, X., Ma, J., Zhang, G.: A linguistic multi-criteria group decision support system for fabric hand evaluation. Fuzzy Optimization and Decision Making 8(4), 395–413 (2009)
13. Ma, J., Lu, J., Zhang, G.: Decider: A fuzzy multi-criteria group decision support system. Knowledge-Based Systems 23(1), 23–31 (2010)
14. Yager, R.: Non-numeric multi-criteria multi-person decision making. Group Decision and Negotiation 2, 81–93 (1993)

15. Chiclana, F., Herrera, F., Herrera-Viedma, E.: Integrating three representation models in fuzzy multipurpose decision making based on fuzzy preference relations. Fuzzy Sets and Systems 97(1), 33–48 (1998)
16. Chiclana, F., Herrera, F., Herrera-Viedma, E.: Integrating multiplicative preference relations in a multiplicative decision making model based on fuzzy preference relations. Fuzzy Sets and Systems 122(2), 277–291 (2001)
17. Alonso, S., Cabrerizo, F., Chiclana, F., Herrera, F., Herrera-Viedma, E.: Group decision-making with incomplete fuzzy linguistic preference relations. International Journal of Intelligent Systems 24(2), 201–222 (2009)
18. Cabrerizo, F., Alonso, S., Herrera-Viedma, E.: A consensus model for group decision making problems with unbalanced fuzzy linguistic information. International Journal of Information Technology & Decision Making 8(1), 109–131 (2009)
19. Herrera, F., Herrera-Viedma, E., Martínez, L.: A fuzzy linguistic methodology to deal with unbalanced linguistic term sets. IEEE Transactions on Fuzzy Systems 16(2), 354–370 (2008)
20. Mata, F., Martínez, L., Herrera-Viedma, E.: An adaptive consensus support model for group decision making problems in a multi-granular fuzzy linguistic context. IEEE Transactions on Fuzzy Systems 17(2), 279–290 (2009)
21. Kacprzyk, J., Zadrozny, S., Ras, Z.: Action rules in consensus reaching process support. In: 9th International Conference on Intelligent Systems Design and Applications (ISDA 2009), pp. 809–814 (2009)
22. Perez, I.J., Cabrerizo, F., Herrera-Viedma, E.: A mobile decision support system for dynamic group decision making problems. IEEE Transactions on Systems, Man and Cybernetics - Part A: Systems and Humans 40(6), 1244–1256 (2010)
23. Zadeh, L.: The concept of a linguistic variable and its applications to approximate reasoning. Information Sciences, Part I, II, III 8, 8, 9, 199–249, 301-357, 43-80 (1975)
24. Herrera, F., Herrera-Viedma, E.: Choice functions and mechanisms for linguistic preference relations. European Journal of Operational Research 120, 144–161 (2000)
25. Xu, Z.: Multiple-attribute group decision making with different formats of preference information on attributes. IEEE Transactions on Systems, Man and Cybernetics Part B-Cybernetics 37(6), 1500–1511 (2007)
26. Xu, Z., Chen, J.: Magdm linear-programming models with distinct uncertain preference structures. IEEE Transactions on Systems, Man and Cybernetics Part B-Cybernetics 38(5), 1356–1370 (2008)
27. Xu, Z.: A method based on the dynamic weighted geometric aggregation operator for dynamic hybrid multi-attribute group decision making. International Journal of Uncertainty, Fuzziness and Knowledge-Based Systems 17(1), 15–33 (2009)
28. Xu, Z.: An interactive approach to multiple attribute group decision making with multigranular uncertain linguistic information. Group Decision and Negotiation 18, 119–145 (2009)
29. Herrera, F., Alonso, S., Chiclana, F., Herrera-Viedma, E.: Computing with words in decision making: Foundations, trends and prospects. Fuzzy Optimization and Decision Making 8(4), 337–364 (2009)
30. García-Lapresta, J., Meneses, L.: Modelling rationality in a linguistic framework. Fuzzy Sets and Systems 160, 3211–3223 (2009)
31. Herrera-Viedma, E., Pasi, G., López-Herrera, A.G., Porcel, C.: Evaluating the information quality of web sites: A methodology based on fuzzy. Journal of the American Society for Information Science and Technology 57(4), 538–549 (2006)
32. Porcel, C., Moreno, J., Herrera-Viedma, E.: A multi-disciplinar recommender system to advice research resources in university digital libraries. Expert Systems with Applications 36(10), 12520–12528 (2009)
33. Herrera, F., Herrera-Viedma, E., Verdegay, J.: Direct approach processes in group decision making using linguistic owa operators. Fuzzy Sets and Systems 79, 175–190 (1996)

34. Chiclana, F., Herrera-Viedma, E., Herrera, F., Alonso, S.: Some induced ordered weighted averaging operators and their use for solving group decision-making problems based on fuzzy preference relations. European Journal of Operational Research 182(1), 383–399 (2007)
35. Yager, R.: Induced aggregation operators. Fuzzy Sets and Systems 137(1), 59–69 (2003)
36. Yager, R., Filev, D.: Induced ordered weighted averaging operators. IEEE Transactions on Systems, Man, and Cybernetics 29(2), 141–150 (1999)
37. Xu, Z.: Eowa and eowg operators for aggregating linguistic labels based on linguistic preference relations. International Journal of Uncertainty, Fuzziness and Knowledge-Based Systems 12, 791–810 (2004)
38. Yager, R.: On ordered weighted averaging aggregation operators in multicriteria decision making. IEEE Transactions on Systems, Man, and Cybernetics 18(1), 183–190 (1988)
39. Kacprzyk, J.: Group decision making with a fuzzy linguistic majority. Fuzzy Sets and Systems 18, 105–118 (1986)
40. Zadeh, L.: A computational approach to fuzzy quantifiers in natural languages. Computer and Mathematics with Applications 9(1), 149–184 (1983)

Fast Mining of Non-derivable Episode Rules in Complex Sequences

Min Gan and Honghua Dai

School of Information Technology, Deakin University,
Burwood, Melbourne, VIC 3125, Australia

Abstract. Researchers have been endeavoring to discover concise sets of episode rules instead of complete sets in sequences. Existing approaches, however, are not able to process complex sequences and can not guarantee the accuracy of resulting sets due to the violation of anti-monotonicity of the frequency metric. In some real applications, episode rules need to be extracted from complex sequences in which multiple items may appear in a time slot. This paper investigates the discovery of concise episode rules in complex sequences. We define a concise representation called non-derivable episode rules and formularize the mining problem. Adopting a novel anti-monotonic frequency metric, we then develop a fast approach to discover non-derivable episode rules in complex sequences. Experimental results demonstrate that the utility of the proposed approach substantially reduces the number of rules and achieves fast processing.

Keywords: Episode rules, complex sequences, sequence data mining.

1 Introduction

Episodes [9] were introduced to model the relative order of different types of events within an event sequence. Episode rule mining is an important problem since episode rules are able to capture associations between the occurrence orders of events. Like traditional association rules [1], episode rules can be discovered in two phases. The first phase is finding frequent episodes. The second phase is generating episode rules from the set of frequent episodes.

Most existing approaches [6,7,8,9,10,11] to frequent episode mining aim to finding all frequent episodes. This may generate a huge number of frequent episodes and episode rules. Although closed frequent episodes [17] substantially reduce the number of generated patterns, they are not sufficiently condensed. Therefore, researchers have been endeavoring to discover concise sets of episode rules instead of complete sets. Harms et al. proposed an algorithm, *Gen-REAR* [5], which is capable of finding representative episode rules in a simple sequence in which no more than one item appears in a time slot (see Fig. 1 (a)). However, *Gen-REAR* suffers from two significant deficiencies. The first deficiency is that it is not able to process complex sequences in which multiple items may appear in one time slot (see Fig. 1 (b)). In some real applications, episode rules need to

V. Torra et al. (Eds.): MDAI 2011, LNAI 6820, pp. 67–78, 2011.

be considered in complex sequences. For example, 10 events need to be considered in each time slot in the analysis of stock prices [6]. We consider complex sequences since episode rules in complex sequences have more extensive applications [6]. *EMMA* in [6] can find episode rules in complex sequences. However, the found complete sets are not easy to put into real utility, as they may contain a huge number of rules including a large portion of redundant rules which can be derived. In order to achieve concise results and better utility, we consider the derivation relationship between episode rules and only extract non-derivable rules. The second deficiency is that the accuracy of the set found by *Gen-REAR* can not be guaranteed since the adopted frequency metric does not satisfy anti-monotonicity [1] [1]. The two deficiencies hinder the application of episode rule discovery in complex sequences. Therefore, this paper investigates the mining of concise episode rules in complex sequences. We define non-derivable episode rules and formularize a problem called mining of non-derivable episode rules in complex sequences. Adopting a novel anti-monotonic frequency metric *T-freq* [7], we then develop an efficient algorithm for discovering non-derivable episodes rules in complex sequences.

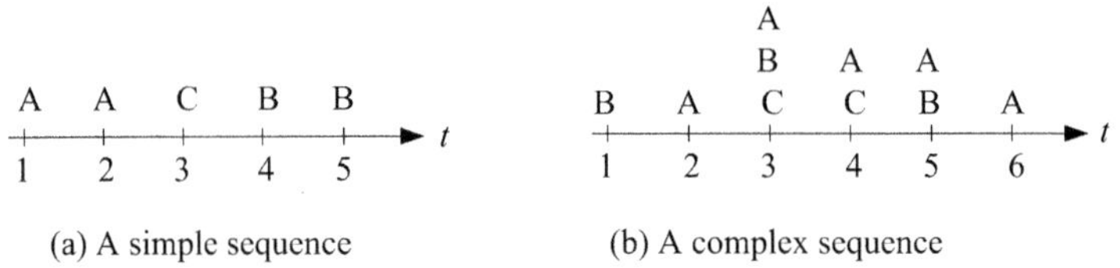

Fig. 1. Single sequences

The rest of this paper is organised as follows. Section 2 presents preliminaries and problem definition. Section 3 addresses the frequency metric and its properties. The mining algorithm is proposed in Section 4 and experimental results are presented in Section 5. Section 6 reviews related works and compares them with our work and Section 7 concludes the paper.

2 Preliminaries and Problem Definition

2.1 Preliminaries

This section presents ordinarily concepts in the literature of episode discovery [6,7,8,9,10,11].

Definition 1 (Complex Sequence). *Let I be a finite set of items. A complex sequence S over I is an ordered list of pairs of elements and timestamps,*

$$S = \langle (E_1, t_1), (E_2, t_2), ..., (E_n, t_n) \rangle \tag{1}$$

[1] A frequency metric is anti-monotonic if under this metric, for any pattern, none of its super-patterns has greater frequency.

where, $E_i \subseteq I$ is called a sequence element, and t_i ($t_1 < t_2 < ... < t_n$) is the timestamp (occurrence time) of E_i in S ($1 \leq i \leq n$). When $E_i \in I$, S is called a simple sequence.

Episodes can de divided into three classes: serial episodes, parallel episodes and composite episodes [9]. In this paper we only consider serial episodes since the other two classes can be constructed from serial episodes.

Definition 2 (Serial Episode). *A serial episode α over I is an ordered list of data elements, denoted as $\alpha = \langle a_1 a_2 ... a_m \rangle$, where $a_i \subseteq I$ ($i = 1, 2, ..., m$). The length of α, denoted as $\alpha.L$, is defined as m. The size of α, denoted as $\alpha.size$, is defined as the number of items contained in α.*

In the rest of the paper, episodes are referred to as serial episodes.

Definition 3 (Sub-episode, Super-episode). *An episode $\beta_{sub} = \langle a_1 a_2 ... a_m \rangle$ is a sub-episode of another episode $\beta = \langle b_1 b_2 ... b_n \rangle$, denoted as $\beta_{sub} \sqsubseteq \beta$, if there exist $1 \leq i_1 < i_2 < ... < i_m \leq n$ such that $a_j \subseteq b_{i_j}$ for all $j = 1, 2, ..., m$. Episode β is a super-episode of β_{sub}.*

For instance, $\langle A(BC) \rangle \sqsubseteq \langle (AD)B(BC) \rangle$. Note that in this paper an episode is written in the form that every element is included in a pair of brackets. The pair of brackets is omitted only when one item is contained in the element. The items in each element are ordered alphabetically.

Definition 4 (Window). *Given $S = \langle (E_1, t_1), (E_2, t_2), ..., (E_n, t_n) \rangle$, a sliding window with width w over S from starting timestamp st, denoted as $win(S, st, w)$, is a sequence segment defined as*

$$win(S, st, w) = \begin{cases} (E_{st})_{st}(E_{st+1})_{st+1} ... (E_{st+w-1})_{st+w-1} & \text{if } st + w - 1 \leq ct \\ (E_{st})_{st}(E_{st+1})_{st+1} ... (E_{ct})_{ct} & \text{otherwise} \end{cases} \quad (2)$$

A window $win(S, st, w)$ contains an episode α if $\alpha \sqsubseteq win(S, st, w)$.

Definition 5 (Frequent Episode, Maximal Frequent Episode). *Given S and min_sup, an episode α is frequent with respect to min_sup if $sup(\alpha) \geq min\_sup$. Episode α is a maximal frequent episode with respect to min_sup if for any $\beta \sqsupseteq \alpha$, $sup(\beta) < min\_sup$.*

Given S, $F(min\_sup)$ is used to denote the set of frequent episodes with respect to min_sup, and $MF(min\_sup)$ is used to denote the set of maximal frequent episodes with respect to min_sup.

Definition 6 (Episode Rule). *An episode rule within an episode α is defined as the implication $\alpha_{sub} \rightarrow \alpha$, where $\alpha_{sub} \sqsubseteq \alpha$ ($\alpha_{sub} \neq \alpha$).*

Definition 7 (Episode Rule Mining). *Given S, min_sup and a minimal threshold of confidence, min_conf, episode rule mining is to discover all episode rules that satisfy both the thresholds of min_sup and min_conf, denoted as $ER(S, min\_sup, min\_conf)$. The confidence of an episode rule $\alpha_{sub} \rightarrow \alpha$ is defined as*

$$conf(\alpha_{sub} \rightarrow \alpha) = sup(\alpha)/sup(\alpha_{sub}) \quad (3)$$

2.2 Problem Definition

This section defines non-derivable episode rules and the mining problem.

Definition 8 (Derivation Relationship). *For two episode rules, $r : \alpha_{sub} \rightarrow \alpha$ and $r' : \alpha'_{sub} \rightarrow \alpha'$, r can be derived from r', denoted as $r' \vdash r$, if $\alpha'_{sub} \sqsubseteq \alpha_{sub}$ and $\alpha \sqsubseteq \alpha'$ ($\alpha'_{sub} \neq \alpha_{sub}$ or $\alpha \neq \alpha'$).*

Definition 9 ((Maximal) Sets of Non-Derivable Episode Rules). *Given S, min_sup and min_conf, we call $\mathcal{R}$ a set of non-derivable episode rules if (1) $R \subseteq ER(S, min\_sup, min\_conf)$ and (2) there exist no $r, r' \in R$ such that $r' \vdash r$. Furthermore, $\mathcal{R}$ is a maximal set of non-derivable episode rules if there exists no larger sets of non-derivable episode rules.*

Definition 10 (The Mining Problem). *Given S, min_sup and min_conf, the problem is discovering the maximal set of non-derivable episode rules, denoted as $\mathcal{MR}$.*

3 Frequency Measurement

The first key issue in episode rule discovery is frequency measurement. This means choosing a frequency metric to measure the frequencies of episodes. To date, several frequency metrics for episodes [6,7,8,9,10,11] have been introduced. Different frequency metrics are adopted in different approaches. Recently, we analysed existing frequency metrics [3], and investigated the impact of these metrics on episode discovery [4]. The metric $fixed\text{-}win\text{-}freq$ is used in *Gen-REAR* [5], and $mo\text{-}freq$ is adopted in $WINEPI$ [11]. However, $fixed\text{-}win\text{-}freq$ does not satisfy anti-monotonicity. Thus, false rules may be included in the result found by *Gen-REAR*. Although $mo\text{-}freq$ is anti-monotonic, it is inconvenient to compute. In this paper, we adopt a novel metric $T\text{-}freq$ [7]. There are two reasons why we choose this metric: (1) it is anti-monotonic and (2) its properties make it convenient to find non-derivable episode rules. This section reviews $T\text{-}freq$ and its basic properties.

Definition 11 (Head Frequency). *Head frequency [7] of episode α in $S = \langle (E_1, t_1), (E_2, t_2), ..., (E_n, t_n) \rangle$ with window width w, denoted as $H\text{-}freq(S, \alpha, w)$, is defined as*

$$H\text{-}freq(S, \alpha, w) = \sum_{i=1}^{n} \delta(win(S, i, w), \alpha) \tag{4}$$

where $\delta(win(S, i, w), \alpha) = 1$ if $a_1 \subseteq E_1$ and $win(S, i, w)$ contains α, otherwise $\delta(win(S, i, w), \alpha) = 0$.

Definition 12 (Total Frequency). *Total frequency [7] of episode α in S with window width w, denoted as $T\text{-}freq(S, \alpha, w)$, is defined as*

$$T\text{-}freq(S, \alpha, w) = \min_{\alpha_{sub} \sqsubseteq \alpha} H\text{-}freq(S, \alpha_{sub}, w) \tag{5}$$

The $T\text{-}freq$ has an incremental property as follows.

Theorem 1 (Incremental Property). *Given sequence* $S = \langle (E_1, t_1), (E_2, t_2), ..., (E_{k-1}, t_{k-1}) \rangle$ *(S is empty when* $k - 1 = 0$*) and min_sup, let* α *be a maximal frequent episode in S (let* α *be an empty episode when S is empty). When a new element is appended, S becomes* $S' = S \circ E_k = \langle (E_1, t_1), (E_2, t_2), ..., (E_k, t_k) \rangle$. *Suppose that* $u \subseteq E_k$ *and u satisfies the following conditions: (1) u is frequent in* S'*, and (2)* $\neg \exists v \subseteq E_k$, $v \supseteq u$ $(u \neq v)$ *and v is frequent in* S'. *Then,* $\alpha \circ u = \langle a_1 a_2 ... a_m u \rangle$ *is a maximal frequent episode in* S'. *[7].*

In this paper, we adopt $T\text{-}freq$ to calculate the frequency of an episode, i.e., $sup(\alpha) = T\text{-}freq(\alpha)$.

4 The Mining Algorithm

In order to find non-derivable episode rules efficiently, we define $\mathcal{MR}$ by multiple layered maximal frequent episodes. Based on the incremental property of $T\text{-}freq$, we then develop an efficient algorithm for mining non-derivable episode rules. The basic idea is to find non-derivable episode rules in two phases.

1. Discovering the set of multiple layered maximal frequent episodes;
2. Extracting non-derivable episode rules from the set of multiple layered maximal frequent episodes.

4.1 Multiple Layered Maximal Frequent Episodes

We divide maximal frequent episodes into different subsets at different layers. Then, non-derivable episodes $\{\alpha_{sub} \rightarrow \alpha\}$ are considered at different layers according to which layer α belongs to. The formal definitions are as follows.

Given S, min_sup and min_conf, let $min\_sup_1 = min\_sup$, $min\_sup_2 = min\_sup_1 + 1 = min\_sup + 1$, ..., $min\_sup_{ml} = min\_sup_{ml-1} + 1 = min\_sup + ml - 1$, where ml is the maximal number of layer. According to the definition of non-derivable episodes, we have the following lemma.

Lemma 1. *Given S, min_sup and min_conf, for any* $\alpha_{sub} \rightarrow \alpha \in \mathcal{MR}$, *if* $sup(\alpha) = min\_sup_r$, *then we have* $\alpha \in MF_r = MF(min\_sup_r)$ *and* $sup(\alpha_{sub}) \in [min\_sup_r, min\_sup_l]$, *where,* $r = 1, 2, ..., ml$, *and* $min\_sup_l = \lfloor \frac{min\_sup_r}{min\_conf} \rfloor$.

Proof. (1) We prove $\alpha \in MF_r$. If $\alpha \notin MF_r$, then $\exists \alpha' \in MF_r$, $\alpha \sqsubseteq \alpha'$ since $sup(\alpha) = min\_sup_r$. Thus, $\alpha_{sub} \rightarrow \alpha' \vdash \alpha_{sub} \rightarrow \alpha$. This deduces that $\alpha_{sub} \rightarrow \alpha \notin \mathcal{MR}$ (a contradiction).

(2) We prove $sup(\alpha_{sub}) \in [min\_sup_r, min\_sup_l]$. We have $sup(\alpha_{sub}) \geq sup(\alpha) = min\_sup_r$ because $T\text{-}freq$ is anti-monotonic. $conf(\alpha_{sub} \rightarrow \alpha) = sup(\alpha)/sup(\alpha_{sub}) = min\_sup_r/sup(\alpha_{sub}) \geq min\_conf$. Therefore, $sup(\alpha_{sub}) \leq min\_sup_r/min\_conf$. Since $sup(\alpha_{sub})$ is an integer, we have the maximal bound $min\_sup_l = \lfloor \frac{min\_sup_r}{min\_conf} \rfloor$. □

We use $\mathcal{MR}_r$ to denote the set of non-derivable episode rules whose consequents have support min_sup_r.

Theorem 2. *Given S, min_sup and min_conf, the maximal set of non-derivable episode rules is $\mathcal{MR} = \cup_{r=1}^{ml} \mathcal{MR}_r$.*

Proof. The theorem can be proven according to Lemma 1 straightaway. □

4.2 Algorithm Description

Non-derivable episode rules are discovered in two phases.

1. Phase 1 (MLMF-Finding) — finding multiple layered maximal frequent episodes.
2. Phase 2 (NR-Extracting) — extracting non-derivable episode rules.

Figure 2 shows the algorithm for mining non-derivable episode rules, MNDER.

Algorithm 1: MNDER(S, min_sup, min_conf)

Input : S, min_sup and min_conf
Output : $\mathcal{MR}$

1 // Phase 1 MLMF-Finding
2 Initialize FR, MF_i and U_i as null for all $i = 1, 2, ..., S.L$;
3 **for** k=1 *to* $S.L$ **do**
4   Read E_k and update FR;
5   Obtain U_i with respect to min_sup_i for all $i = 1, 2, ..., S.L$;
6     $MF_i \leftarrow MF_i \times U_i$ for all $i = 1, 2, ..., S.L$;
7 Obtain ml by reading FR;
8 // Phase 2 NR-Extracting;
9 **for** r=1 *to* ml **do**
10     $l \leftarrow \lfloor min\_sup_r / min\_conf \rfloor$;
11   **for** k=1 *to* $s_{max} - 1$ **do**
12   Obtain $SE_k = \{\alpha | \alpha.size = k \wedge \alpha \sqsubseteq \beta \wedge \alpha \neq \beta \wedge \beta \in MF_r\}$;
13   **for each** $\alpha \in SE_k$ **do**
14   **if** $sup(\alpha) \in [min\_sup_l, min\_sup_r]$ **then**
15   **for each** $\beta \in MF_r$ **do**
16   **if** $\alpha \sqsubseteq \beta$ $(\alpha \neq \beta)$ *and* $\neg \exists r' \in \mathcal{MR}$ *s.t.* $r' \vdash \alpha \rightarrow \beta$ **then**
17   Insert $\alpha \rightarrow \beta$ into $\mathcal{MR}_r$;
18 Return ($\mathcal{MR}$);

Fig. 2. The MNDER algorithm

In Fig. 2, Phase 1 is extended from the algorithm in [7]. We use FR to record the frequency of length-1 episodes. From Line 3 to Line 6, whenever an element E_k is read, FR is updated (Line 4), U_i and MF_i are obtained (Lines 5 and 6) according to Theorem 1. In Line 5, U_i refers to the set of length-1 episodes $\{u \subseteq E_k\}$ that satisfies the two conditions in Theorem 1 with respect to min_sup_i (min_sup is replaced with min_sup_i). In Line 7, maximal number of layer ml is obtained.

In Phase 2 (Lines 9-17), $\mathcal{MR}_r$ is extracted from MF. In Line 10, we compute the maximal bound of layer (support) for the left hand side with respect to the right hand side from MF_r. From Line 11 to Line 17, we consider episode rules with the right hand side at the r-th layer. In line 11, s_{max} refers to the maximal size of episodes in MF_r. In Line 12, sub-episodes with size k are extracted from MF_r. From Line 13 to Line 17, for each size-k episode α, consider $\beta \in MF_r$, and insert $\alpha \rightarrow \beta$ into $\mathcal{MR}_r$ if it satisfies the conditions in Line 16.

It is important to note that (1) $sup(\alpha)$ is embedded in the layered maximal frequent episodes; $sup(\alpha) = min\_sup_m$ if the super-episode of α with the maximal support is at Layer m; and (2) if $\alpha \rightarrow \beta$ has been inserted into $\mathcal{MR}$, any rule $\alpha' \rightarrow \beta$ is not considered if $\alpha' \sqsupseteq \alpha$.

4.3 A Running Example

We use an example to illustrate how non-derivable episodes are discovered by the algorithm.

Example 1. Given S as shown in Fig. 1 (b), $min\_sup = 2$, $min\_conf = 0.5$, MNDER is used to discover $\mathcal{MR}$ from S.

The process of finding $\mathcal{MR}$ is shown in Fig. 3. Figure 3 (a) shows FR, U and MF when an element is read, and Fig. 3 (b) shows the final MF. The found $\mathcal{MR}$ is shown in Fig. 3 (c), where the support and confidence are behind every rule. Only eight rules are included in $\mathcal{MR}$. In contrast, more than 100 rules are included in the resulting set if all rules are found. Note that other rules outside $\mathcal{MR}$ can be derived, and the confidence of each derived rule can be obtained from the found MF. So, $\mathcal{MR}$ is a highly condensed and information lostless set.

5 Experimental Results

The proposed algorithm was performed on synthetic data. Comparisons of condensation and time efficiency were conducted between our method and three previous approaches, *WINEPI* [11], *Gen-REAR* [5] and *EMMA* [6]. The algorithms were implemented in Java. All experiments were performed on a computer with 2.0Ghz CPU and 1GB memory, running on Windows XP.

The synthetic data was created by an IBM synthetic sequence generator [2]. Short sequences are created by the generator first, and the generated short sequences are connected to form long sequences. The generator involves 5 major parameters [2]: C (average number of elements per sequence), T (average number of items per sequence element), N (number of different items), S (average length of maximal potential large sequences) and I (average size of elements in maximal potentially large sequences). We use L (in 000s) to represent the length of a long sequence. Since *WINEPI* [11] and *Gen-REAR* [5] only process simple sequences, we generated two kinds of sequences; simple sequences and complex sequences. Four experiments were conducted to evaluate the performance.

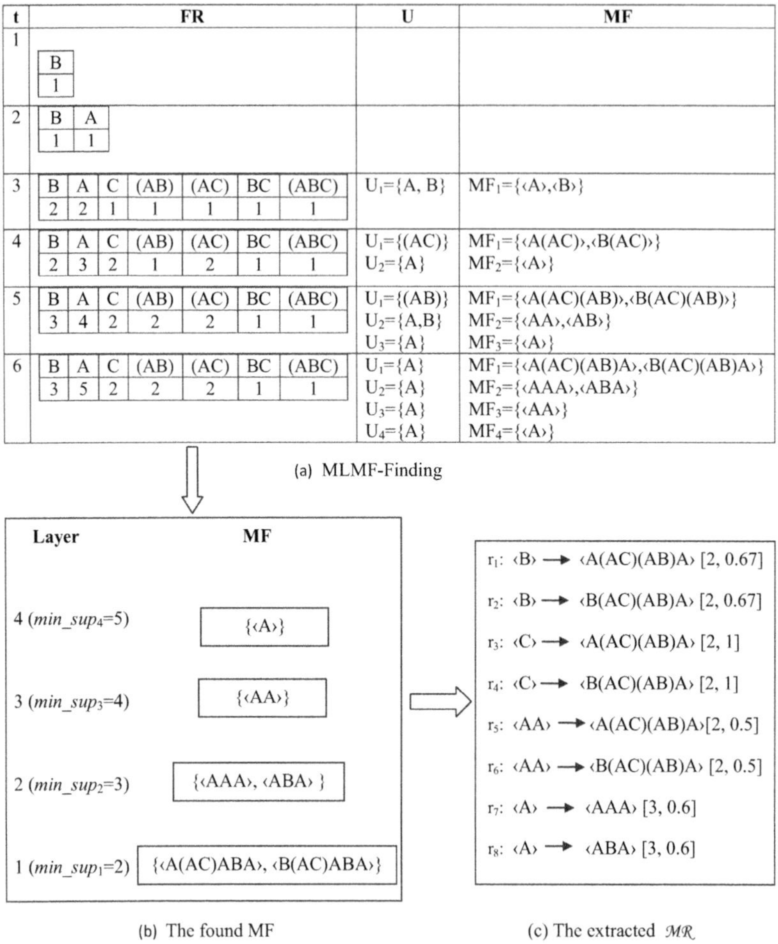

Fig. 3. The process of the example

Experiments 1 and 2 evaluate the performance on simple sequences. Experiments 3 and 4 evaluate the performance on complex sequences.

In Experiment 1, three algorithms ($MNDER$, $WINEPI$ [11] and Gen-$REAR$ [5]) were performed on a simple sequence L10C10T1N50S6I4 and the number of found rules was compared when $min\_sup = 3\%$ and min_conf varies. As shown in Table 1, $\mathcal{MR}$ found by $MNDER$ substantially reduces the size of the complete set found by $WINEPI$ by 300-400 times, and is smaller than the set of representative rules found by Gen-$REAR$. The value in the bracket behind every number is the ratio of this number over the corresponding number in Column 2, e.g., in Column 3, 2.96=548/185. In addition, the higher min_conf is, the lower the number of rules generated.

Table 1. The number of rules found from simple sequences

min_conf	*MNDER*	*Gen-REAR*	*WINEPI*
0.25	185	548 (2.96)	64432 (348.28)
0.3	94	324 (3.45)	38926 (414.11)
0.35	53	195 (3.68)	20814 (392.72)
0.4	26	105 (4.04)	9843 (378.58)

Table 2. A comparison of run-time on simple sequences (in seconds)

min_sup	*MNDER*	*Gen-REAR*	*WINEPI*
2%	6.3	113.8 (18.06)	6735.4 (1069.11)
3%	5.1	78.3 (15.35)	5053.7 (990.92)
4%	3.2	52.6 (16.44)	2626.4 (820.75)
5%	2.6	28.5 (10.96)	693.8 (266.85)

In Experiment 2, we compared the run-time on L10C10T1N50S6I4 with varying *min_sup* and *min_conf* = 0.35. Column 2 in Table 2 shows that the proposed algorithm can be finished in several seconds. The values in brackets demonstrate that *MNDER* is one order of magnitude faster than Gen-REAR and 2-3 orders of magnitude faster than *WINEPI*. The high efficiency benefits from one scan of the sequence and the generation of fewer candidates.

In Experiment 3, we evaluated the number of rules generated by *MNDER* from complex sequences when T and N vary, *min_sup* = 3% and *min_conf* = 0.35. Figure 4 (a) shows the number of rules found in L10C10T2-20N50S6I4 (T varies from 2 to 20). Figure 4 (b) shows the number of rules found in L10C10T5N10-100S6I4 (N varies from 10 to 100). The results in Fig. 4 show that sets of non-derivable episode rules compress complete sets by two orders of magnitude, and more rules are generated when either T or N increases.

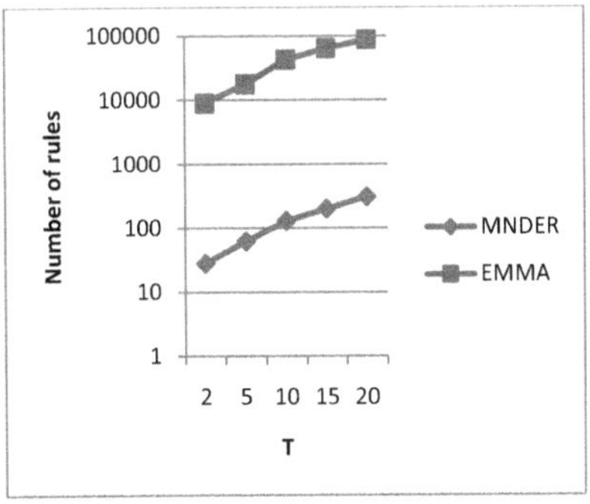

(a) Number of rules vs. T

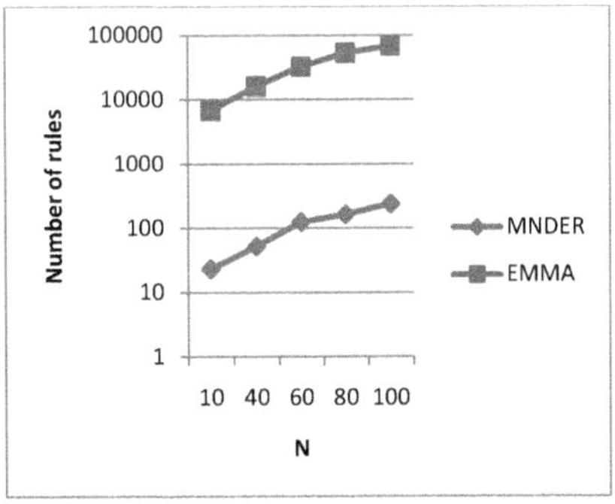

(b) Number of rules vs. N

Fig. 4. Number of rules found from complex sequences

Experiment 4 evaluates run-time of $MNDER$ and $EMMA$ on complex sequences L10C10T5N10-100S6I4 (N varies from 10 to 100) and L1-100C10T5N50S 6I4 (L varies from 1k to 100k) when $min\_sup = 3\%$ and $min\_conf = 0.35$. Figure 5 (a) shows that more time is needed when N increases. This is because more items and episodes need to be considered when N increases. Figure 5 (b) shows that MNDER spends more time on longer sequences. It can be seen from Fig. 5 that $MNDER$ runs faster than $EMMA$.

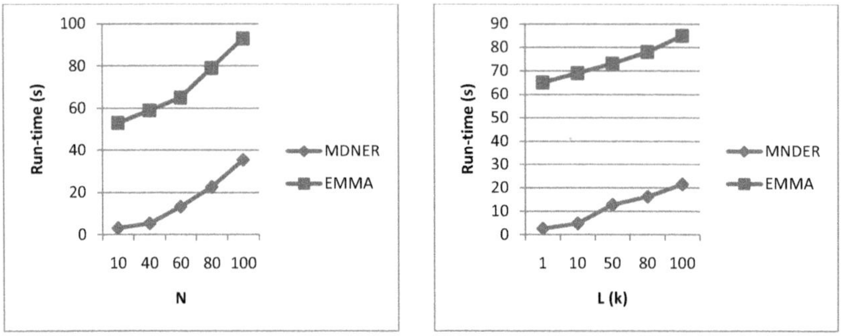

(a) Run-time vs. N (b) Run-time vs. L

Fig. 5. Run-time on complex sequences

6 Related Work

Traditional association rule mining in transactional databases [1] has been well developed. However, without considerations of relative order of occurrence, approaches to traditional association rule mining are not applicable to frequent episode mining and episode rule discovery [9]. Consequently, episode rule discovery has been treated and investigated as a separate topic.

The original framework for episode rule mining [9] was introduced by Mannila et al.. Since then, Mannila further improved the framework by introducing a new frequency metric $mo\text{-}freq$ and more efficient search strategies [10,11]. Besides Mannila's work [9,10,11], there have been other studies. The studies in [7,8] focus on episode discovery on data streams. Zhou et al. investigated the discovery of closed frequent episodes [17]. Huang et al. [6] considered complex sequences and proposed an efficient algorithm for mining frequent episodes in complex sequences. All these studies [6,7,8,9,10,11] concentrate on finding complete sets of episode rules other than concise sets. Less attention has been paid to the discovery of concise episode rules. Harms et al. introduced representative episode rules and developed an efficient mining algorithm $Gen\text{-}REAR$ [5]. Nevertheless, $Gen\text{-}REAR$ can not process complex sequences.

The existing approaches [5,6,7,8,9,10,11] can be classified according to four major angles: input (simple sequences vs. complex sequences), frequency metrics (anti-monotonic or not), output (complete sets vs. concise sets) and efficiency. In Table 3, a comparison from four angles shows the advantages of our method against three previous approaches, *WINEPI* [11], *Gen-REAR* [5] and *EMMA* [6].

Table 3. A comparison between our method and previous approaches

Input	**Frequency metric** (Anti-monotonic?)	**Output** Rule set	Size	**Efficiency**	**Approach**
A simple sequence	mo-freq (Y)	complete	large	slow	*WINEPI*
	fixed-win-freq (N)	representative	small	fast	*Gen-REAR*
A complex sequence	Distinct-bound-st (N)	complete	large	fast	*EMMA*
	T-freq (Y)	non-derivable	smaller	faster	*MNDER*

Multiple layered maximal frequent episodes used in our approach are essentially closed frequent episodes. So, closed subsequence mining is related to the problem considered in this paper. Closed subsequence mining has been developed in two streams. The first stream is closed sequential pattern mining (CSPM) [12,13,14,15,16] and the other is closed frequent episode mining (CFEM) [17]. Mining techniques for CSPM is different from our approach because CSPM discover closed frequent subsequences from sequence databases, while our approach extracts closed episodes from a single long sequence. The problem of mining non-derivable episodes is defined based on CSPM [17]. However, Clo-episode in [17] is not applicable to our problem as the adopted frequency metrics are different.

7 Conclusion

This paper proposed and investigated a new problem: the discovery of non-derivable episode rules in complex sequences. We developed an efficient mining algorithm *MNDER* for discovering non-derivable episode rules in complex sequences. The found sets are not only highly condensed but also information lostless. Experimental results on synthetic data showed that the proposed method outperforms previous approaches with the advantages of processing complex sequences, adopting an anti-monotonic frequency metric and achieving more condensed results and faster processing.

References

1. Agrawal, R., Imielinski, T., Swami, A.: Mining Association Rules between Sets of Items in Large Databases. In: ACM-SIGMOD International Conference on Management of Data, Washington, USA, pp. 207–216 (1993)

2. Agrawal, R., Srikant, R.: Fast Algorithms for Mining Association Rules. In: 20th International Conference on Very Large Data Bases, pp. 487–499 (1994)
3. Gan, M., Dai, H.: A Study on the Accuracy of Frequency Measures and its Impact on Knowledge Discovery in Single Sequences. In: Workshops at IEEE 10th Int. Conf. on Data Mining, Sydney, Australia, pp. 859–866 (2010)
4. Gan, M., Dai, H.: Obtaining Accurate Frequencies of Sequential Patterns over a Single Sequence. ICIC Express Letters 5(4) (2011) (in press)
5. Harms, S.K., Saquer, J., Tadesse, T.: Discovering Representative Episodal Association Rules from Event Sequences Using Frequent Closed Episode Sets and Event Constraints. In: IEEE International Conference on Data Mining (2001)
6. Huang, K., Chang, C.: Efficient Mining of Frequent Episodes from Complex Sequences. Information Systems 33(1), 96–114 (2008)
7. Iwanuma, K., Ishihara, R., Takano, Y., Nabeshima, H.: Extracting Frequent Subsequences from a Single Long Data Sequence: a Novel Anti-monotonic Measure and a Simple On-line Algorithm. In: 3rd IEEE International Conference on Data Mining, Texas, USA, pp. 186–193 (2005)
8. Laxman, S., Sastry, P., Unnikrishnan, K.: A Fast Algorithm for Finding Frequent Episodes in Event Streams. In: 13th International Conference on Knowledge Discovery and Data Mining, California, USA, pp. 410–419 (2007)
9. Mannila, H., Toivonen, H., Verkamo, A.: Discovering Frequent Episodes in Sequences. In: 1st International Conference on Knowledge Discovery and Data Mining, Montreal, Canada, pp. 210–215 (1995)
10. Mannila, H., Toivonen, H.: Discovering Generalized Episodes Using Minimal Occurrences. In: 2nd International Conference on Knowledge Discovery and Data Mining, Oregon, USA, pp. 146–151 (1996)
11. Mannila, H., Toivonen, H., Verkamo, A.I.: Discovery of Frequent Episodes in Event Sequences. Data Mining and Knowledge Discovery 1(3), 259–289 (1997)
12. Pei, J., Liu, J., Wang, H., Wang, K., Yu, P., Wang, J.: Efficiently Mining Frequent Closed Partial Orders. In: 5th IEEE International Conference on Data Mining, pp. 753–756 (2005)
13. Tzvetkov, P., Yan, X., Han, J.: TSP: Mining Ttop-k Closed Sequential Patterns. Knowl. Inf. Syst. 7(4), 438–457 (2005)
14. Wang, J., Han, J.: BIDE: Efficient Mining of Frequent Closed Sequences. In: 20th International Conference on Data Engineering, Boston, MA, USA, pp. 79–90 (2004)
15. Wang, J., Han, J., Li, C.: Frequent Closed Sequence Mining without Candidate Maintenance. IEEE Trans. Knowl. Data Eng. 19(8), 1042–1056 (2007)
16. Yan, X., Han, J., Afshar, R.: CloSpan: Mining Closed Sequential Patterns in Large Databases. In: SIAM International Conference on Data Mining (2003)
17. Zhou, W., Liu, H., Cheng, H.: Mining Closed Episodes from Event Sequences Efficiently. In: 14th Pacific-Asia Conference on Knowledge Discovery and Data Mining, Hyderabad, India, pp. 310–318 (2010)

Hybridizing Data Stream Mining and Technical Indicators in Automated Trading Systems

Michael Mayo

Department of Computer Science
University of Waikato
Hamilton, New Zealand
mmayo@waikato.ac.nz

Abstract. Automated trading systems for financial markets can use data mining techniques for future price movement prediction. However, classifier accuracy is only one important component in such a system: the other is a decision procedure utilizing the prediction in order to be long, short or out of the market. In this paper, we investigate the use of technical indicators as a means of deciding when to trade in the direction of a classifier's prediction. We compare this "hybrid" technical/data stream mining-based system with a naive system that always trades in the direction of predicted price movement. We are able to show via evaluations across five financial market datasets that our novel hybrid technique frequently outperforms the naive system. To strengthen our conclusions, we also include in our evaluation several "simple" trading strategies without any data mining component that provide a much stronger baseline for comparison than traditional buy-and-hold or sell-and-hold strategies.

1 Introduction

Analysing a financial market is a necessary precursor to the development of any trading strategy for that market. The type of analysis can vary greatly. For example, *fundamental* analysis is concerned with the broad economic factors and sweeping long term trends of a market [1]; *technical* analysis is concerned with finding clues to future price movements in historic market data and other variables [2]; and *sentiment* analysis involves gauging the opinion of market participants as to overall market direction [3]. Trading strategies may involve one, two or all of these methods of analysis.

Our research falls squarely into the technical analysis camp. Over the past hundred years or so, numerous technical indicators and technical charting methods (such as trend lines) have been developed for so-called "price chart reading" (e.g. [4]). These indicators and methods are now so firmly entrenched in the psychology of market participants that they often become self-fulfilling prophecies rather than independent predictors. With the advent of computers, these traditional indicators are now considerably easier to compute, and literally every trader can have a hundred or so different indicators available at her fingertips.

V. Torra et al. (Eds.): MDAI 2011, LNAI 6820, pp. 79–90, 2011.

In terms of research, academics routinely apply new computerised methods such as data mining (e.g. [6], [8]), neural networks (e.g. [5], [7], and [10]), evolutionary algorithms (e.g. [9]), and recently data stream mining ([11]) to the markets in order to develop newer and better trading techniques, but also to better understand how the markets work.

In this paper, we describe one such new technique which fuses the predictions made by a data mining classifier with a decision procedure based on technical analysis. The simple rule is that both types of analysis must agree before a trade in the predicted direction is made; if they disagree, no action is taken regardless of the classifier's prediction.

Our results show that in most cases, performance using this rule increases significantly compared to a trading system that *only* follows the classifier's recommendations. Furthermore, the number of trades (and this applies even to the situations where there is no significant improvement in trading performance) is considerably reduced – to around 50% in many cases – leading therefore to much reduced transaction costs.

We also provide a much more solid baseline for our experimental evaluations. Often in this field, it is considered "standard" to compare new strategies to buy-and-hold (whereby a long position is established at the beginning of the evaluation period and held to the end) or sell-and-hold (in which a short position is established and held to the end). The returns of the buy-and-hold or sell-and-hold strategies can then be compared to that of the new method under consideration. However, in modern markets, overly simplistic strategies such as buy-and-hold frequently underperform as Figure 1 illustrates.

Fig. 1. Daily closing prices for the EURUSD market, 23 May 2003 - 3 Dec 2010

This figure shows the daily closing prices for the EURUSD or "Eurodollar" market over the period from 23 May 2003 to 3 Dec 2010. The first closing price at the start of the period is \$1.16 and the final closing price is \$1.32 – representing a paltry 13.7% return (not annualized!) for a buy-and-hold strategy over a nearly 7 year period. Despite this, an inspection of the price series shows that price swung greatly several times in amounts far exceeding this net 13.7% movement. In fact, the highest recorded price is

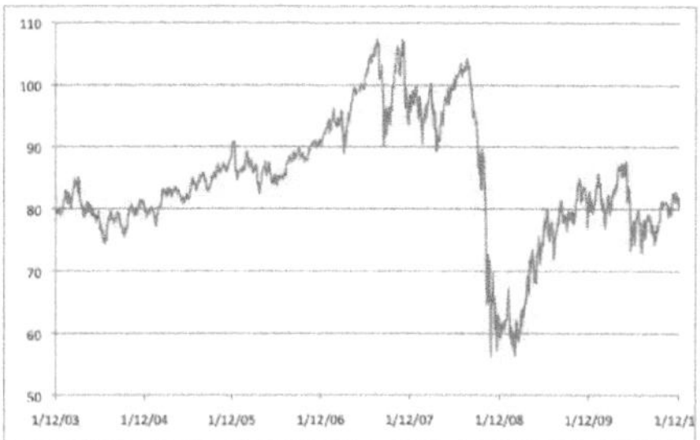

Fig. 2. Daily closing prices for the AUDJPY forex market, 1 Dec 2003 - 3 Dec 2010

around $1.60 and the lowest below $1.10. Clearly, any strategy just a little more intelligent than buy-and-hold could capture vastly more profit. Yet many papers compare their new "intelligent" strategy to buy-and-hold or sell-and-hold. A similar argument can be made for Figure 2, which shows prices for the Australian Dollar/Japanese Yen (AUDJPY) market.

We advocate significantly more challenging baseline strategies inspired by (and including) the simple strategies first proposed by Tiňo [12], which are designed specifically to be conducive for statistical significance testing.

In the next section, we outline our new method in more detail, discussing the technical and classifier components of the system as well as the strategy execution on a price series. In Section 3 we detail the experimental setup, in particular focussing on the baseline simple strategies (superior to buy-and-hold) that were used for comparison, as well as the evaluation measures usede. Section 4 describes the actual evaluation itself, with the datasets, and then the results. Finally, Section 5 concludes the paper.

2 Proposed New Trading Strategy Framework

Our hybrid framework for trading strategy design consists of two main components: a *technical component* based on standard technical indicators, and a *data stream mining component*, which is an abstaining classifier trained on a stream of historic price data. Besides price, the values of various indicators and other indexes may also be included in the stream.

2.1 The Technical Trading Rule (or Filtering) Component

A technical trading rule generally involves the computation of one or more technical indicators from historic price data. Because technical indicators are often designed to gauge a market's price trend direction, a trading rule is essentially a filter for trading actions, for example to rule out buy trades when the market is trending down.

One of the simplest technical indicators is the Simple Moving Average (SMA) [4]. Two instances of this indicator are depicted in Figure 3 where they are overlaid on the closing price series for the USDJPY market from the period 15 May 2003 to 3 Dec 2010. The dark, slower-moving line is the 200-period SMA while the medium-grey, faster-moving line is the 20-period SMA. Because the 200-period SMA lags behind the 20-period SMA, a good technical trading rule (and the one adopted in this paper) is to go long (buy) only if the 20 SMA is above the 200 SMA; and to go short (sell) only if the 20 SMA is below the 200 SMA.

We can see that using this rule would have resulted in mostly buying in the approximate period May '05 to May '07 because the 20 SMA is mostly above the 200 SMA during this period. Thereafter, the 20 SMA is mostly below the 200 SMA and therefore most trades would have been short (selling).

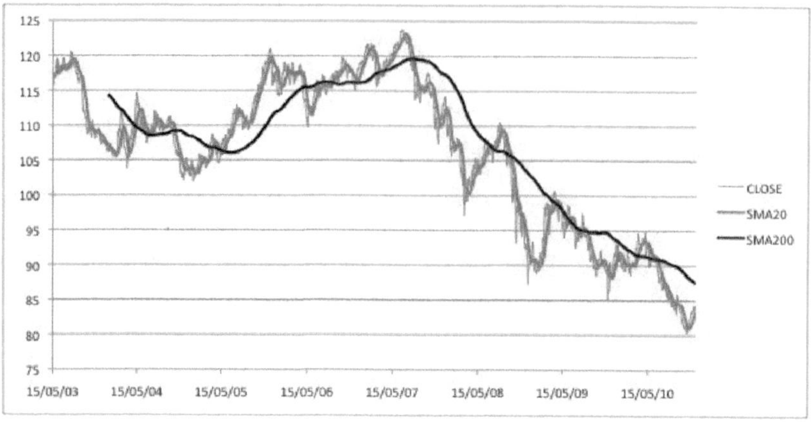

Fig. 3. Daily closing prices for the USDJPY forex market, 15 May 2003 - 3 Dec 2010, with the 20 and 200 period SMAs overlaid

Note that the direction of the technical trading rules does not force a trade to be made; rather it is applied as a filter to eliminate potentially incorrect predictions made by the abstaining classifier component described next.

2.2 The Abstaining-Classifier Component

The abstaining-classifier component is a machine learning classifier capable of abstaining from a prediction if the uncertainty is too high. The simplest way to achieve this is to have the classifier predict not a binary direction (e.g. up or down) for the market over the next period, but a probability distribution over market directions. If the probabilities are within a small deviation of 0.5 (which in our case is 0.0001), then the classifier abstains from making a prediction and there is no trade.

2.3 Strategy Execution

The basic rule is that in order for a trade to occur, the most likely market direction (up or down) as predicted by the classifier must agree with the technical trading trade. In other words, the 20-period SMA must exceed the 200-period SMA *and* the classifier must predict an upwards price movement in order for a long trade to happen; vice-versa for a short trade. If the classifier abstains or the classifier's prediction conflicts with the technical trading rule, then no trade is made.

We also use a standard "sliding window" method for executing our strategy. The basic idea is that (as opposed to performing a single train/test split for an entire dataset), a new classifier is instead trained for every single prediction that needs to be made. The training data for the classifier is obtained by sliding a 200-day fixed-size window along the price stream, so that only the most recent data (up to and excluding the test instance) is used for prediction. Using this method, older data is gradually discarded. Each instance in our data stream consists of 10 price points leading up to the day to be predicted.

Also, it should be noted that instead of raw prices, we use the *log-return* values:

$$r_n = sgn(c_n - o_n) \times log(K|c_n - o_n|) \tag{1}$$

where r_n is the log return, o_n is the opening price of the nth day, c_n is the same day's closing price, $sgn(.)$ is the sign function, and K is an abitrary constant. This feature proved far superior to raw price during initial testing.

The strategy assumes that trades are held only during market opening hours, and that they can only be initiated at the market open (i.e.. a buy at price o_n), and closed at the end of the day (at price c_n). No positions are allowed to be held overnight or over weekends, which eliminates the effects of gap ups and gap downs. No stops are used, which means that we do not need to be concerned with the order that prices were visited during the day – only o_n and c_n are significant.

Finally, the decision to trade and the direction of the trade for the next day are made at the immediate close of the current day, as soon as the SMAs and the classifier can be updated.

3 Methodology

In this section, we present four different experimental conditions that we were concerned with, and briefly describe the trading strategy evaluation measures used.

3.1 The Four Experimental Conditions

Simple, Non-Filtered. In the simple, non-filtered case, we adopt Tiňo's [12] four proposed baseline strategies. They are Simple$_L$, a strategy that goes long every day; Simple$_S$, a strategy that goes short every day; Simple$_{TR}$, a trend following strategy that buys if the previous day's close was higher than its open, and sells whenever yesterday's close was below its open; and Simple$_{CT}$, a counter-trend strategy that does the opposite of Simple$_{TR}$.

Note that while Simple$_L$ and Simple$_S$ are superficially similar to buy-and-hold and sell-and-hold, they exit the market at the close of each day, and re-enter the next day. Buy-and-hold and sell-and-hold on the other hand enter the market once at the period beginning and exit once at the end.

Simple, Filtered. The simple, filtered strategies are four additional strategies that are introduced in this paper. The basic idea is to take Tiňo's four baseline strategies described above and apply the technical trading rule described in Section 2.1. This generates four new strategies which are filtered – that is, they are only in the market if the trade direction agrees with the technical trading rule, and they are out of the market (flat) otherwise.

Machine Learning, Non-Filtered. In the Machine Learning (ML) non-filtered set of strategies, we use an abstaining classifier to predict market direction and trade whenever the classifier makes a prediction. The classifiers we use are Naive Bayes (NB) [13], Support Vector Machines (SVMs) [15] and Random Forest (RF) [16]. We also add a simple classifier, ZeroR (0R) which only ever predicts the majority class from the 200-day training dataset. This serves as an additional baseline for the classifiers. The implementations of the classifiers are those found in Weka 3.6.6 [17] with all default parameters, bar the Random Forest classifier which consists of 100 instead of 10 random trees.

Machine Learning, Filtered. Finally, the set of strategies in this group represent our target group: they are a full implementation of the system described in Section 2 in which an abstaining classifier's predictions are combined with a technical trading rule. They vary only in the choice of classifier.

3.2 Evaluation Measures

In this section, we briefly outline the evaluation measures we used.

Accuracy. The accuracy measures we report give the percentage of times that the strategy correctly predicts the market direction (up or down). We exclude situations where there is no trade (for example, because the classifier disagrees with the technical rule).

Net Profit Ratio. Most trading strategies are concerned with maximising net profit whilst minimising risk. This corresponds to having winning trades that return as much profit as possible, and losing trades that make minimal losses. One way to measure this is the Net Profit Ratio (NPR), in which total Net Profit (NP, i.e. sum of all wins from all winning trades less sum of losses from all losing trades) divided by Maximum Drawdown (MDD):

$$NPR = \frac{NP}{MDD} \tag{2}$$

In this ratio, the MDD is defined as the maximum drop in NP that a trading strategy experiences over a particular period. For example, if a strategy starts at \$0 NP, then reaches \$100 NP after some wins, then drops to \$50 NP after some losses, and finally ends the testing period (after further wins and losses) with \$120 net profit, then the MDD is \$50 which corresponds to the largest drop of profits from \$100 to \$50. The NPR therefore would be $\frac{\$120}{\$50} = 2.4$.

Ideally, we want to find trading strategies with NPRs as high as possible. This will tell us that the strategy has a high NP relative to its MDD. Strategies that have a NPR of 1.0 or less are undesirable for actual live trading, because such a low NPR implies that the MDD is greater than (or at least equal to) the NP, which may make the strategy a riskier proposition.

Statistical Significance. We also assess each trading strategy's performance statistically using Monte Carlo Permutation Testing (MCPT) [18] [19]. MCPT takes the daily positions (long, short or flat) made by a strategy, and randomly permutes them M times to produce M randomized trading strategies or "samples". It then computes the total NP of each sample and compares these using a conservative right-tailed test to the total NP achieved by the strategy. MCPT is useful for evaluating the statistical significance of a trading strategy because it makes no assumptions about the performances of other possible trading strategies – i.e. NPs achieved by the random strategies need not have a normal distribution, nor do they need to have a zero mean (which is a highly unlikely assumption in a strongly bullish or bearish market).

Fig. 4. Daily closing prices for the GOOGLE stock market, 25 Oct 2004 - 3 Dec 2010

Significance is reported for each strategy as a p value, where a smaller p value indicates greater significance. Values less than 0.05 are significant at 95% confidence.

4 Evaluation

We now describe the evaluation and our results in detail.

4.1 Datasets

We acquired five daily streaming datasets from Dukascopy [20]. They are the EURUSD, AUDJPY and USDJPY datasets already discussed and depicted in Figures 1-3, along with two stock market datasets, one for Google (Figure 4) and the other for Boeing (Figure 5).

Fig. 5. Daily closing prices for the BOEING stock market, 17 Mar 2003 - 3 Dec 2010

The data sets each comprise open, close, minimum and maximum prices for each trading day. We further added the 20 and 200 SMAs to the streams. In all cases, there are 2000 days worth of data, except for AUDJPY which has only 1832 days, and Google, which has 1583 days. EURUSD and USDJPY were chosen because they are the most commonly traded forex markets; AUDJPY was chosen because it is an interesting market with a high volume of carry trades; and Google and Boeing were selected because they represent two different but popular companies in the stock market.

In each dataset, predictions were not made for the first 210 days because these days were the minimum needed to construct a full dataset for training the classifiers given the window size of 200 and the instance size of 10.

4.2 Results

We executed each of the strategies in each of the four conditions (Filtered vs. non-filtered, simple vs. ML-based) on each of the five datasets. This gave a total of $16 \times 5 = 80$ experiments that were performed. To evaluate the effect of the classifier abstentions and technical filtering, we first of all counted the number of trades that were actually executed. They are given in Table 1. A key point from this table is that the effect of filtering varies massively. In some cases, the number of trades is reduced only somewhat, for example from 1790 to 1360 in the case of filtered 0R applied to EURUSD. However, in other cases the trade reduction is huge, such as the drop from 1790 trades to 476 trades in the case of Boeing with filtered Simple_L. This corresponds to trading about once every three or four days instead of every day.

Table 1. Number of trades by condition (row) and dataset (column)

Strategy	EURUSD	USDJPY	AUDJPY	GOOGLE	BOEING
NON-Simple_L	1790	1790	1622	1373	1790
NON-Simple_S	1790	1790	1622	1373	1790
NON-Simple_{TR}	1790	1790	1622	1373	1790
NON-Simple_{CT}	1790	1790	1622	1373	1790
NON-0R	1716	1694	1597	1318	1689
NON-NB	1790	1790	1622	1373	1789
NON-SVM	1790	1790	1622	1373	1790
NON-RF	1750	1749	1597	1336	1736
FIL-Simple_L	1121	794	1097	910	1314
FIL-Simple_S	669	996	525	463	476
FIL-Simple_{TR}	924	896	850	742	904
FIL-Simple_{CT}	866	894	772	631	886
FIL-0R	1360	957	1152	896	1207
FIL-NB	1019	925	937	807	987
FIL-SVM	977	917	977	890	984
FIL-RF	971	918	909	794	934

Table 2 gives the overall accuracies. In most cases the accuracy is around 50%, with the exception of AUDJPY in which filtered Simple_L achieves about 55%. This can be most likely explained as long-bias due to the carry trade. The near-random degree of accuracy concurs with previous results such as [21] and [8] where only small gains in accuracy (about 1-2%) above random were achievable when new methods were tested.

The NPRs for each strategy and each dataset are given in Table 3, which shows considerable variation.

About half of the strategies fail to make any profit at all, ending the testing period with a net loss (negative NPR). Of those remaining, many have a NPR below 1.0, which suggests that these strategies tend to make large losses in comparison to their final net profit.

Table 2. Strategy directional accuracy by condition (row) and dataset (column)

Strategy	EURUSD	USDJPY	AUDJPY	GOOGLE	BOEING
NON-Simple$_L$	49.4%	50.6%	**55.1%**	47.9%	49.5%
NON-Simple$_S$	48.9%	48.7%	43.8%	47.6%	48.6%
NON-Simple$_{TR}$	46.4%	46.8%	49.7%	48.0%	45.8%
NON-Simple$_{CT}$	**51.9%**	**52.5%**	49.2%	47.5%	**52.3%**
NON-0R	49.2%	51.1%	54.5%	47.4%	48.2%
NON-NB	49.7%	50.6%	52.0%	47.3%	48.6%
NON-SVM	50.6%	51.7%	51.7%	**48.4%**	49.1%
NON-RF	48.9%	49.8%	51.8%	47.2%	49.8%
FIL-Simple$_L$	50.6%	50.0%	**55.2%**	51.3%	50.0%
FIL-Simple$_S$	51.4%	48.3%	44.2%	49.2%	50.2%
FIL-Simple$_{TR}$	48.2%	46.2%	51.9%	49.3%	46.7%
FIL-Simple$_{CT}$	**53.8%**	51.0%	51.4%	**52.1%**	**53.5%**
FIL-0R	50.3%	50.1%	54.5%	49.3%	48.6%
FIL-NB	51.0%	49.9%	53.6%	49.3%	49.2%
FIL-SVM	51.7%	51.0%	53.2%	50.0%	50.0%
FIL-RF	50.3%	49.7%	53.7%	48.7%	50.9%

On the other hand, there are a few strategies that are big winners in NPR terms. For example, the filtered Simple$_{CT}$ strategy on EURUSD achieves a NPR of 2.142 – implying that more than $2 profit were made for each $1 of loss. However, this strategy does not include a classifier, and the strategies that did tended to perform not as well on the EURUSD dataset.

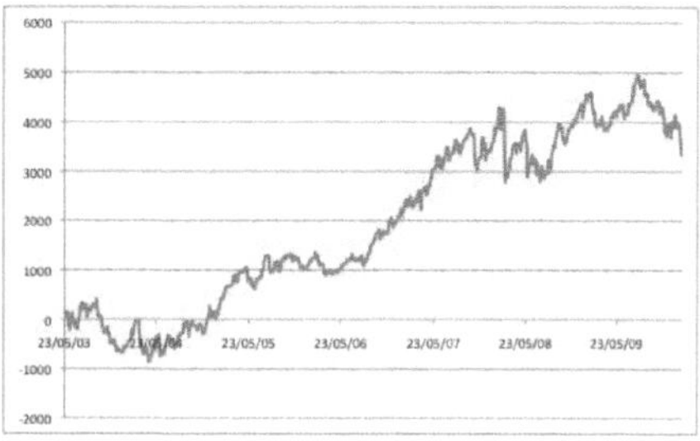

Fig. 6. Daily equity curve for the Filtered Simple$_{CT}$ strategy on EURUSD. The axes are day (x) vs. profit (y, in points, 1 point=0.0001 dollars).

The opposite is true however for the AUDJPY and BOEING datasets. In these experiments, filtered classifier-based strategies achieve NPRs of 1.444 and 3.909 respectively, with the classifiers being Naive Bayes in the first case and Random Forest in the second case. These two cases represent markets on which our new approach works exceedingly well.

Table 3. Strategy net profit ratio by condition (row) and dataset (column)

Strategy	EURUSD	USDJPY	AUDJPY	GOOGLE	BOEING
NON-Simple$_L$	-0.058	-0.595	0.004	0.522	0.184
NON-Simple$_S$	0.068	1.280	-0.008	-0.571	-0.228
NON-Simple$_{TR}$	-0.706	-0.755	-0.235	**0.394**	-0.436
NON-Simple$_{CT}$	**1.487**	**2.485**	0.506	-0.218	0.746
NON-0R	0.397	-0.818	-0.390	-0.661	-0.401
NON-NB	0.918	0.041	**1.186**	-0.853	-0.372
NON-SVM	1.408	1.204	0.400	-0.697	-0.261
NON-RF	-0.448	-0.128	-0.029	-0.853	**1.889**
FIL-Simple$_L$	0.670	-0.917	-0.052	1.243	2.043
FIL-Simple$_S$	0.794	-0.267	-0.067	0.305	1.084
FIL-Simple$_{TR}$	-0.299	-0.887	-0.280	**1.475**	0.560
FIL-Simple$_{CT}$	**2.142**	**0.382**	0.344	0.830	1.961
FIL-0R	0.871	-0.864	-0.361	-0.008	0.668
FIL-NB	1.229	-0.769	**1.444**	-0.535	0.585
FIL-SVM	1.362	-0.269	0.238	-0.157	1.012
FIL-RF	0.144	-0.598	0.044	-0.409	**3.909**

Table 4. Strategy statistical significance by condition (row) and dataset (column)

Strategy	EURUSD	USDJPY	AUDJPY	GOOGLE	BOEING
NON-Simple$_L$	1.000	1.000	1.000	1.000	1.000
NON-Simple$_S$	1.000	1.000	1.000	1.000	1.000
NON-Simple$_{TR}$	0.868	0.976	0.701	**0.378**	0.781
NON-Simple$_{CT}$	**0.132**	0.025	0.300	0.622	0.220
NON-0R	0.369	0.903	0.808	0.864	0.666
NON-NB	0.238	0.480	**0.102**	0.978	0.739
NON-SVM	0.138	**0.018**	0.352	0.978	0.626
NON-RF	0.719	0.560	0.504	0.966	**0.060**
FIL-Simple$_L$	0.209	0.916	0.526	**0.201**	0.060
FIL-Simple$_S$	0.209	0.916	0.526	**0.201**	0.060
FIL-Simple$_{TR}$	0.593	0.991	0.665	0.229	0.326
FIL-Simple$_{CT}$	**0.085**	**0.339**	0.367	0.347	0.056
FIL-0R	0.249	0.974	0.680	0.530	0.328
FIL-NB	0.158	0.820	**0.205**	0.788	0.296
FIL-SVM	0.103	0.625	0.410	0.611	0.231
FIL-RF	0.455	0.779	0.511	0.703	**0.017**

Table 4 gives the statistical significance values. In this table, a lower value indicates greater significance. Comparing the two tables, we see that in many cases, a low p value correlates to a high NPR. Note that there are only a handful of strategies that are significant at 95% level: they are non-filtered Simple$_{CT}$ and SVM strategies applied to the USDJPY market (the SVM strategy has greater significance); and the Random Forest-based strategies applied to BOEING. Some of the other high-NPR also have low p values, but they are not quite significant, such as the non-filtered Naive Bayes strategy with a p value of 0.102 which is nearly significant at a level of 90%.

The equity curves of some of the better-performing strategies are presented next. Figure 6 shows the reasonably good performance of the filtered simple countertrend strategy on the EURUSD market. Note that the strategy actually loses money for the first year or so before profits start to increase.

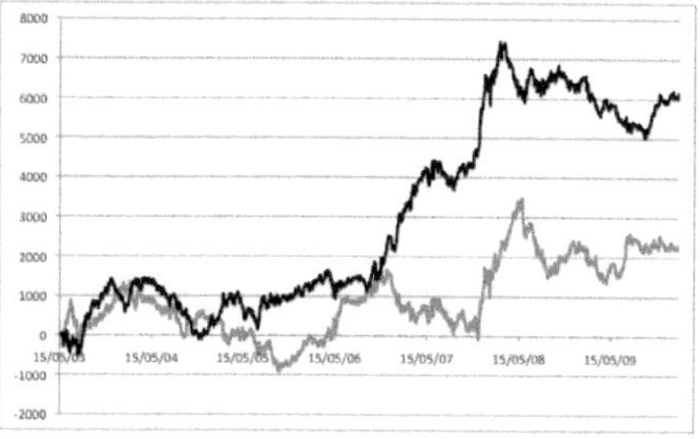

Fig. 7. Daily equity curve for the Non-filtered Simple$_{CT}$ (black line, upper) and SVM (grey line, lower) strategies on USDJPY. The axes are day (x) vs. profit (y, in points, 1 point=0.01 Yen).

Figure 7 shows the equity curves for the two highly-performing strategies applied to the USDJPY market, specifically the non-filtered countertrend strategy and the non-filtered SVM strategy. Although the latter strategy has a higher statistical significance according to permutation testing, the former strategy actually produces a greater NPR over time. This emphasizes one of the key points of the permutation test for trading strategies, which is that the test does not rank strategies according to profitability: instead, it ranks them according to how unlikely it would be for a random strategy to produce the same result.

Figures 8 and 9 show further equity curves, this time for two filtered strategies, specifically Naive Bayes on AUDJPY and Random Forest on BOEING. Note than in all cases, the equity curves show the *points* or *cents* won. This measurement is independent of position or account size. If the curves were depicted with account size on the y axis instead and compounding of position sizes was employed, it would be expected that the curves would be much steeper.

5 Conclusion

To conclude, we have demonstrated that a novel hybridized data mining/technical trading rule strategy can perform effectively and significantly in some markets. However, there is no single optimal or "holy grail" strategy that fits all five of our test datasets. Rather, each market appears to have its own dynamics and character, and therefore requires its own unique investigation. It is also known that markets change gradually over time (i.e. the distribution of price changes is non-stationary), so the process of optimizing the hybrid strategy is likely to be continuous rather than a one-off event.

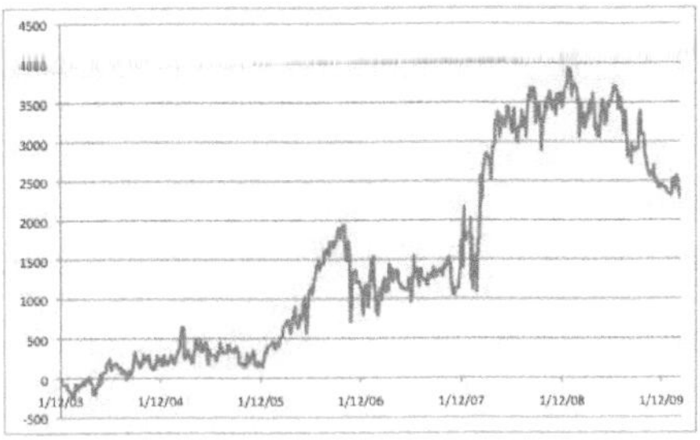

Fig. 8. Daily equity curve for the Filtered Naive Bayes strategy on AUDJPY. The axes are day (x) vs. profit (y, in points, 1 point=0.01 Yen).

We have also compared our "intelligent" strategies to a set of very strong simplistic strategies which can sometimes themselves yield high profits and near-statistical significance. In this respect, our research here differs considerably from that of prior literature where the baseline strategy, if one is proposed, is most often an easily out-performed buy-and-hold strategy. We feel the more rigorous evaluations performed here give a more realistic view of the performance of our approach.

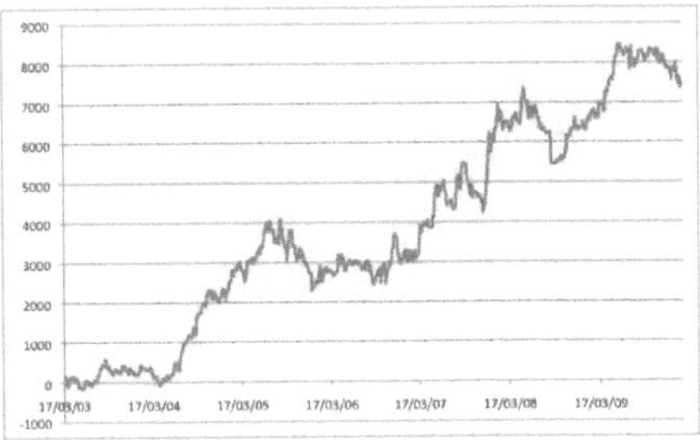

Fig. 9. Daily equity curve for the Filtered Random Forests strategy on BOEING. The axes are day (x) vs. profit (y, in cents).

There is also one caveat that should be made concerning this research: we have not included transaction and slippage costs in our simulations. For the forex markets, the transaction costs are very low compared to the stock market, but costs are changing rapidly with time. Slippage and costs are difficult to model because they are dependent on the broker as well as market conditions not available in the price data stream. Individuals constructing a live implementation of an automated trading system such as the one introduced here should make appropriate assumptions about their own costs when they evaluate potential strategies.

References

1. Schwager, J.: Futures: Fundamental Analysis. Wiley, Chichester (1995)
2. Pring, M.: Market Momentum. McGraw-Hill Companies, New York (1997)
3. Saettele, J.: Sentiment in the Forex Market. Wiley, Chichester (2008)
4. Pring, M.: Technical Analysis Explained. McGraw-Hill, New York (2002)
5. Lean, Y., Lai, K.: Foreign Exchange Rate Forecasting with Artificial Neural Networks. Springer, Heidelberg (2007)
6. Liu, Z., Xiu, D.: An automated trading system with multi-indicator fusion based on D-S evidence theory in forex market. In: Proc. Sixth International Conference on Fuzzy Systems and Knowledge Discovery, pp. 239–243. IEEE, Los Alamitos (2009)
7. Ni, H., Yin, H.: Exchange rate prediction using hybrid neural networks and trading indicators. Neurocomputing 72, 2815–2832 (2009)
8. Barbosa, R., Belo, O.: Autonomous Forex Trading Agents. In: Perner, P. (ed.) ICDM 2008. LNCS (LNAI), vol. 5077, pp. 389–403. Springer, Heidelberg (2008)
9. Hirabayashi, A., Aranha, C., Iba, H.: Optimization of the Trading Rule in Foreign Exchange using Genetic Algorithm. In: Proc. GECCO 2009, pp. 1529–1536 (2009)
10. Walczak, S.: An Empiracal Analysis of Data Requirements for Financial Forecasting with Neural Networks. Journal of Management Information Systems 17(4), 203–222 (2001)
11. Montana, G., Trianafyllopoulos, K., Tsagaris, T.: Data stream mining for market-neutral algorithmic trading. In: Proc. Symposium on Applied Computing (SAC 2008), pp. 966–970 (2008)
12. Tiňo, P., Schittenkopf, C., Dorffner, G.: Financial Volatility Trading using Recurrent Neural Networks. IEEE Trans. on Neural Networks 12(4), 865–874 (2002)
13. John, G., Langley, P.: Estimating Continuous Distributions in Bayesian Classifiers. In: Eleventh Conference on Uncertainty in Artificial Intelligence, San Mateo, pp. 338–345 (1995)
14. le Cessie, S., van Houwelingen, J.C.: Ridge estimators in logistic regression. Applied Statistics 41(1), 191–201
15. Platt, J.C.: Fast training of support vector machines using sequential minimal optimization. In: Schölkopf, B., Burges, C., Smola, A. (eds.) Advances in Kernel Methods – Support Vector Learning. MIT Press, Cambridge (1998)
16. Breiman, L.: Random Forests. Machine Learning 45(1), 5–32
17. Hall, M., Frank, E., Holmes, G., Pfahringer, B., Reutemann, P., Witten, I.H.: The WEKA data mining software: An update. SIGKDD Explorations 11(1) (2009)
18. Aronson, D.: Evidence-Based Technical Analysis: Applying the Scientific Method and Statistical Inference to Trading Signals
19. Masters, T.: Monte-Carlo Evaluation of Trading Systems (2006), `http://www.evidencebasedta.com/MonteDoc12.15.06.pdf` (retrieved December 13, 2010)
20. Dukascopy Swiss Forex Bank & Marketplace, `http://www.dukascopy.com` (Data retrieved December 4 2010)
21. Krause, A.: Evaluating the performance of adapting trading strategies with different memory lengths. In: Corchado, E., Yin, H. (eds.) IDEAL 2009. LNCS, vol. 5788, pp. 711–718. Springer, Heidelberg (2009)

Semi-supervised Dimensionality Reduction via Harmonic Functions

Chenping Hou[1,⋆], Feiping Nie[2], and Yi Wu[1]

[1] Department of Mathematics and System Science, National University of Defense Technology, Changsha, 410073, China
hcpnudt@gmail.com
[2] Department of Computer Science and Engineering, University of Texas, Arlington, 76019, USA

Abstract. Traditional unsupervised dimensionality reduction techniques are widely used in many learning tasks, such as text classification and face recognition. However, in many applications, a few labeled examples are readily available. Thus, semi-supervised dimensionality reduction(SSDR), which could incorporate the label information, has aroused considerable research interests. In this paper, a novel SSDR approach, which employs the harmonic function in a gaussian random field to compute the states of all points, is proposed. It constructs a complete weighted graph, whose edge weights are assigned by the computed states. The linear projection matrix is then derived to maximize the separation of points in different classes. For illustration, we provide some deep theoretical analyses and promising classification results on different kinds of data sets. Compared with other dimensionality reduction approaches, it is more beneficial for classification. Comparing with the transductive harmonic function method, it is inductive and able to deal with new coming data directly.

Keywords: semi-supervised dimensionality reduction, harmonic function, soft label, weighted complete graph.

1 Introduction

Dimensionality reduction is a big challenge in many areas, such as pattern recognition and machine learning. It is a frequently used preprocessing technique for learning tasks. Reducing dimensions may improve the classifier performance since it can suppress noise in the data and act as a form of regularization. Moreover, a meaningful low dimensional representations can help in visualizing the data. It is an important tool in exploratory data analysis.

However, in many practical applications, we usually face the semi-supervised learning problem. One often has a few prior information, since obtaining prior knowledge, such as labeling points or constructing links often requires expensive

⋆ Corresponding author. Thanks to the NSFC China, No.60975038, 61005003 for their supports.

V. Torra et al. (Eds.): MDAI 2011, LNAI 6820, pp. 91–102, 2011.

human labor and much time. On the contrary, a large number of unlabeled points can be much easier to obtain. For example, in text classification, one can easily access to a plenty of documents by crawling the Web, but only a small percent of them are classified by hand. Thus, how to design an effective semi-supervised dimensionality reduction (SSDR) approach to reduce the dimensionality of this kind of data is a challenging problem.

Traditional dimensionality reduction approaches are not suitable to solve this problem since 1) Unsupervised methods, such as Principle component analysis(PCA) [1], [2], Locally linear embedding (LLE) [3] and Maximum variance unfolding (MVU) [4], often suffer from a low discriminant power due to its unsupervised nature. 2) Supervised methods, e.g., Linear discriminant analysis (LDA) [5], Generalized additive model [6], often require a large number of points.

There is little work concerning about SSDR. Considering the types of prior knowledge, we can classify previous SSDR approaches into three categories. 1) The first kind of approaches adopt pre-defined low dimensional representations of several points. Typical method is proposed by Yang et al[7]. They modified several typical nonlinear dimensionality reduction techniques by taking into account prior information on exact mapping of certain data points. 2) The second type methods employ domain knowledge in the form of pairwise constraints. Zhang et al first specified whether a pair of instances belong to the same class (must-link constraints) or different classes (cannot-link constraints) and then computed linear transformations[8]. These two kinds of links have been used in previous for improving the performance of K-means firstly[9]. 3) The third type of SSDR methods use label information directly. They commonly employ labeled and unlabeled data points to construct a weighted graph. Typical method may include Semi-supervised Discriminant Analysis (SDA) [10], which uses labeled data points to maximize the separability between different classes and the unlabeled data points to estimate the intrinsic geometric structure of the data.

Despite the success of applying semi-supervised dimensionality reduction approaches to many fields[11], there are still some problems that are not properly addressed till now. The performance of different kinds of SSDR approaches can also be improved. The first kind of approaches are not suitable for real applications, since the required prior information is too strict in practice. The second kind of approaches require plenty of links to guarantee their validity. Moreover, labeling a few points is more realistic than constructing a large number of links. The third type of methods do not perform well if the labeled points are not sufficient.

To address the above issues, we propose a novel method called Semi-supervised dimensionality reduction via harmonic functions (SSDR *via* HF). The SSDR *via* HF algorithm first computes the states of the whole data set using harmonic functions in Gaussian fields[12], [13]. We show that the state of a point can also be regarded as its soft label. Then, we construct a weighted complete graph. For each pair of vertexes, the weight measures the similarity of two linked points. After that, a linear transformation matrix can be derived by maximizing all dissimilarities and simultaneously, minimizing all similarities. The transformation matrix is constrained to be orthogonal or uncorrelated respectively. We prove theoretically that the state computation can be solved in a closed form and SSDR approach [8] can be regarded as a special case of our method. Finally, the experiments on image, digit and text are presented to show the effectiveness.

The remainder of this paper is organized as follows. Section 2 will review the related work and Section 3 will show the SSDR *via* HF algorithm in detail. The analysis of SSDR *via* HF will be proposed in Section 4. Section 5 presents the experimental results on synthetic and real-world data sets, followed by the conclusions and future works in Section 6.

2 Notations and Harmonic Function

In this section, we will briefly review the label propagation technique based on the harmonic function[13]. First, let us introduce some notations. A set of n data points in $\mathbb{R}^{D_1}$ is represented by $\mathcal{X} = \{x_1, x_2, \cdots, x_l, x_{l+1}, \cdots, x_n\}$, and $\mathcal{L} = \{1, 2, \cdots, c\}$ is the label set. There are c classes in total. The first l points $\mathcal{X}_L = \{x_i\}_{i=1}^{l}$ are labeled as $\mathcal{T}_L = \{t_i \in \mathcal{L}\}_{i=1}^{l}$ and the remaining points $\mathcal{X}_U = \{x_i\}_{i=l+1}^{n}$ are unlabeled. Commonly, for an inductive method, we may also have t unseen points $\mathcal{X}_T = \{x_i\}_{i=n+1}^{n+t}$ for testing. These unseen points are not available for training. Our goal is to find a suitable mapping matrix P, which is computed base on $\mathcal{X}$ and projects the whole data to a low dimensional space, i.e., $y_i = Px_i$ for $i = 1, 2, \cdots, n+t$. Here $y_i \in \mathbb{R}^{D_2}$ and $D_2 \ll D_1$.

The harmonic function aims to predict states of unlabeled points and employ values on several known points as the constraints. In essence, this method plays the role to propagate labels from labeled points to unlabeled data.

For training with c classes, we need to compute a vector-based function $f : \mathcal{X} \to \mathbb{R}^c$, where the jth element of $f(x_i)$ (a row vector) corresponds to the probability that x_i belongs to the jth class (we will show this result in Section 4.). Therefore, for a labeled data x_i, if it is in the jth class, the jth element of $f(x_i)$ equals to one and other elements are zeros.

First, a weighted neighborhood graph is constructed on the whole data set. An $n \times n$ symmetric weight matrix W on the edges of the graph can be computed by gaussian function, i.e.,

$$w_{ij} = exp(- \parallel x_i - x_j \parallel^2 / (2\sigma^2)). \tag{1}$$

Here σ is the scale hyperparameter.

We constrain $f(x_i) = t_i$ on the labeled data set $\mathcal{X}_L$ and choose the quadratic energy function:

$$E(f) = \frac{1}{2} \sum_{i,j} w_{ij} (f(x_i) - f(x_j))(f(x_i) - f(x_j))^T. \tag{2}$$

The minimum energy function $f = argmin_{\{f(x_i)=t_i, i=1,\cdots,l\}} E(f)$ is *harmonic*, namely, it satisfies $\Delta f = 0$ on unlabeled data $\mathcal{X}_U$ and is equal to $\mathcal{T}_L$ on the labeled data points $\mathcal{X}_L$. Here $\Delta = D - W$, $D = diag(d_i)$ is the diagonal matrix with entries $d_i = \sum_j w_{ij}$. "0" is a matrix with all zeros. The harmonic property means that the value of f at each unlabeled data point is the average of f values at its neighboring points.

$$f(x_j) = \frac{1}{d_j}\sum_{i\sim j} w_{ij} f(x_i), \text{for } j = l+1, \cdots, n. \tag{3}$$

Denote

$$f_l = \begin{bmatrix} f(x_1) \\ f(x_2) \\ \cdots \\ f(x_l) \end{bmatrix}, f_u = \begin{bmatrix} f(x_{l+1}) \\ f(x_{l+2}) \\ \cdots \\ f(x_n) \end{bmatrix}, f = \begin{bmatrix} f_l \\ f_u \end{bmatrix}. \tag{4}$$

$Q = D^{-1}W$, Eq. (3) can be expressed slightly differently: $f = Qf$. Let $f(x_i) \triangleq [f_1^{(i)}, f_2^{(i)}, \cdots, f_c^{(i)}]$. Since the harmonic functions have maximum principle, f is unique and is either a constant or the jth element of $f(x_i)$ (row vector) satisfies $0 < f_j^{(i)} < 1$ for $i = l+1, \cdots, n$, $j = 1, \cdots, c$.

To compute the harmonic solution in a closed form, we split Δ as follows:

$$\Delta = \begin{bmatrix} \Delta_{ll} & \Delta_{lu} \\ \Delta_{ul} & \Delta_{uu} \end{bmatrix}.$$

Considering the harmonic properties, we can formulate the following equation based on $\Delta f = 0$ on unlabeled data.

$$\Delta_{ul} f_l + \Delta_{uu} f_u = 0. \tag{5}$$

Since D is a diagonal matrix, $\Delta_{ul} = -W_{ul}$. We compute f_u by the following equation in a closed form.

$$f_u = -\Delta_{uu}^{-1}\Delta_{ul} f_l = (D_{uu} - W_{uu})^{-1} W_{ul} f_l = (I - Q_{uu})^{-1} Q_{ul} f_l. \tag{6}$$

As shown in Eq. (6), the harmonic function can be automatically computed in a closed form. More importantly, label information is propagated and thus it is beneficial for the following dimensionality reduction.

3 The Algorithm

In this section, we will formally present our SSDR *via* HF algorithm, which aims to discover the low dimensional representations of original points.

3.1 Computing the State by Harmonic Function

We directly apply Harmonic Function[13], [14] to compute the states f_u of unlabeled points, i.e., the possibilities that one point belongs to the c classes. Through this way, we can propagate the label information from labeled points to unlabeled samples. More concretely, we integrate the geometry characters W from all points to enlarge the label information f_l. Thus, the label information is effectively used.

There are some aspects that should be highlighted for this step:

(1) Since the points, whose states are known as labels, are represented by probability matrix f_l, each row of f_u also corresponds to the soft label of an unlabeled point. The input label information f_l is formulated as a $l \times c$ matrix, therefore, the output f_u is a $(n-l) \times c$ matrix. More accurately, they are probabilities of the $(n-l)$ points belonging to c different classes, we will give the proof in Section 4. In other words, we can propagate label information from labeled points to the whole data set by harmonic functions.

(2) Without iteration, we can directly compute f_u by Eq. (6), the adding computational requirement in this step is limited.

3.2 Constructing a Weighted Complete Graph

In this section, we will explain how to construct a weighted complete graph based on the soft label matrix f.

For any two points, since we have known their soft labels, i.e., the probability that they belong to the c classes, it is direct to construct a complete graph by connecting every two points. The soft labels can be employed to measure the similarity between every two points. We will show how to define these weights.

Take two points x_i and x_j as examples, their corresponding harmonic function values are denoted by $f(x_i)$ and $f(x_j)$. Recall the intuition of classification, i.e., two points in the same class are much more similar than two points belonging to different classes. Note that $f(x_i) = [f_1^{(i)}, f_2^{(i)}, \cdots, f_c^{(i)}]$ for $i = 1, 2, \cdots, n$. Since $f_k^{(i)}$ is the probability of x_i belonging to the kth class, it is directly to define the similarity S_{ij} between x_i and x_j by

$$S_{ij} = \sum_{k=1}^{c} f_k^{(i)} f_k^{(j)} = f(x_i) f(x_j)^T. \tag{7}$$

It is the probability that x_i and x_j belonging to the same class. More concretely, if x_i and x_j are in the same class k, $f_k^{(i)}$ and $f_k^{(j)}$ are much larger than other elements of $f(x_i)$ and $f(x_j)$, therefore, S_{ij} is relatively large. Comparing with SSDR, S_{ij} can also be regarded as the probability that x_i and x_j has a must-link connecting them.

To compute the similarity matrix in terms of matrix operations, we can directly formulate similarity matrix S in the following form.

$$S = ff^T. \tag{8}$$

Here f is a $n \times c$ matrix that are defined by Eq. (4).

There is another simple strategy to use the soft label matrix to construct similarity matrix. We can first change the soft label of a point to the hard label by assigning the point to the class, which corresponds to the largest probability. Take a particular point x_i as an example, if $f_k^{(i)}$ is the largest element of $f(x_i)$, then

$$f_m^{(i)} = \begin{cases} 1 & \text{if } m = k, \\ 0 & \text{if } m \neq k. \end{cases}$$

After the changing, we apply the same method to construct similarity matrix based on the new f by Eq.(8). In essential, the two strategies have no significant difference in deriving low dimensional embeddings since we will take a parameter to balance the effects of two items (see Eq. (9)). In the following experiments, we will simply employ the first strategy.

In summary, we have construct a complete graph with a weight measuring the similarity for two connected points. The larger similarity is, the more likely two linked points belong to the same class and vice versa.

3.3 Deriving Projection Matrix

The final step of SSDR *via* HF is to derive projection matrix P on the whole data such that $y_i = Px_i$, for $i = 1, 2, \cdots, n+t$. Denote $X \triangleq [x_1, x_2, \cdots, x_l, x_{l+1}, \cdots, x_n]$, $X_T \triangleq [x_{n+1}, x_{n+2}, \cdots, x_{n+t}]$, $Y \triangleq [y_1, y_2, \cdots, y_l, y_{l+1}, \cdots, y_n]$ and $Y_T \triangleq [y_{n+1}, y_{n+2}, \cdots, y_{n+t}]$.

In the second step, we have constructed a complete graph whose edge weights can measure the similarity of two connected points. When we refer to the dimensionality reduction, it is intuitive that if two points are in the same class, their low dimensional representations are expected to be near. On the contrary, if two points belong to two different classes, their expected representations should be far away. Since S_{ij} is the probability that x_i and x_j belong to the same class, we expect to find the embedding that maximizes

$$E(Y) = \sum_{i=1}^{n}\sum_{j=1}^{n}(1 - S_{ij})\|y_i - y_j\|^2 - \lambda\sum_{i=1}^{n}\sum_{j=1}^{n} S_{ij}\|y_i - y_j\|^2 \tag{9}$$

Here $1 - S_{ij}$ is the probability of x_i and x_j belonging to two different class, the first item measures the dissimilarities for all points belonging to different classes. On the contrary, the second item measures the similarities. More concretely, comparing with SSDR, the first item is the sum of weighted distances for cannot link and the second item is for must-link. λ is a tradeoff parameter which can balance the effects of two items.

Intuitively, if two points are in the same class, S_{ij} is comparatively larger than $1 - S_{ij}$, thus, the maximization of $E(Y)$ is mainly focus on minimizing the last item, i.e., compressing distance of two points in the same class. On the contrary, if two points belong to different classes, S_{ij} is much smaller than $1 - S_{ij}$. The maximization of $E(Y)$ is approximately equivalent to enlarging the first item, which is the sum of weight distances between points of different classes.

If we replace y_i by Px_i, $E(Y)$ is rewritten as

$$E(Y) = \sum_{i=1}^{n}\sum_{j=1}^{n}(1 - S_{ij})\|Px_i - Px_j\|^2 - \lambda\sum_{i=1}^{n}\sum_{j=1}^{n} S_{ij}\|Px_i - Px_j\|^2. \tag{10}$$

To compute the optimal solution in a closed form, we rewrite $E(Y)$ in the form of matrix.

$$E(Y) = tr(PX(\Phi - \Gamma)X^T P^T). \tag{11}$$

Here, Γ is a matrix whose element $\Gamma_{ij} = 1 - (1+\lambda)S_{ij}$. $\Phi = diag(\phi_i)$ is a diagonal matrix with entries $\phi_i = \sum_j \Gamma_{ij}$.

Clearly, to guarantee that Eq.(11) has an optimal solution, we should add some constraints on the projection matrix P. There are two commonly used constraints: orthogonal and uncorrelated.

Orthogonal constraint. If we expect that the low dimensional representations are just a rotation on original data, the orthogonal constraint, i.e. $PP^T = I$ is a good choice. This constraint is commonly used in unsupervised dimensionality reduction approaches, such as PCA. SSDR *via* HF with orthogonal constraint can be regarded as solution to the following problem.

$$\begin{aligned} argmax \;\; & tr(PX(\Phi - \Gamma)X^T P^T), \\ \text{s.t.} \;\; & PP^T = I. \end{aligned} \tag{12}$$

This problem can be easily solved by eigen-decomposition of $X(\Phi - \Gamma)X^T$. We call our method Orthogonal semi-supervised dimensionality reduction via harmonic function (OSSDR *via* HF) in this situation.

Uncorrelated constraint. Since it has been pointed out that uncorrelated constraint is more reasonable than orthogonal constraint in some cases[15], we also directly employ uncorrelated constraint, i.e. $PS_tP^T = I$ in our method. Here S_t is the covariance matrix of all data points in the original space. Uncorrelated constraint has been successfully used in supervised dimensionality reduction methods, such as LDA. SSDR *via* HF with uncorrelated constraint is computed by

$$\begin{aligned} argmax \;\; & tr(PX(\Phi - \Gamma)X^T P^T), \\ \text{s.t.} \;\; & PS_tP^T = I. \end{aligned} \tag{13}$$

This problem can also be simply solved[15] in the close form by generalized eigen-decomposition of $X(\Phi - \Gamma)X^T$ and S_t. We call our method Uncorrelated Semi supervised dimensionality reduction via harmonic function (USSDR *via* HF) with this kind of constraint.

For an unseen data, e.g, $x_i \in \mathcal{X}_T$, we can directly reduce its dimensionality by employing the projection matrix P, i.e., $y_i = Px_i$ for $i = n+1, \cdots, n+t$.

4 Analysis and Extensions

Performance analysis. First, we will show the reason why we can directly employ Eq. (6) to compute the states of all points. More concretely, we will prove that the minimum energy function f shown in Section 2 is harmonic.

Theorem 1. *The optimal function f as defined by Eq. (3) that minimizes energy function shown in Eq. (2) with constraint $f\mid_{\mathcal{X}_L} = f_l$ is harmonic.*

Then, we will show why we can apply Eq. (8) to construct the similarity matrix. The reason is that f_u is a probability matrix in common cases, i.e., the sum of each row's elements of f_u is equal to one and all the items of f_u is non-negative. The following theorem shows this conclusion.

Theorem 2. *Assume that $\rho(Q_{uu}) < 1$, then the optimal function f_u computed by Eq. (6) is a probability matrix, i.e., $f_u \geq 0$ and $f_u \mathbf{1}_{c\times 1} = \mathbf{1}_{(n-l)\times 1}$. Here $\rho(Q_{uu})$ represents the spectral radius of Q_{uu}, $f_u \geq 0$ means that all elements of f_u is non-negative. $\mathbf{1}_{c\times 1}$ represents a $c \times 1$ vector with all ones.*

Finally, since we assume that $\rho(Q_{uu}) < 1$ in Theorem 2. We will explain what this assumption actually means. In essential, $\rho(Q) = 1$ since Q is a probability matrix. Q_{uu} is only a block of Q and commonly, $\rho(Q_{uu}) < 1$. If $\rho(Q_{uu}) = 1$, it means that the original graph is disconnected and some connected components have no labeled data. The harmonic function can not be used in this situation. More concretely, we have the following theorem.

Theorem 3. *Assume that $\mathcal{C}_1, \mathcal{C}_2, \cdots, \mathcal{C}_m$ are m connected components formulated by unlabeled data points of neighborhood graph. For each component, there is at least one labeled point that connects with at least one point of this component. Then, $\rho(Q_{uu}) < 1$.*

For conciseness, we would like to omit the proofs. In real applications, it is usually assumed that the points in the same class are in one component of the neighborhood graph and simultaneously, there is at least one labeled point related to this component. This guarantees that the connected component is not isolated.

In summary, the above three theorems guarantee that we can employ S in Eq.(8) to measure the similarities between every two points.

Relations to other approaches. SSDR *via* HF has close relationship to other approaches, such as SSDR method[8]. More concretely, for every two labeled points x_i and x_j, if they are of the same label, they have a must-link connecting them and $S_{ij} = 1$. On the contrary, they have a cannot-not link and $S_{ij} = 0$. Denote n_M and n_C are the numbers of cannot links and must links. As in SSDR, if we introduce two parameters α, β and assume

$$\lambda = \frac{(\beta/n_M - 1/n^2)}{(1/n^2 + \alpha/n_C)}, S_{ij} = \frac{1}{\lambda+1} - \frac{1/n^2}{(1/n^2+\alpha/n_C)(\lambda+1)}, \tag{14}$$

then, Eq. (9) becomes

$$\begin{aligned}(\frac{1}{n^2} + \frac{\alpha}{n_C})E(Y) = & \sum_{x_i,x_j\in\mathcal{X}} \frac{1}{n^2}\|Px_i - Px_j\|^2 + \sum_{x_i,x_j\in\mathcal{C}} \frac{\alpha}{n_C}\|Px_i - Px_j\|^2 - \\ & \sum_{x_i,x_j\in\mathcal{M}} \frac{\beta}{n_M}\|Px_i - Px_j\|^2\end{aligned} \tag{15}$$

Here $\mathcal{C}$ and $\mathcal{M}$ contain points pairs who have cannot-link and must-link. The right side of Eq. (15) is just the objective function of SSDR[8]. Thus, SSDR can be considered as a special case of our method. Moreover, the element S_{ij} in our approach is learned via label propagation and that in SSDR is predefined.

Moreover, SSDR *via* HF has some relationships with SDA[10]. SDA employs XLX^T as the regularizer, which is added to the total scatter matrix S_t. The objective function of one-dimensional SDA is as follows

$$max \frac{A^T S_b A}{A^T(S_t + \alpha XLX^T)A}. \tag{16}$$

Here L is the Laplacian matrix. S_b is the between-class scatter matrix and S_t is the total scatter matrix. Comparing the optimization problem shown in Eq. (13) and Eq. (16), it is clear that our method has a similar formulation with SDA, except that they have different numerators and denominators, i.e., they use the prior information in different way. However, they can all be solved by the general spectral decompositions[10].

5 Experiments and Discussions

In this section, several experiments are performed to test our algorithm. The experiments mainly include image classification, digits recognition and text categorization.

In following numerical comparisons, since there is no common metric to measure the performance of dimensionality reduction approaches, the classification accuracy is employed as our metric. *Nearest Neighbor* classifier (NN) is applied on the embeddings of unlabeled points (transductive methods) or unseen sample (inductive approaches) to compute the classification accuracies. The results are all averaged over 50 independent trails. There are totally two different kinds of experiments. The first is to compare the performances of different transductive methods. The second type of experiments is inductive.

Image classification. In this case study, we will focus on the problem of classifying images of different rotated objects. The Umist [16] data set is adopted. It consists of 575 face images of 20 people. Each covers a range of poses from profile to frontal views. Subjects cover a range of race/sex/appearance. The precropped images are rescaled to 23 × 28 and hence $D_1 = 644$. In each run of the following experiments, we split the points in each class into three parts. We randomly choose $l/20$ (l is the number of labeled points in total) label points, 2 unseen data. The rest are considered as unlabeled points. The number of nearest neighbors, i.e., k, is set to ten manually. Meanwhile, dimensionality of embedding space D_2 is set to $c - 1$.

We employ our methods with two different kinds of constraints: orthogonal (OSSDR *via* HF) and uncorrelated (USSDR *via* HF). With different numbers of labeled points per class, the transductive and inductive classification accuracies averaged over 50 independently trails are summarized in Table 1 and Table 2 respectively. The standard derivations are within the brackets. "-" indicates that the corresponding method can not be used in that situation.

Table 1. Classification accuracy on the **Umist** data set for different **Transductive** methods with different number of labeled points

	l=20	l=40	l=60	l=80
PCA	0.4574(0.0367)	0.6219(0.0343)	0.7231(0.0270)	0.7991(0.0289)
LDA	—	0.7175(0.0479)	0.8251(0.0331)	0.9049(0.0347)
SSDR	0.4857(0.0405)	0.7036(0.0479)	0.8114(0.0306)	0.8894(0.0345)
SDA	0.5505(0.0459)	0.7512(0.0524)	0.8505(0.0309)	0.9087(0.0261)
TSVM	0.6010(0.0465)	0.8124(0.0524)	0.8777(0.0321)	0.9132(0.0159)
OSSDR *via* HF	0.8048(0.0410)	0.8629(0.0393)	0.8949(0.0212)	0.9267(0.0214)
USSDR *via* HF	0.8046(0.0386)	0.8533(0.0331)	0.8870(0.0164)	0.9221(0.0200)
HF	0.8095(0.0387)	0.8568(0.0340)	0.8910(0.0164)	0.9241(0.0210)

Table 2. Classification accuracy on the **Umist** data set for different **Inductive** methods with different number of labeled points

	l=20	l=40	l=60	l=80
NN	0.4400(0.0362)	0.6083(0.0473)	0.7033(0.0647)	0.7833(0.0304)
PCA	0.4467(0.0560)	0.6083(0.0568)	0.6967(0.0680)	0.7850(0.0288)
LDA	—	0.7133(0.0618)	0.8183(0.0574)	0.9067(0.0410)
SSDR	0.4700(0.0362)	0.6983(0.0726)	0.7917(0.0486)	0.8933(0.0362)
SDA	0.5300(0.0571)	0.7450(0.0624)	0.8383(0.0599)	0.9083(0.0364)
TSVM	0.5900(0.0425)	0.8133(0.0745)	0.8633(0.0522)	0.9050(0.0317)
OSSDR *via* HF	**0.7600**(0.0479)	**0.8400**(0.0399)	**0.8950**(0.0497)	**0.9211**(0.0360)
USSDR *via* HF	0.7000(0.0609)	0.7983(0.0352)	0.8500(0.0383)	0.9083(0.0352)

As seen from Table 1 and Table 2, NN is not included in Table 2 for Transductive and HF is not included in Table 1 for inductive methods. In Table 1 we can see that our methods (OSSDR via HP and USSDR via HP) achieve the same accuracy as HF and perform better than other Transductive methods. Also, in Table 2 we see that for the inductive approach, our methods perform the best. This is because that our methods could integrate the geometry structure of the data to propagate the label.

Text categorization. In this section, we validate our methods on test categorization based on a subset of the Newsgroup data, which is preprocessed by Yu et al [17]. It contains 8014 dimensional TFIDF features. There are totally 4 different classes, covering *autos*, *motorcycles*, *baseball* and *hockey*. We bring the first 200 points in each class. The labeled points varies from 5, 10, 15, 20 and the test points are 50 per class. All of them are randomly selected in each run. The number of nearest neighbors, i.e., k, is set to 12 and D_2 is set to 100 manually. Similarly, we also compare our algorithms with above-mentioned methods with different number of labeled points in transductive and inductive situations. The results averaged over 50 trails are listed in Table 3 and Table 4.

Comparing with the corresponding results in Table 1 and Table 2, we have the same conclusions. Our methods, i.e., OSSDR *via* HF and USSDR *via* HF, perform as well as HF and better than other transductive methods, especially

Table 3. Classification accuracy on a subset of the **Newsgroup** data for different **Transductive** methods with different number of labeled points

	l=20	l=40	l=60	l=80
PCA	0.4163(0.0610)	0.4722(0.0353)	0.5529(0.0321)	0.5613(0.0335)
LDA	0.6058(0.0324)	0.6981(0.0321)	0.7671(0.0284)	0.8000(0.0268)
SSDR	0.6132(0.0376)	0.7022(0.0327)	0.7709(0.0297)	0.8016(0.0223)
SDA	0.6068(0.0329)	0.7025(0.0332)	0.7676(0.0306)	0.8019(0.0271)
TSVM	0.5789(0.0521)	0.6894(0.0345)	0.7547(0.0246)	0.7975(0.0215)
OSSDR *via* HF	0.6300(0.0558)	0.7537(0.0215)	0.8203(0.0275)	0.8509(0.02315)
USSDR *via* HF	0.6210(0.0322)	0.7503(0.0325)	0.8101(0.0327)	0.8409(0.0124)
HF	0.6382(0.0381)	0.7557(0.0195)	0.8236(0.0315)	0.8533(0.0225)

Table 4. Classification accuracy on a subset of the **Newsgroup** data for different **Inductive** methods with different number of labeled points

	l=20	l=40	l=60	l=80
NN	0.5845(0.0207)	0.6125(0.0300)	0.6685(0.0381)	0.6815(0.0248)
PCA	0.3460(0.0904)	0.3710(0.0719)	0.4050(0.0762)	0.4260(0.0884)
LDA	0.6135(0.0328)	0.6980(0.0458)	0.7575(0.0207)	0.7905(0.0278)
SSDR	0.6100(0.0330)	0.7055(0.0564)	0.7620(0.0268)	0.7985(0.0368)
SDA	0.6095(0.0300)	0.7020(0.0501)	0.7540(0.0248)	0.7930(0.0316)
TSVM	0.5690(0.0313)	0.6450(0.0424)	0.7115(0.0332)	0.7665(0.0302)
OSSDR *via* HF	**0.6085**(0.0567)	**0.7385**(0.0591)	**0.8150**(0.0303)	**0.8470**(0.0327)
USSDR *via* HF	0.5995(0.0610)	0.7245(0.0619)	0.8065(0.0313)	0.8395(0.0332)

Table 5. Classification accuracy on a subset of the **USPS** data for different **Transductive** methods with different number of labeled points

	l=20	l=40	l=60	l=80
PCA	0.7384(0.0440)	0.8094(0.0334)	0.8263(0.0307)	0.8667(0.0210)
LDA	0.7703(0.0509)	0.8349(0.0372)	0.8722(0.0285)	0.8931(0.0170)
SSDR	0.7672(0.0468)	0.8346(0.0337)	0.8680(0.0310)	0.8880(0.0225)
SDA	0.7941(0.0487)	0.8505(0.0223)	0.8988(0.0290)	0.9274(0.0179)
TSVM	0.8599(0.0510)	0.9095(0.0272)	0.9151(0.0108)	0.9217(0.0095)
OSSDR *via* HF	0.8592(0.0306)	0.8683(0.0252)	0.8862(0.0109)	0.9109(0.0137)
USSDR *via* HF	0.9164(0.0357)	0.9459(0.0330)	0.9602(0.0196)	0.9706(0.0229)
HF	0.9316(0.0389)	0.9549(0.0210)	0.9633(0.0213)	0.9735(0.0060)

Table 6. Classification accuracy on a subset of the **USPS** data for different **Inductive** methods with different number of labeled points

	l=20	l=40	l=60	l=80
NN	0.7270(0.0512)	0.7880(0.0452)	0.8100(0.0535)	0.8430(0.0479)
PCA	0.7310(0.0506)	0.8040(0.0430)	0.8230(0.0591)	0.8600(0.0389)
LDA	0.7560(0.0693)	0.8290(0.0477)	0.8430(0.0362)	0.8660(0.0347)
SSDR	0.7670(0.0696)	0.8260(0.0513)	0.8460(0.0493)	0.8620(0.0394)
SDA	0.7930(0.0668)	0.8490(0.0412)	0.8750(0.0284)	0.9170(0.0279)
TSVM	0.8570(0.0427)	0.8750(0.0530)	0.9100(0.0292)	0.9300(0.0194)
OSSDR *via* HF	0.8510(0.0374)	0.8530(0.0365)	0.8860(0.0406)	0.8940(0.0263)
USSDR *via* HF	**0.9120**(0.0352)	**0.9250**(0.0321)	**0.9440**(0.0259)	**0.9670**(0.0263)

when the labeled points are rare. Moreover, OSSDR *via* HF performs the best among all the inductive methods. Additionally, all the methods have higher accuracies in transductive condition since we have integrated unlabeled point for training.

Digit recognition. The final experiment has been performed on the handwritten digits. The data set that we adopt is the USPS handwritten 16×16 digit data set[18]. We choose 100 images for each category in these experiments. Since $c = 10$, there are totally $n = 1000$ points. In each run, we randomly split the samples in each class into three parts: labeled, unlabeled and unseen. The labeled points number varies from 2 to 6 and the number of unseen points is fixed to 10. Other parameters are as follows: $k = 12$ and $d = 15$. With the same setting, the results are shown in Table 6 and Table 7.

Based on these results, we have the same conclusions. Moreover, as seen from these results, the uncorrelated constraint is more suitable for this kind of data set.

6 Conclusions and Future Works

In this paper, a novel semi-supervised dimensionality reduction approach: SSDR *via* HF is proposed. It aims to use the prior information in a more effective way. We use harmonic function in the Gaussian random field to predict states of unlabeled points and then construct a weighted complete graph. Finally, the linear transformation is computed according to these weights. We provide many experiments to show the effectiveness of our method. In our future work, we will focus on the kernel extensions and computational cost issues of our SSDR *via* HF algorithm.

References

1. Collins, M., Dasgupta, S., Schapire, R.: A generalization of principal component analysis to the exponential family. In: NIPS, vol. 13 (2001)
2. Wang, H., Wang, Z., et al.: Pca plus F-LDA: a new approach for face recognition. IJPRAI 21(6), 1059–1068 (2007)
3. Saul, L., Roweis, S.: Think globally, fit locally: unsupervised learning of low dimensional manifolds. Journal of Machine Learning Research 4, 119–155 (2003)
4. Weinberger, K., Saul, L.: Unsupervised learning of image manifolds by semidefinite programming. In: Proceedings of the IEEE Conference on Computer Vision and Pattern Recognition (CVPR 2004), vol. 2, pp. 988–995 (2004)
5. Fukunaga, K.: Introduction to Statistical Pattern Recognition, 2nd edn. Academic Press, London (1990)
6. Stone, C.: The Dimensionality Reduction Principle for Generalized Additive Models. The Annals of Statistics 14(2), 590–606 (1986)
7. Yang, X., Fu, H., Zha, H., Barlow, J.L.: Semisupervised nonlinear dimensionality reduction. In: ICML-2006, Pittsburgh, PA, pp. 1065–1072 (2006)
8. Zhang, D., Zhou, Z., Chen, S.: Semi-supervised dimensionality reduction. In: SIAM Conference on Data Mining, SDM (2007)
9. Wagstaff, K., Cardie, C., Rogers, S., Schroedl, S.: Constrained k-means clustering with back- ground knowledge. In: Proc. 18th International Conference on Machine Learning (2001)
10. Cai, D., He, X., Han, J.: Semi-supervised discriminant analysis. In: Proceedings of the International Conference on Computer Vision (2007)
11. Song, Y., Nie, F., Zhang, C., Xiang, S.: A unified framework for semi-supervised dimensionality reduction. Pattern Recognition 41(9), 2789–2799 (2008)
12. Zhu, X.: Semi-supervised learning literature survey, Computer Sciences Technical report 1530, Univ. of Wisconsin, Madison (2007)
13. Zhu, X., Ghahramani, Z., Lafferty, J.: Semi-supervised learning using Gaussian fields and harmonic functions. In: Proceedings of the 20th International Conference on Machine Learning (2003)
14. Zhu, X.: Semi-Supervised Learning with Graphs. PhD thesis, Carnegie Mellon University, CMU-LTI-05-192 (2005)
15. Ye, J., Janardan, R., Li, Q., Park, H.: Feature Reduction via Generalized Uncorrelated Linear Discriminant Analysis. IEEE Transactions on Knowledge and Data Engineering 18(10), 1321–1322 (2006)
16. UMIST face database, `http://images.ee.umist.ac.uk/danny/database.html`
17. Yu, K., Bi, J., Tresp, V.: Active learning via transductive experimental design. In: ICML 2006, pp. 1081–1088 (2006)
18. USPS database, `http://www.cs.toronto.edu/~roweis/data.html`

Semi-supervised Agglomerative Hierarchical Clustering with Ward Method Using Clusterwise Tolerance

Yukihiro Hamasuna[1], Yasunori Endo[2], and Sadaaki Miyamoto[2]

[1] Department of Informatics, School of Science and Engineering,
Kinki University,
Kowakae 3-4-1, Higashi Osaka, Osaka, 577-8502, Japan
yhama@info.kindai.ac.jp

[2] Department of Risk Engineering, Systems and Information Engineering,
University of Tsukuba, Tennodai 1-1-1, Tsukuba, Ibaraki, 305-8573, Japan
{endo,miyamoto}@risk.tsukuba.ac.jp

Abstract. This paper presents a new semi-supervised agglomerative hierarchical clustering algorithm with ward method using clusterwise tolerance. Recently, semi-supervised clustering has been remarked and studied in many research fields. In semi-supervised clustering, must-link and cannot-link called pairwise constraints are frequently used in order to improve clustering properties. First, a clusterwise tolerance based pairwise constraints is introduced in order to handle must-link and cannot-link constraints. Next, a new semi-supervised agglomerative hierarchical clustering algorithm with ward method is constructed based on above discussions. Moreover, the effectiveness of proposed algorithms is verified through numerical examples.

Keywords: semi-supervised clustering, pairwise constraints, agglomerative hierarchical clustering, ward method, clusterwise tolerance.

1 Introduction

The aim of data analysis is to discover important properties or structures from massive and complex databases. Recently, semi-supervised learning has also been remarked and studied in many research fields [1]. In the field of clustering [2], pairwise constraints are frequently introduced in order to improve clustering properties [3,4]. Also, pairwise constraints problems are formulated with probabilistic model [5], fuzzy clustering model [6] and regularization terms [7]. These semi-supervised clustering methods are divided into two groups. One is hard constraints based methods, and the other is soft ones. In hard constraints based methods, pairwise constraints are always satisfied, while they are not always satisfied in soft constraints based ones. These hard and soft constraints are typical way to handle prior knowledges in semi-supervised learning.

In recent years, semi-supervised clustering which are based on k-means and fuzzy c-means clustering have been widely studied [3,6,7]. Semi-supervised clustering methods which are based on agglomerative hierarchical clustering (AHC)

V. Torra et al. (Eds.): MDAI 2011, LNAI 6820, pp. 103–113, 2011.

are also discussed [8,9,10]. In these methods, pairwise constraints referred to must-link and cannot-link are used as prior knowledge about which data should be in the same or different cluster [3]. However, because of the squared L_2-norm which is used as dissimilarity, it is difficult to introduce pairwise constraints in the L_2-space. In Constrained Complete–Link (CCL) [9], cannot-link constraint is handled as $d(G, G') = +\infty$. This means that a point is at the infinity, which generally breaks the L_2-space. In order to avoid such situations, the methods with kernel function have been proposed [6,7]. In these methods with kernel function, pairwise constraints are considered not input space but high-dimensional feature space.

By the way, we have proposed the concept of clusterwise tolerance in order to handle different sizes or shapes of clusters [11,12]. This clusterwise tolerance is based on the concept of tolerance [13]. In the proposed clustering methods for data with clusterwise tolerance, the squared L_2-norm is rewritten as the dissimilarity between data with clusterwise tolerance vector and cluster center. By using the concept of clusterwise tolerance, we can handle different sizes or shapes of clusters in the L_2-space. From that sense, we propose clusterwise tolerance based pairwise constraints in order to introduce pairwise constraints into the L_2-space in natural way. By introducing the concept of clusterwise tolerance based pairwise constraints into fuzzy c-means clustering, we have proposed new semi-supervised fuzzy c-means clustering algorithms [14,15]. In addition to those methods, we have proposed semi-supervised agglomerative hierarchical clustering algorithm with centroid method by using clusterwise tolerance based pairwise constraints (AHCCTP$_{cm}$) [16]. In particular, centroid and ward methods are based on the L_2-space. Therefore, we will consider the semi-supervised agglomerative hierarchical clustering algorithm with ward method by using clusterwise tolerance based pairwise constraints.

In this paper, we will propose semi-supervised agglomerative hierarchical clustering algorithm with ward method by using clusterwise tolerance based pairwise constraints. The contents of this paper are the followings. In the second section, we introduce some symbols, agglomerative hierarchical clustering algorithm (AHC) and pairwise constraints. In the third section, we propose clusterwise tolerance based pairwise constraints. In the forth section, we propose semi-supervised agglomerative hierarchical clustering with ward method using clusterwise tolerance based pairwise constraints (AHCCTP$_{wm}$). In the fifth section, we show the effectiveness of proposed methods through numerical examples. In the last section, we conclude this paper.

2 Preparation

First, a set of data to be clustered is given. A data set is denoted by $X = \{x_1, \ldots, x_n\}$ in which x_k, $(k = 1, \ldots, n)$ is a data. In most cases, $x_1, \ldots, x_n$ are vectors of real p-dimensional space $\Re^p$, that is, a data $x_k \in \Re^p$. Generally, a hard cluster is denoted by G_i is a subset of X. A set of clusters is denoted as follows:

$$\mathcal{G} = \{G_1, G_2, \ldots, G_C\},$$

where the clusters are disjoint and their union is a set of data as follows:

$$\bigcup_{i=1}^{C} G_i = X, \quad G_i \cap G_j = \emptyset \quad (i \neq j).$$

2.1 Agglomerative Hierarchical Clustering

In agglomerative hierarchical clustering (AHC), the dissimilarity denoted by $d(G, G')$ $(G, G' \in \mathcal{G})$ is used for measuring nearness between two clusters.

First, we describe a general algorithm of AHC [17].

Algorithm 1. AHC

AHC 1 Assume that initial clusters are given by

$$\mathcal{G} = \left\{\hat{G}_1, \hat{G}_2, \ldots, \hat{G}_{N_0}\right\}.$$

Set $C := N_0$. (C is the number of clusters and N_0 is the initial number of clusters)

$$G_i := \hat{G}_i (i = 1, \ldots, C).$$

Calculate $d(G, G')$ for all pairs $G, G' \in \mathcal{G}$.

AHC2 Search the pair of minimum dissimilarity:

$$(G_p, G_q) = \arg \min_{G, G' \in \mathcal{G}} d(G, G').$$

Merge: $G_r =: G_p \cup G_q$.

Add G_r to $\mathcal{G}$ and delete G_p, G_q from $\mathcal{G}$.

$C := C - 1$.

If $C = 1$ then stop and output the dendrogram. Otherwise, go to **AHC 3**.

AHC 3 Update $d(G_r, G'')$ for all $G'' \in \mathcal{G}$.

Go to **AHC 2**.

End AHC.

2.2 Centroid Method

In AHC procedure, there are five methods for updating dissimilarity, that is, single linkage, complete linkage, average linkage, centroid method, and ward method. In particular, centroid and ward methods are based on the L_2-space.

First, we note two definitions of centroid method, that is, the centroid of cluster and the dissimilarity between two clusters.

Let the centroid of a cluster G be

$$M(G) = \frac{1}{|G|} \sum_{x_k \in G} x_k, \tag{1}$$

and let the squared L_2-norm used as dissimilarity be

$$d(G, G') = \|M(G) - M(G')\|^2. \tag{2}$$

2.3 Ward Method

Assume $M(G)$ is the centroid of cluster G is the same as (1), and let

$$E(G) = \sum_{x_i \in G} \|x_i - M(G)\|^2. \tag{3}$$

Define $d(G, G')$ as follows:

$$d(G, G') = E(G \cup G') - E(G) - E(G'). \tag{4}$$

2.4 Pairwise Constraints

Typical examples of pairwise constraints are must-link and cannot-link [3]. These constraints are considered as prior knowledge about which data should be in the same or different cluster. A set $ML = \{(x_k, x_l)\} \subset X \times X$ consists of must-link pairs so that x_k and x_l should be in the same cluster, while another set $CL = \{(x_i, x_j)\} \subset X \times X$ consists of cannot-link pairs so that x_i and x_j should be in different cluster. Obviously, ML and CL are assumed to be symmetric, that is, if $(x_k, x_l) \in ML$ then $(x_l, x_k) \in ML$, and if $(x_i, x_j) \in CL$ then $(x_j, x_i) \in CL$.

In semi-supervised clustering, these pairwise constraints are considered as hard or soft constraints. In case of hard constraints, pairwise constraints ML and CL are always satisfied in clustering procedures and results, while they are not always satisfied in case of soft constraints. Many semi-supervised clustering methods have been proposed in order to improve clustering results by using prior knowledge of data sets [3,4,6,7,9,10].

3 Clusterwise Tolerance Based Pairwise Constraints

3.1 Clusterwise Tolerance

Each data has the tolerance κ_k which means the upper bound of clusterwise tolerance vectors. A clusterwise tolerance vector is the vector within the range of tolerance. A set of clusterwise tolerance vector is defined as $\Delta = \{\delta_{11}, \ldots, \delta_{kl}, \ldots, \delta_{nn}\}$ in which δ_{kl} is a clusterwise tolerance vector of p-dimensional real space $\Re^p$.

If $(x_k, x_l) \in ML$, δ_{kl} and δ_{lk} are calculated to be near each other, while $(x_i, x_j) \in CL$, δ_{ij} and δ_{ji} are calculated to be distant each other.

A constraint for clusterwise tolerance vector is as follows:

$$\|\delta_{kl}\|^2 \leq (\kappa_k)^2 \quad (\kappa_k \geq 0), {}^\forall k, \ l. \tag{5}$$

Figure 1 shows a clusterwise tolerance in $\Re^2$.

In this example, $(x_1, x_2) \in ML$ and $(x_1, x_3) \in CL$. Also, each data has tolerance. Therefore, the dissimilarity between each data are calculated as follows:

$$d(x_1, x_2) = (\|x_1 - x_2\| - \kappa_1 - \kappa_2)^2,$$
$$d(x_1, x_3) = (\|x_1 - x_3\| + \kappa_1 + \kappa_3)^2,$$
$$d(x_2, x_3) = \|x_2 - x_3\|^2.$$

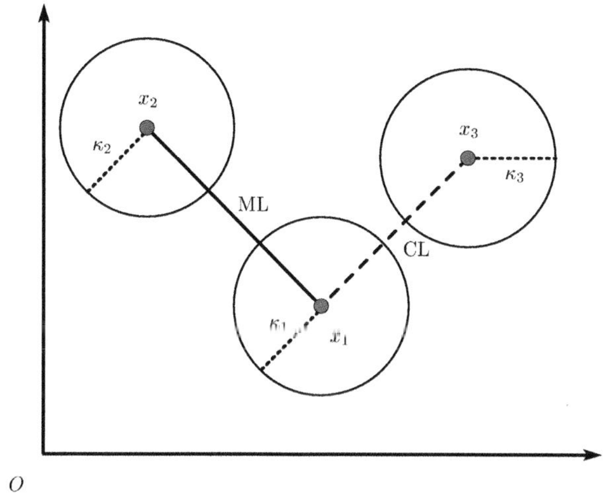

Fig. 1. An illustrative example of the concept of clusterwise tolerance

3.2 Clusterwise Tolerance Based Pairwise Constraints

First, a set of must or cannot-linked objects are defined. A set $ML(x; G')$ consists of must-linked objects which in cluster G' with a data x, while $CL(x; G')$ consists of cannot-linked objects which in cluster G' with a data x.

$$ML(x; G') = \{\xi \mid \xi \in G', (x, \xi) \in ML\}, \tag{6}$$

$$CL(x; G') = \{\xi \mid \xi \in G', (x, \xi) \in CL\}. \tag{7}$$

In addition, $ML(G; G')$ is defined as a union of sets $ML(x; G')$, while $CL(G; G')$ is defined as a union of sets $CL(x; G')$ as follows:

$$ML(G; G') = \bigcup_{x \in G} ML(x; G'), \tag{8}$$

$$CL(G; G') = \bigcup_{x \in G} CL(x; G'). \tag{9}$$

A concept of clusterwise tolerance based pairwise constraints uses these sets in order to calculate the clusterwise tolerance which is defined between clusters.

Here, we propose clusterwise tolerance based pairwise constraints. A value of $K(x; G')$ and $K(G; G')$ are the sum of tolerance κ_k which in a set of must or cannot-linked data.

$$K(x; G') = \sum_{x_k \in ML(x; G')} \kappa_k - \sum_{x_l \in CL(x; G')} \kappa_l. \tag{10}$$

If $K(x; G') > 0$, x is considered must-linked data with G', while $K(x; G') < 0$, x is considered cannot-linked data with G'. The upper bound of clusterwise tolerance is defined as $|K(x;G')|$. Obviously, it is depended on the value of κ_k whether x is must or cannot-linked with G'.

$$K(G;G') = \sum_{x_k \in ML(G;G')} \kappa_k - \sum_{x_l \in CL(G;G')} \kappa_l. \tag{11}$$

If $K(G;G') > 0$, G is considered must-linked cluster with G', while $K(G;G') < 0$, G is considered cannot-linked cluster with G'. The upper bound of clusterwise tolerance is defined as $|K(G;G')|$. Obviously, it is depended on the value of κ_k whether G is must or cannot-linked with G'. Therefore, $K(G;G')$ and $K(G';G)$ are asymmetric.

Next, we show an illustrative example of clusterwise tolerance based pairwise constraints. Figure 2 is a simple example of proposed method.

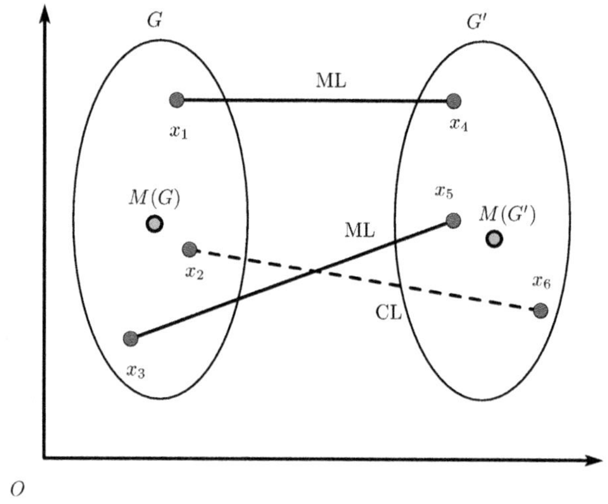

Fig. 2. An illustrative example of clusterwise tolerance based pairwise constraints

In this example, $(x_1, x_4), (x_3, x_5) \in ML$ and $(x_2, x_6) \in CL$. Therefore, $ML(x;G')$, $CL(x;G')$ $ML(G;G')$ and $CL(G;G')$ are as follows:

$$\begin{aligned}
&ML(x_1;G') = \{x_4\},\ ML(x_2;G') = \emptyset,\ ML(x_3;G') = \{x_5\},\\
&CL(x_1;G') = \emptyset,\ CL(x_2;G') = \{x_6\},\ CL(x_3;G') = \emptyset,\\
&ML(G;G') = \{x_4, x_5\},\ CL(G;G') = \{x_6\}.
\end{aligned}$$

Also, $ML(x;G)$, $CL(x;G)$ $ML(G';G)$ and $CL(G';G)$ are as follows:

$$\begin{aligned}
&ML(x_4;G) = \{x_1\},\ ML(x_5;G) = \{x_3\},\ ML(x_6;G) = \emptyset,\\
&CL(x_4;G) = \emptyset,\ CL(x_5;G) = \emptyset,\ CL(x_6;G) = \{x_2\},\\
&ML(G';G) = \{x_1, x_3\},\ CL(G';G) = \{x_2\}.
\end{aligned}$$

If all κ_k are the same value, $K(G;G')$ and $K(G';G)$ are both positive. This means that G and G' are must-linked clusters each other.

4 Semi-supervised Agglomerative Hierarchical Clustering Using Clusterwise Tolerance Based Pairwise Constraints

In this section, we propose semi-supervised AHC using clusterwise tolerance based pairwise constraints (AHCCTP). First, we introduce AHCCTP with centroid method (AHCCTP$_{cm}$) [16]. Next, we propose AHCCTP with ward method (AHCCTP$_{wm}$). In proposed methods, the centroid of each cluster is calculated as the same procedure (1), while the dissimilarity between two clusters are different.

4.1 Centroid Method

First, we introduce AHCCTP with centroid method (AHCCTP$_{cm}$) [16]. Assume $M(G)$ is the centroid of cluster G is the same as (1) and let the squared L_2-norm between clusters as follows:

$$d(G,G') = \begin{cases} (\|M(G) - M(G')\| - K(G;G') - K(G';G))^2 \\ \qquad (\|M(G) - M(G')\| > K(G;G') + K(G';G)), \\ 0 \quad \text{(otherwise)}. \end{cases} \tag{12}$$

4.2 Ward Method

Next, we propose AHCCTP with ward method (AHCCTP$_{wm}$). In this method, $M(G)$ is also the same as (1), and let

$$E(G) = \sum_{x_i \in G} (\max\{\|x_i - M(G)\| - K(x_i;G), 0\})^2. \tag{13}$$

Define $d(G,G')$ as follows:

$$d(G,G') = E(G \cup G') - E(G) - E(G').$$

4.3 Algorithm

Next, we describe an algorithm of AHCCTP.

5 Numerical Examples

In this section, we show numerical examples with Iris data set published in UCI machine learning repository (http://archive.ics.uci.edu/ml/index.html). This data set consists of 150 data with four attributes and should be classified into

Algorithm 2. AHCCTP

AHCCTP 1 Assume that initial clusters are given by

$$\mathcal{G} = \left\{ \hat{G}_1, \hat{G}_2, \dots, \hat{G}_{N_0} \right\}.$$

Set $C = N_0$. (C is the number of clusters and N_0 is the initial number of clusters)

$$G_i = \hat{G}_i (i = 1, \dots, C).$$

Set ML, CL, κ_k and $M(G)$.
Calculate $K(G; G')$ and $d(G, G')$ for all pairs $G, G' \in \mathcal{G}$.

AHCCTP 2 Search the pair of minimum dissimilarity:

$$(G_p, G_q) = \arg \min_{G, G' \in \mathcal{G}} d(G, G').$$

Merge: $G_r = G_p \cup G_q$.
Add G_r to $\mathcal{G}$ and delete G_p, G_q from $\mathcal{G}$.
$C := C - 1$.
If $C = 1$ then stop and output the dendrogram. Otherwise, go to **AHCCTP 3**.

AHCCTP 3 Update $ML(G_r; G'')$, $CL(G_r; G'')$, $K(G_r; G'')$, $M(G_r)$ and $d(G_r, G'')$ for all $G'' \in \mathcal{G}$.
Go to **AHCCTP 2**.

End AHCCTP.

Table 1. The average of misclassified data out of 100 trials by AHCCTP$_{cm}$ with ML

κ_k \ $\lvert ML \rvert$	100	300	500
0.05	26.06	0.67	0.0
0.10	7.38	0.11	0.0
0.30	1.81	0.09	0.0

Table 2. The average of misclassified data out of 100 trials by AHCCTP$_{wm}$ with ML

κ_k \ $\lvert ML \rvert$	100	300	500
0.05	6.56	2.20	0.17
0.10	5.50	0.23	12.25
0.30	8.39	39.87	76.56

three clusters. Each attribute is normalized between 0 and 1. We show the average of misclassified data and the number of violated constraints out of 100 trials by AHCCTP$_{cm}$ and AHCCTP$_{wm}$. The number of misclassified data is 6 by conventional centroid method, while the one is 5 by conventional ward method.

Here, we consider three cases as follows. In each cases, the number of ML and CL are denoted as $|ML|$ and $|CL|$. First, we set $|ML| = \{100, 300, 500\}$. Second, we set $|CL| = \{100, 300, 500\}$ as well as ML. Third, we set ML and CL at the same time $|ML| = \{100, 300, 500\}$ and $|CL| = \{100, 300, 500\}$. Thus, the sum of ML and CL are $\{200, 600, 1000\}$ in all. Tables 1, 2, 5, 6, 9 and 10 show the results of misclassified data. Also, Tables 3, 4, 7, 8, 11 and 12 show the results of the number of violated constraints.

We can see that must-link constraints more affect than cannot-link constraints from these tables. In particular, the difference of ML and CL are significant in case of $|ML|$, $|CL|$ or the value of κ_k is small. If the value of κ_k is small, large $|ML|$ and $|CL|$ are required to take large $K(G; G')$ and $K(x; G)$. For must-link constraints, $|ML|$ and κ_k are both required, while $|CL|$ is required for cannot-link

Table 3. The average of the number of violated constraints out of 100 trials by AHCCTP$_{cm}$ with ML

κ_k \ $\lvert ML \rvert$	100	300	500
0.05	0.22	0.11	0.0
0.10	0.06	0.0	0.0
0.30	0.0	0.0	0.0

Table 4. The average of the number of violated constraints out of 100 trials by AHCCTP$_{wm}$ with ML

κ_k \ $\lvert ML \rvert$	100	300	500
0.05	5.82	5.37	0.47
0.10	3.15	0.20	0.50
0.30	1.04	1.16	17.26

Table 5. The average of misclassified data out of 100 trials by AHCCTP$_{cm}$ with CL

κ_k \ $\lvert CL \rvert$	100	300	500
0.05	6.02	5.43	5.35
0.10	6.10	5.62	5.42
0.30	6.21	5.59	5.49

Table 6. The average of misclassified data out of 100 trials by AHCCTP$_{wm}$ with CL

κ_k \ $\lvert CL \rvert$	100	300	500
0.05	6.26	5.57	5.58
0.10	6.00	5.66	5.42
0.30	5.43	5.54	5.38

Table 7. The average of the number of violated constraints out of 100 trials by AHCCTP$_{cm}$ with CL

κ_k \ $\lvert CL \rvert$	100	300	500
0.05	3.09	9.61	15.60
0.10	3.04	9.28	16.07
0.30	2.77	9.44	16.18

Table 8. The average of the number of violated constraints out of 100 trials by AHCCTP$_{wm}$ with CL

κ_k \ $\lvert CL \rvert$	100	300	500
0.05	3.27	9.70	10.05
0.10	3.35	9.50	16.38
0.30	2.86	9.87	16.18

Table 9. The average of misclassified data out of 100 trials by AHCCTP$_{cm}$ with ML and CL

κ_k \ $\lvert ML \rvert, \lvert CL \rvert$	100	300	500
0.05	2.85	0.11	0.0
0.10	1.96	0.01	0.0
0.30	1.63	0.02	0.0

Table 10. The average of misclassified data out of 100 trials by AHCCTP$_{wm}$ with ML and CL

κ_k \ $\lvert ML \rvert, \lvert CL \rvert$	100	300	500
0.05	5.64	1.78	0.03
0.10	3.82	0.15	0.0
0.30	2.91	0.04	0.0

Table 11. The average of the number of violated constraints out of 100 trials by AHCCTP$_{cm}$ with ML and CL

κ_k \ $\lvert ML \rvert, \lvert CL \rvert$	100	300	500
0.05	2.42	0.13	0.0
0.10	0.91	0.0	0.0
0.30	0.59	0.01	0.0

Table 12. The average of the number of violated constraints out of 100 trials by AHCCTP$_{wm}$ with ML and CL

κ_k \ $\lvert ML \rvert, \lvert CL \rvert$	100	300	500
0.05	8.64	7.81	0.17
0.10	3.80	0.37	0.0
0.30	1.53	0.02	0.0

constraints. In Table 2 and 4, the average of misclassified data is much larger than other results. Thus, determining the adequate κ_k and $|ML|$ in AHCCTP$_{wm}$ is quite important problem. In our proposed methods, the pairwise constraints are considered soft constraints in case of small κ_k, while they are considered hard constraints in case of large κ_k.

6 Conclusions

In this paper, we introduced semi-supervised agglomerative hierarchical clustering using clusterwise tolerance based pairwise constraints (AHCCTP$_{cm}$) and proposed AHCCTP$_{wm}$. The proposed method can handle the pairwise constraints without breaking the L_2-space by using the concept of clusterwise tolerance. Moreover, we showed the effectiveness of proposed methods through numerical examples.

In future works, we will show the effectiveness and difference of proposed methods through numerical examples and dendrogram with various kinds of data sets. Moreover, we will consider the mathematical discussions about updating dissimilarity process and reversal in the dendrogram.

Acknowledgments. This study is partly supported by Research Fellowships of the Japan Society for the Promotion of Science for Young Scientists and the Grant-in-Aid for Scientific Research (C) (Project No.21500212) from the Ministry of Education, Culture, Sports, Science and Technology, Japan.

References

1. Chapelle, O., Schoölkopf, B., Zien, A. (eds.): Semi-Supervised Learning. MIT Press, Cambridge (2006)
2. Miyamoto, S., Ichihashi, H., Honda, K.: Algorithms for Fuzzy Clustering. Springer, Heidelberg (2008)
3. Wagstaff, K., Cardie, C., Rogers, S., Schroedl, S.: Constrained k-means clustering with background knowledge. In: Proc. of the 18th International Conference on Machine Learning (ICML 2001), pp. 577–584 (2001)
4. Basu, S., Banerjee, A., Mooney, R.J.: Active semi-supervision for pairwise constrained clustering. In: Proc. of the SIAM International Conference on Data Mining (SDM 2004), pp. 333–344 (2004)
5. Basu, S., Bilenko, M., Mooney, R.J.: A probabilistic framework for semi-supervised clustering. In: Proc. of the 10th ACM SIGKDD International Conference on Knowledge Discovery and Data Mining (KDD 2004), pp. 59–68 (2004)
6. Miyamoto, S., Yamazaki, M., Terami, A.: On semi-supervised clustering with pairwise constraints. In: Proc. of The 7th International Conference on Modeling Decisions for Artificial Intelligence (MDAI 2009), (CD-ROM), pp. 245–254 (2009)
7. Kulis, B., Basu, S., Dhillon, I., Mooney, R.: Semi-supervised graph clustering: a kernel approach. Machine Learning 74(1), 1–22 (2009)
8. Talavera, L., Béjar, J.: Integrating declarative knowledge in hierarchical clustering tasks. In: Hand, D.J., Kok, J.N., Berthold, M.R. (eds.) IDA 1999. LNCS, vol. 1642, pp. 211–222. Springer, Heidelberg (1999)

9. Klein, D., Kamvar, S., Manning, C.: From instance-level constraints to space-level constraints: making the most of prior knowledge in data clustering. In: Proc. of the 19th International Conference on Machine Learning (ICML 2002), pp. 307–314 (2002)
10. Davidson, I., Ravi, S.S.: Agglomerative hierarchical clustering with constraints: theoretical and empirical results. In: Proc. of 9th European Conference on Principles and Practice of Knowledge Discovery in Databases (KDD 2005), pp. 59–70 (2005)
11. Hamasuna, Y., Endo, Y., Miyamoto, S.: On Tolerant Fuzzy c-Means. Journal of Advanced Computational Intelligence and Intelligent Informatics (JACIII) 13(4), 421–427 (2009)
12. Hamasuna, Y., Endo, Y., Miyamoto, S.: Fuzzy c-Means Clustering for Data with Clusterwise Tolerance Based on L_2- and L_1-Regularization. Journal of Advanced Computational Intelligence and Intelligent Informatics (JACIII) 15(1), 68–75 (2011)
13. Endo, Y., Murata, R., Haruyama, H., Miyamoto, S.: Fuzzy c-Means for Data with Tolerance. In: Proc. of International Symposium on Nonlinear Theory and Its Applications (Nolta 2005), pp. 345–348 (2005)
14. Hamasuna, Y., Endo, Y., Miyamoto, S.: Semi-supervised fuzzy c-means clustering using clusterwise tolerance based pairwise constraints. In: Proc. of 2010 IEEE International Conference on Granular Computing (GrC 2010), pp. 188–193 (2010)
15. Hamasuna, Y., Endo, Y.: Semi-supervised fuzzy c-means clustering for data with clusterwise tolerance with pairwise constraints. In: Joint 5th International Conference on Soft Computing and Intelligent Systems and 11th International Symposium on Advanced Intelligent Systems (SCIS & ISIS 2010), pp. 397–400 (2010)
16. Hamasuna, Y., Endo, Y., Miyamoto, S.: Semi-supervised agglomerative hierarchical clustering using clusterwise tolerance based pairwise constraints. In: Torra, V., Narukawa, Y., Daumas, M. (eds.) MDAI 2010. LNCS (LNAI), vol. 6408, pp. 152–162. Springer, Heidelberg (2010)
17. Miyamoto, S.: Introduction to Cluster Analysis: Theory and Applications of Fuzzy Clustering. Morikita-Shuppan, Tokyo (1999) (in Japanese)

Agglomerative Clustering Using Asymmetric Similarities

Satoshi Takumi[1] and Sadaaki Miyamoto[2]

[1] Graduate School of Systems and Information Engineering
University of Tsukuba, 1-1-1 Tennodai, Tsukuba, Ibaraki 305-8573, Japan
e0711227@edu.esys.tsukuba.ac.jp
[2] Department of Risk Engineering, Faculty of Systems and Information Engineering
University of Tsukuba, 1-1-1 Tennodai, Tsukuba, Ibaraki 305-8573, Japan
miyamoto@risk.tsukuba.ac.jp

Abstract. Algorithms of agglomerative hierarchical clustering using asymmetric similarity measures are studied. Two different measures between two clusters are proposed, one of which generalizes the average linkage for symmetric similarity measures. Asymmetric dendrogram representation is considered after foregoing studies. It is proved that the proposed linkage methods for asymmetric measures have no reversals in the dendrograms. Examples based on real data show how the methods work.

Keywords: agglomerative clustering, asymmetric similarity, asymmetric dendrogram.

1 Introduction

Cluster analysis alias clustering has now become a standard tool in modern data mining and data analysis. Clustering techniques are divided into two classes of hierarchical and non-hierarchical methods. The major technique in the first class is the well-known agglomerative hierarchical clustering [1,2] which is old but has been found useful in a variety of applications.

Agglomerative hierarchical clustering uses a similarity or dissimilarity measure between a pair of objects to be clustered, and the similarity/dissimilarity is assumed to have symmetric property. In some real applications, however, relation between objects are asymmetric, e.g., citation counts between journals and import of goods between two countries. In such cases we have a motivation to analyze asymmetric measures and obtain clusters having asymmetric features.

Not many but several studies have been done on clustering based on asymmetric similarity measures. Hubert [3] defined clusters using the concept of the connectivity of asymmetric weighted graphs. Okada and Teramoto [6] used the mean of asymmetric measures with an asymmetric dendrogram. Yadohisa [10] studied the generalized linkage method of asymmetric measures with a variation of asymmetric dendrogram representing two levels on a branch.

We propose two new linkage methods for asymmetric similarity measures in this paper. A method is a generalization of the average linkage for symmetric similarity and another is a model-dependent method having the concept of

V. Torra et al. (Eds.): MDAI 2011, LNAI 6820, pp. 114–125, 2011.

average citation probability from a cluster to another cluster. As the asymmetric dendrogram herein, we use a variation of that by Yadohisa [10]. We also prove that the proposed methods have no reversals in the dendrogram.

To see how the proposed methods work, we show three examples based on real data.

2 Agglomerative Hierarchical Clustering

We first review the general procedure of agglomerative hierarchical clustering and then introduce asymmetric similarity measures.

2.1 Preliminaries

Let the set of objects for clustering be $X = \{x_1, \dots, x_N\}$. Generally a cluster denoted by G_i is a subset of X. The family of clusters is denoted by

$$\mathcal{G} = \{G_1, G_2, \dots, G_K\},$$

where the clusters form a crisp partition of X:

$$\bigcup_{i=1}^{K} G_i = X, \qquad G_i \cap G_j = \emptyset \quad (i \neq j).$$

Moreover the number of objects in G is denoted by $|G|$.

Agglomerative hierarchical clustering uses a similarity or dissimilarity measure. We use similarity here: similarity between two objects $x, y \in X$ is assumed to be given and denoted by $s(x, y)$. Similarity between two clusters is also used, which is denoted by $s(G, G')$ $(G, G' \in \mathcal{G})$ which also is called an inter-cluster similarity.

In the classical setting a similarity measure is assumed to be symmetric:

$$s(G, G') = s(G', G).$$

Let us first describe a general procedure of agglomerative hierarchical clustering [4,5].

AHC (Agglomerative Hierarchical Clustering) Algorithm:

AHC1: Assume that initial clusters are given by
$\mathcal{G} = \{\hat{G}_1, \hat{G}_2, \dots, \hat{G}_{N_0}\}$. where $\hat{G}_1, \hat{G}_2, \dots, \hat{G}_N$ are given initial clusters.
Generally $\hat{G}_j = \{x_j\} \subset X$, hence $N_0 = N$.
Set $K = N_0$.
(K is the number of clusters and N_0 is the initial number of clusters)
$G_i = \hat{G}_i$ $(i = 1, \dots, K)$.
Calculate $s(G, G')$ for all pairs $G, G' \in \mathcal{G}$.

AHC2: Search the pair of maximum similarity:

$$(G_p, G_q) = \arg \max_{G_i, G_j \in \mathcal{G}} s(G_i, G_j), \tag{1}$$

and let

$$m_K = s(G_p, G_q) = \max_{G_i, G_j \in \mathcal{G}} s(G_i, G_j). \tag{2}$$

Merge: $G_r = G_p \cup G_q$.
Add G_r to $\mathcal{G}$ and delete G_p, G_q from $\mathcal{G}$.
$K = K - 1$.
if $K = 1$ then stop and output the *dendrogram*.

AHC3: Update similarity $s(G_r, G'')$ and $s(G'', G_r)$ for all $G'' \in \mathcal{G}$.
Go to **AHC2**.

End AHC.

Note 1. The calculation of $s(G'', G_r)$ in **AHC3** is unnecessary when the measure is symmetric: $s(G_r, G'') = s(G'', G_r)$.

Well-known linkage methods such as the single link, complete link, and average link all assume symmetric dissimilarity measures [1,2,4]. In particular, the single link uses the following inter-cluster similarity definition:

$$s(G, G') = \max_{x \in G, y \in G'} s(x, y), \tag{3}$$

When G_p and G_q are merged into G_r, the updating formula in **AHC3** by the single link is:

$$s(G_r, G'') = s(G_p \cup G_q, G'') = \max\{s(G_p, G''), s(G_q, G'')\}. \tag{4}$$

The average link defines the next inter-cluster similarity:

$$s(G, G') = \frac{1}{|G||G'|} \sum_{x \in G, y \in G'} s(x, y). \tag{5}$$

and the updating formula in **AHC3** by the average link is:

$$s(G_r, G'') = s(G_p \cup G_q, G'') = \frac{|G_p|}{|G_r|} s(G_p, G'') + \frac{|G_q|}{|G_r|} s(G_q, G''). \tag{6}$$

There are two more linkage methods of the centroid link and the Ward method that assume objects are points in the Euclidean space. They use dissimilarity measures related to the Euclidean distance. For example, the centroid link uses the square of the Euclidean distance between two centroids of the clusters. Anyway, the above mentioned five linkage methods all assume the symmetric property of similarity and dissimilarity measures.

For the single link, complete link, and average link, it is known that we have the *monotonicity* of m_K:

$$m_N \geq m_{K-1} \geq \cdots \geq m_2 \geq m_1. \tag{7}$$

If the monotonicity does not hold, we have a *reversal* in a dendrogram: it means that G and G' are merged into $\hat{G} = G \cup G'$ at level $m = s(G, G')$ and after that $\hat{G}$ and G'' are merged at the level $\hat{m} = s(\hat{G}, G'')$, and $\hat{m} > m$ occurs.

Reversals in a dendrogram is observed for the centroid method. Consider the next example [4,5]:

Example 1. If three points A, B, C in a plane are near equilateral triangle but two points A, B are nearer, these two are made into a cluster, and then the distance between the mid point (centroid) of AB and C will be smaller than the distance between A and B. We thus have a reversal.

Apparently, if the monotonicity always holds for a linkage method, no reversals in the dendrogram will occur. A simple example of a reversal is shown in Fig. 1.

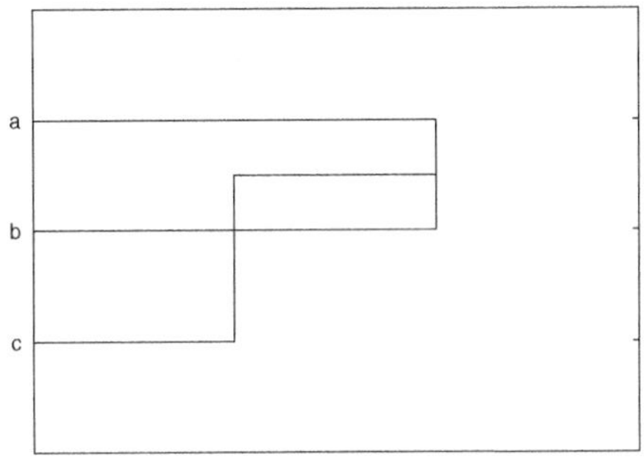

Fig. 1. A simple example of reversal

By reviewing the above, the way how we calculate asymmetric similarity is given in the next section.

3 Asymmetric Similarity Measures

We assume hereafter that similarity measures are asymmetric in general:

$$s(G, G') \neq s(G', G).$$

We moreover use other symbols such as $p(G, G')$ and $r(G, G')$ instead of $s(G, G')$ for the sake of convenience.

First, we use **AHC** algorithm in the previous section, which means that two clusters (G_p, G_q) with

$$s(G_p, G_q) = \max_{G_i, G_j \in \mathcal{G}} s(G_i, G_j) \tag{8}$$

will be merged regardless of asymmetric property. The above equation can be rewritten as

$$s(G_p, G_q) = \max_{i<j} \max\{s(G_i, G_j), s(G_j, G_i)\}. \tag{9}$$

Let us consider concrete linkage methods.

3.1 Asymmetric Average Link

In this section we use symbol $p(x, y)$ and $p(G, G')$ instead of $s(G, G')$ to emphasize the asymmetric property:

$$p(x,y) \neq p(y,x), \qquad p(G,G') \neq p(G',G).$$

Before introducing asymmetric average link, let us review the variation of the single link which has already been studied in Hubert [3]. That is, we define

$$p(G,G') = \max_{x \in G, y \in G'} p(x,y), \tag{10}$$

which is the same as (3) with the replacement of $p(x,y)$ into $s(x,y)$. As the measure is asymmetric, we have $p(G,G') \neq p(G',G)$ but we still use **AHC** algorithm. It is then straightforward to see that the resulting clusters imply weak connectivity of the weighted graph [3]. This method is thus already known, but the same idea leads to an average link.

It is now natural to introduce a method of average link using the same equation as (5).

Definition 1. *Assume that an asymmetric similarity measure $p(x,y)$ is given for every pair of objects $x, y \in X$. The inter-cluster similarity of an average link is defined by*

$$p(G,G') = \frac{1}{|G||G'|} \sum_{x \in G, y \in G'} p(G,G'). \tag{11}$$

Unlike the symmetric measure, we have $p(G,G') \neq p(G',G)$ in general.

We immediately have the following formula for updating similarities in **AHC3**. Note again that the same **AHC** algorithm is used for asymmetric measures.

Proposition 1. *When G_p and G_q are merged into G_r ($G_r = G_p \cup G_q$), the updating formula in* **AHC3** *is:*

$$p(G_r, G'') = p(G_p \cup G_q, G'') = \frac{|G_p|}{|G_r|} p(G_p, G'') + \frac{|G_q|}{|G_r|} p(G_q, G''), \tag{12}$$

$$p(G'', G_r) = p(G'', G_p \cup G_q) = \frac{|G_p|}{|G_r|} p(G'', G_p) + \frac{|G_q|}{|G_r|} p(G'', G_q). \tag{13}$$

The proof is straightforward and omitted. Note that we need to calculate both $p(G_r, G'')$ and $p(G'', G_r)$ for the updating.

This proposition shows that the method to use $p(G,G')$ is the same as the group average method in [7,9], which means that this method itself is not new.

3.2 A Probabilistic Model

We assume a specific example of handling citations between journals in this section. This example seems very specific, but the proposed model can easily be extended to a wide class of applications. This specification for citation is thus for the sake of simplicity.

We hence call objects in X *journals*. Assume that $n(x,y)$ is the number of citations from x to y: journal x cites y for $n(x,y)$ times. Moreover $\bar{n}(x)$ is the total number of citations of x, i.e., the number of citations from x to all journals. We have

$$\bar{n}(x) \geq \sum_{y \in X} n(x,y). \tag{14}$$

Note that $\bar{n}(x) = \sum_{y \in X} n(x,y)$ does not hold in general, since X does not generally exhaust all journals in the world.

We can define the estimate of citation probability from x to y:

$$\pi(x,y) = \frac{n(x,y)}{\bar{n}(x)} \tag{15}$$

which may be generalized to inter-cluster similarity:

$$\pi(G,G') = \frac{\sum_{x \in G, y \in G'} n(x,y)}{\sum_{x \in G} \bar{n}(x)}. \tag{16}$$

This measure $\pi(G,G')$ is, however, inconvenient for clustering, as we discuss in the next section. Hence we define asymmetric similarity as follows.

Definition 2. *Assume that $n(x,y)$ and $\bar{n}(x)$ are given as above, and G, G' are arbitrary two clusters of X. Then an average citation probability from G to G' is defined by*

$$r(G,G') = \frac{\pi(G,G')}{|G'|} = \frac{\sum_{x \in G, y \in G'} n(x,y)}{|G'| \sum_{x \in G} \bar{n}(x)}. \tag{17}$$

We also define

$$n(G,G') = \sum_{x \in G, y \in G'} n(x,y), \tag{18}$$

$$\bar{n}(G) = \sum_{x \in G} \bar{n}(x). \tag{19}$$

We then have

$$r(G,G') = \frac{n(G,G')}{|G'|\bar{n}(G)}. \tag{20}$$

Note that if $G = \{x\}$ and $G' = \{y\}$, we have

$$r(G,G') = r(\{x\},\{y\}) = \pi(x,y).$$

Hence this measure is based on the citation probability from x to y.

We have the following formula for the updating in **AHC3**.

Proposition 2. *When G_p and G_q are merged into G_r ($G_r = G_p \cup G_q$), the updating formula in* **AHC3** *is:*

$$r(G_r,G'') = r(G_p \cup G_q, G'') = \frac{n(G_p,G'') + n(G_q,G'')}{|G''|(\bar{n}(G_p) + \bar{n}(G_q))}, \tag{21}$$

$$r(G'',G_r) = r(G'', G_p \cup G_q) = \frac{n(G'',G_p) + n(G'',G_q)}{(|G_p| + |G_q|)\bar{n}(G'')}. \tag{22}$$

The proof is straightforward and omitted.

4 Dendrogram without Reversals

As noted earlier, the average link for symmetric measures have no reversals in the dendrograms [1,4,5]. We will show that this property of *no reversals* also holds for the proposed methods.

First, we define

$$\mathcal{S}(K) = \{s(G, G') : \forall (G, G') \in \mathcal{G} \times \mathcal{G},\ G \neq G'\} \tag{23}$$

where K is the index in **AHC** and $\mathcal{G}$ changes as K varies, e.g., $|\mathcal{G}| = K$. Hence $\mathcal{S}(K)$ is the set of all values of similarity for K. Note that $s(G, G')$ is rewritten as $p(G, G')$ and $r(G, G')$ when asymmetric measures are discussed.

We also assume

$$\max \mathcal{S}(K)$$

is the maximum value of $\mathcal{S}(K)$: it exactly is m_K given by (2). We have the following lemma.

Lemma 1. *If* $\max \mathcal{S}(K)$ *is monotonically non-increasing with respect to* K*:*

$$\max \mathcal{S}(N) \geq \max \mathcal{S}(N-1) \geq \cdots \geq \max \mathcal{S}(2) \geq \max \mathcal{S}(1), \tag{24}$$

then there is no reversal in the dendrogram.

Proof. The proof is almost trivial, since $\max \mathcal{S}(K) = m_K$. Thus (24) is exactly the same as (7), thus we have the conclusion. □

We have the next two propositions regarding $p(G, G')$ and $r(G, G')$.

Proposition 3. *Assume that* $p(G, G')$ *are used. For* $G, G', G'' \in \mathcal{G}$*, we have*

$$p(G \cup G', G'') \leq \max\{p(G, G''), p(G', G'')\}, \tag{25}$$

$$p(G'', G \cup G') \leq \max\{p(G'', G), p(G'', G')\}. \tag{26}$$

Proposition 4. *Assume that* $r(G, G')$ *are used. For* $G, G', G'' \in \mathcal{G}$*, we have*

$$r(G \cup G', G'') \leq \max\{r(G, G''), r(G', G'')\}, \tag{27}$$

$$r(G'', G \cup G') \leq \max\{r(G'', G), r(G'', G')\}. \tag{28}$$

The proofs of these two propositions are omitted, as straightforward calculations are sufficient for the proof.

We finally have the following propositions.

Proposition 5. *Assume that* X *and* $p(x, y), x, y \in X$*, are arbitrarily given. We use the definition* $p(G, G')$ *by (11) and perform* **AHC** *algorithm. Then we have a dendrogram without any reversal.*

Proposition 6. *Assume that* X*,* $n(x, y), x, y \in X$*,* $\bar{n}(x)$*, are arbitrarily given as a citation model. We use the definition* $r(G, G')$ *by (17) and perform* **AHC** *algorithm. Then we have a dendrogram without any reversal.*

Proof. The proof of the last two propositions is now easy. Propositions 3 and 4 imply that $\max \mathcal{S}(K)$ is monotonically non-increasing. Hence Lemma 1 is applied and we have the desired conclusions. □

4.1 Asymmetric Dendrogram

Foregoing studies propose asymmetric dendrograms [6,10]. Note again that (8) is equivalent to (9). When G_p and G_q are merged at the level $s(G_p, G_q)$, we have $s(G_p, G_q) \leq s(G_q, G_p)$. Yadohisa [10] proposed to show the value $s(G_q, G_p)$ in addition to the merged level $s(G_p, G_q)$ in the dendrogram using another lines. This idea is used here, which is shown as Fig. 2. The merged level $s(G_p, G_q)$ is shown with the solid lines with an arrow, while the red and thin line shows the other level $s(G_q, G_p)$. Note that Yadohisa did not use an arrow, whereas the arrows are adopted here to show the direction $G_p \rightarrow G_q$. Thus the difference between $s(G_p, G_q)$ and $s(G_q, G_p)$ shows the degree of asymmetricity.

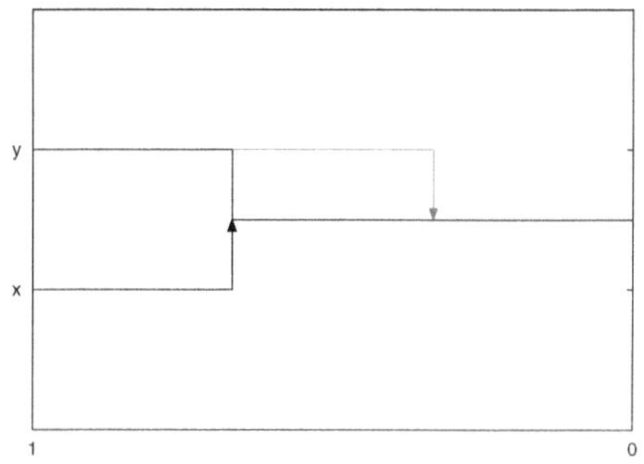

Fig. 2. Asymmetric dendrogram: a variation of Yadohisa's [10]

5 Application to Real Data

Three data sets were used. They are as follows:

1. First data set is the numbers of citations among eight journals on statistics. The original data are omitted here, which are given in [8]. We call this data set *citation data.*
2. Second data set is the total amount of foreign trade among nine countries. The original data are omitted here, which are given in [11]. We call this data *trade data.*
3. Third data set is the input-output table among 13 sectors in economics in Japan, 2005. The original data are omitted here, which are given in [12]. We call this data set *input-output data.*

The details of the description of these data sets are omitted, since our purpose here is to show the way how the methods work and not to discuss their semantics in detail.

We used the both methods: the method using $p(G,G')$ is called the *asymmetric average link*; the method using $r(G,G')$ is called the *probability model*.

Note that the probability model can be applied to all the three models. For the second data set $r(G,G')$ is interpreted as the ratio of the trade from the group of countries G to G', to the total trade of G. For the third data set $r(G,G')$ is interpreted as the ratio of input from G to G', to the total input of G.

The method of the asymmetric average link can also be applied to all these three examples. The number or amount themselves, and not the ratio, is dealt with in this method.

Hence we show two (asymmetric) dendrograms for each data set. The results are briefly commented below.

Citation data: Figures 3 and 4 were respectively obtained from the asymmetric average link and the probability model. In the both figures a cluster of three

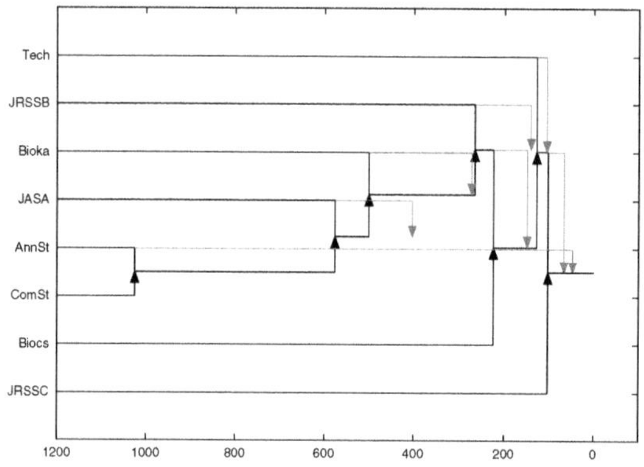

Fig. 3. Dendrogram of journal citation data using the asymmetric average link

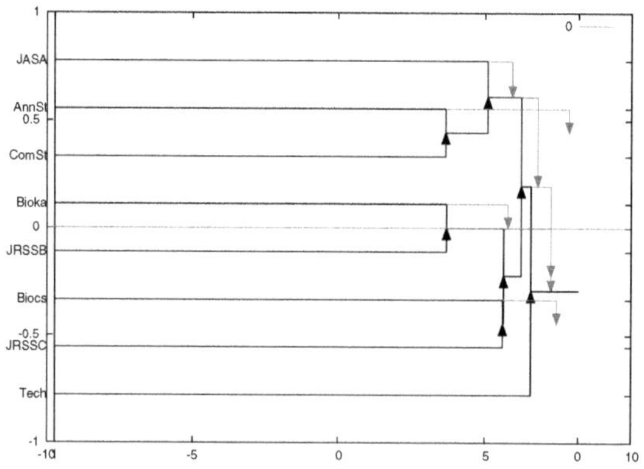

Fig. 4. Dendrogram of journal citation data using the probability model

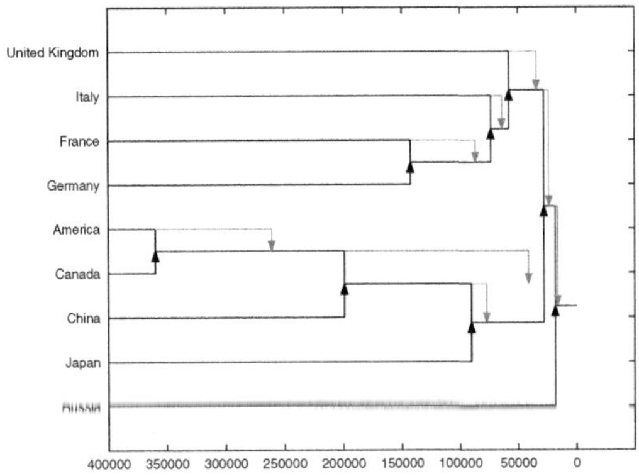

Fig. 5. Dendrogram of the trade data using the asymmetric average link

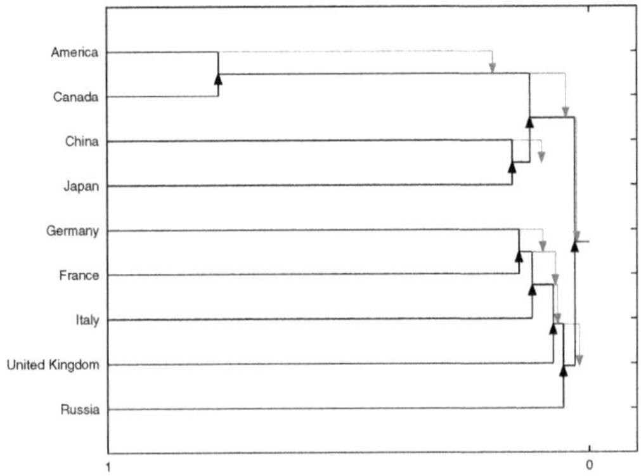

Fig. 6. Dendrogram of the trade data using the probability model

journals 'JASA', 'AnnSt', and 'ComSt' are formed. They are frequently citing one another. An example of asymmetry of citation is found between 'AnnSt' and 'ComSt'; the citation from 'ComSt' to 'AnnSt' is stronger than the reverse direction in both the asymmetric average link and the probability model.

Trade data: Figures 5 and 6 were respectively obtained from the asymmetric average link and the probability model. We observe two clusters in the both figures: one cluster consists of 'America', 'Canada', 'China', and 'Japan'; another consists of European countries. A strong asymmetry is observed, for example, between 'America' and 'Canada': 'Canada' exports more to 'America' than the reverse direction.

Input-output data: Figures 7 and 8 were respectively obtained from the asymmetric average link and the probability model. The asymmetric average link

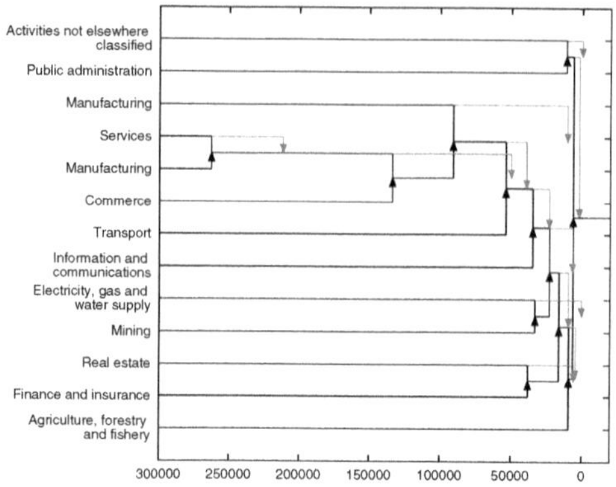

Fig. 7. Dendrogram of the input-output table Japan 2005 using the asymmetric average link

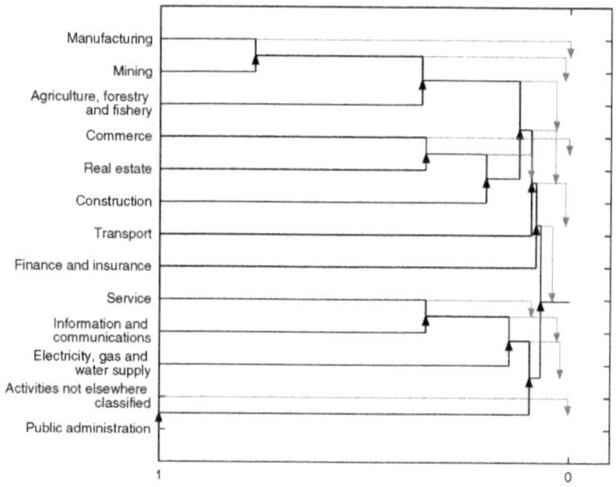

Fig. 8. Dendrogram of the input-output table Japan 2005 using the probability model

model handled the monetary amount, while the probability model uses the ratios and the quantities are normalized. We have different cluster structures, since the similarity between objects are different. Which structures are useful for adequate interpretation needs expert opinions and it is a subject for a future study.

6 Conclusion

We have developed two linkage methods for asymmetric measures of similarity. One is an asymmetric version of the average link, and another is based on the ratio or probability. The second method has been described in terms of citations, but the method is not restricted to bibliographic citations, as we have seen in the applications.

The both methods are useful: the first uses the amount of interaction directly, and the second normalizes the amount using the concept of probability.

The theory of reversals in dendrograms has also been described and it has been proved that the two methods do not have reversals. They are useful properties, as a method with reversals is inconvenient in applications.

We have shown applications of small scales. As a future study, larger scale applications should be studied. Moreover the theory of reversals should be studied for other classes of linkage methods.

Acknowledgement. We thank Dr. Keiichi Okajima for giving us useful advices on applications. We also thank anonymous reviewers for their useful comments.

References

1. Anderberg, M.R.: Cluster Analysis for Applications. Academic Press, New York (1960)
2. Everitt, B.S.: Cluster Analysis, 3rd edn. Arnold, London (1993)
3. Hubert, L.: Min and max hierarchical clustering using asymmetric similarity measures. Psychometrika 38(1), 63–72 (1973)
4. Miyamoto, S.: Fuzzy Sets in Information Retrieval and Cluster Analysis. Kluwer, Dordrecht (1990)
5. Miyamoto, S.: Introduction to Cluster Analysis. Morikita-Shuppan, Tokyo (1999) (in Japanese)
6. Okada, A., Iwamoto, T.: A Comparison before and after the Joint First Stage Achievement Test by Asymmetric Cluster Analysis. Behaviormetrika 23(2), 169–185 (1996)
7. Saito, T., Yadohisa, H.: Data Analysis of Asymmetric Structures. Marcel Dekker, New York (2005)
8. Stigler, S.M.: Citation Patterns in the Journals of Statistics and Probability. Statistical Science 9, 94–108 (1994)
9. Takeuchi, A., Saito, T., Yadohisa, H.: Asymmetric agglomerative hierarchical clustering algorithms and their evaluations. Journal of Classification 24, 123–143 (2007)
10. Yadohisa, H.: Formulation of Asymmetric Agglomerative Clustering and Graphical Representation of Its Result. J. of Japanese Society of Computational Statistics 15(2), 309–316 (2002) (in Japanese)
11. `http://www.jetro.go.jp/indexj.html`
12. `http://www.stat.go.jp/index.htm`

On Hard c-Means Using Quadratic Penalty-Vector Regularization for Uncertain Data

Yasunori Endo[1], Arisa Taniguchi[2], Aoi Takahashi[2], and Yukihiro Hamasuna[3]

[1] Department of Risk Engineering, University of Tsukuba, Tennodai 1-1-1, Tsukuba, Ibaraki, 305-8573, Japan
endo@risk.tsukuba.ac.jp

[2] Graduate School of Systems and Information Engineering, University of Tsukuba, Tennodai 1-1-1, Tsukuba, Ibaraki, 305-8573, Japan
{taniguchi,aoi}@soft.risk.tsukuba.ac.jp

[3] Department of Informatics, Kinki University, 3-4-1, Kowakae, Higashiosaka, Osaka 577-8502, Japan
yhama@info.kindai.ac.jp

Abstract. Clustering is one of the unsupervised classification techniques of the data analysis. Data are transformed from a real space into a pattern space to apply clustering methods. However, the data cannot be often represented by a point because of uncertainty of the data, e.g., measurement error margin and missing values in data. In this paper, we introduce quadratic penalty-vector regularization to handle such uncertain data into hard c-means (HCM) which is one of the most typical clustering algorithms. First, we propose a new clustering algorithm called hard c-means using quadratic penalty-vector regularization for uncertain data (HCMP). Second, we propose sequential extraction hard c-means using quadratic penalty-vector regularization (SHCMP) to handle datasets whose cluster number is unknown. Moreover, we verify the effectiveness of our propose algorithms through some numerical examples.

1 Introduction

Clustering methods are known as very useful tools in many fields for data mining and we can find the construction of datasets through the clustering methods.

As computers have become sophisticated, the more studies for uncertainty are done. In the past, each datum handled by the computers was approximately represented as one point or value because of poor ability of the computers. However, the ability is now enough to handle the data with uncertainty called uncertain data and a lot of researchers have tried to handle original data from the viewpoint that the datum should be represented as not one point approximately but certain distribution exactly in a data space.

V. Torra et al. (Eds.): MDAI 2011, LNAI 6820, pp. 126–138, 2011.

Whenever we construct the clustering methods for the uncertain data, we have one problem, that is, how should we represent the uncertainty of data ?

To solve the above problems, we have proposed "tolerance" as a convenient tool to handle uncertain data and applied some of clustering algorithms [3–8]. In our proposed tolerance, tolerance vectors [3] and penalty ones [8, 2] play main role. Each uncertain datum is allowed to allocate any position by those vectors as far as the constraints for those vectors are satisfied and the position is derived as an optimal solution of a given objective function. Hence, we can say that this concept is in the framework of methodology of soft computing. Penalty vectors are similar to tolerance ones and the methods using penalty vectors become more flexible than tolerance vectors because no constraint for the vectors is needed. Then, we consider the penalty vectors in this paper.

By the way, sequential extraction hard c-means is proposed in Ref [9] which is based on noise clustering in Ref [11, 12]. The clustering does not need the initial number of clusters. The whole dataset is classified into one cluster and one noise dataset and data of the cluster are removed. The sequential extraction HCM classifies the dataset by repeating the above procedure and it can thus handle datasets whose cluster number is unknown.

The goal of this paper is to propose two new clustering algorithms for uncertain data based on hard c-means (HCM) [10], that is, hard c-means using quadratic penalty-vector regularization (HCMP), and sequential extraction hard c-means using quadratic penalty-vector regularization (SHCMP). We believe the proposed algorithms can classify the datasets which consists of uncertain data and whose cluster number is unknown.

2 Preliminaries

In this section, we explain the basic concept of tolerance and penalty, and hard c-means (HCM) clustering. First of all, we define some symbols. Each data is denoted $x_k = (x_{k1}, \ldots, x_{kp})^T \in \Re^p$ and the dataset $X = \{x_1, \ldots, x_n\}$ is given. Each cluster $C_i (i = 1, \ldots, c)$ has a cluster center $v_i = (v_{i1}, \ldots, v_{ip})^T \in \Re^p$. V means a set of cluster centers $\{v_1, \ldots, v_c\}$. A membership grade for x_k to C_i which means belongingness of x_k to C_i is denoted by u_{ki}. U means a partition matrix $(u_{ki})_{1 \le k \le n, 1 \le i \le c}$.

2.1 Tolerance and Penalty Vectors

In this paragraph, we explain two basic concepts, tolerance and penalty as the tools to handle uncertain data in the framework of optimization.

First, we describe the basic concept of tolerance. In general, a datum $x \in \Re^p$ with uncertainty is presented by some interval, i.e.,

$$[\underline{x}, \overline{x}] = [(\underline{x}_1, \ldots, \underline{x}_p)^T, (\overline{x}_1, \ldots, \overline{x}_p)^T] \subset \Re^p.$$

In our proposed tolerance, such a datum is represented by

$$\begin{aligned} x+\varepsilon &= (x_1,\dots,x_p)^T + (\varepsilon_1,\dots,\varepsilon_p)^T \in \Re^p \\ &= (x_1+\varepsilon_1,\dots,x_p+\varepsilon_p)^T \end{aligned}$$

and a constraint for ε_j like that

$$|\varepsilon_j| \leq \xi_j.$$

A vector $\varepsilon = (\varepsilon_1,\dots,\varepsilon_p)^T \in \Re^p$ is called tolerance vector. If we assume that

$$\begin{cases} x_j = \dfrac{\overline{x}_j + \underline{x}_j}{2}, \\ \xi_j = \dfrac{|\overline{x}_j - \underline{x}_j|}{2}, \end{cases}$$

the formulation is equivalent to the above interval.

This concept of tolerance is very useful because we can handle uncertain data in the framework of optimization to use the concept, without introducing some particular measure between intervals, e.g., minimum, maximum, or Hausdorff distance. If we use tolerance, we don't need any particular distance, that is, a distance $d(X,Y)$ between $X = x + \varepsilon_x$ ($\|\varepsilon_x\| \leq \xi_x$) and $Y = y + \varepsilon_y$ ($\|\varepsilon_y\| \leq \xi_y$) can be calculated as $\|(x-y)+(\varepsilon_x-\varepsilon_y)\|$. From the above, we know that this tool is useful when we handle the data, especially data with missing values of their attributes, in the framework of optimization like as fuzzy c-means clustering [6].

We can introduce the concept of penalty based on the tolerance. The concept is similar to the concept of tolerance but it differs from the tolerance in that there is no constraint for penalty vectors.

We define some symbols at the beginning. In addition to the symbols in the above section, we define penalty vector $\delta_k = (\delta_{k1},\dots,\delta_{kp})^T \in \Re^p$, and a set of penalty vectors $\Delta = \{\delta_1,\dots,\delta_n\}$. The uncertain datum is represented as $x_k + \delta_k$. In addition, we define weighting coefficient w_{klj} ($w_{klj} \geq 0$) and weighting matrix as follows:

$$W_k = \begin{pmatrix} w_{k11} & \cdots & w_{k1p} \\ \vdots & \ddots & \vdots \\ w_{kp1} & \cdots & w_{kpp} \end{pmatrix}.$$

One of the simplest form of the matrix is as follows:

$$W_k = \begin{pmatrix} w_{k1} & & 0 \\ & \ddots & \\ 0 & & w_{kp} \end{pmatrix}. \tag{1}$$

This form is not only simple but useful.

Now, we introduce the following quadratic penalty-vector regularization term:

$$\sum_{k=1}^{n} {\delta_k}^T W_k \delta_k = \sum_{j=1}^{p}\sum_{l=1}^{p} w_{klj}\delta_{kl}\delta_{kj}.$$

We assume that W_k is a symmetric matrix, i.e., $w_{kj} = w_{jk}$. In case that W_k is a diagonal matrix, the regularization term is represented as follows:

$$\sum_{k=1}^{n} {\delta_k}^T W_k \delta_k = \sum_{k=1}^{n} \sum_{j=1}^{p} w_{kjj} \left(\delta_{kj}\right)^2 .$$

2.2 Hard c-Means

We describe here the conventional hard c-means (HCM) clustering.

The objective function of HCM J_{HCM} and the constraints are defined as follows:

$$J_{\mathrm{HCM}}(U,V) = \sum_{k=1}^{n} \sum_{i=1}^{c} u_{ki} \|x_k - v_i\|^2,$$

$$\sum_{i=1}^{c} u_{ki} = 1, \quad \forall k.$$

The optimal solutions u_{ki} and v_i which minimize J_{HCM} are as follows:

$$u_{ki} = \begin{cases} 1, & (v_i = \arg\min_l \|x_k - v_l\|^2) \\ 0, & (\text{otherwise}) \end{cases}$$

$$v_i = \frac{\sum_{k=1}^{n} u_{ki} x_k}{\sum_{k=1}^{n} u_{ki}}.$$

The algorithm of HCM is as follows:

Algorithm 1. HCM

Give X.
$c \leftarrow$ a fixed value
$V \leftarrow$ initial values
while The stop criterion does not satisfy **do**
  for all k such that $1 \le k \le n$ **do**
    for all i such that $1 \le i \le c$ **do**
      $u_{ki} \leftarrow \begin{cases} 1 & (v_i = \arg\min_l \|x_k - v_l\|^2) \\ 0 & (\text{otherwise}) \end{cases}$
    end for
  end for
  for all i such that $1 \le i \le c$ **do**
    $v_i \leftarrow \frac{\sum_{k=1}^{n} u_{ki} x_k}{\sum_{k=1}^{n} u_{ki}}$
  end for
end while

3 Hard c-Means Using Quadratic Penalty-Vector Regularization

In this section, we propose a new clustering algorithm, hard c-means using quadratic penalty-vector regularization (HCMP) to handle uncertain data. HCMP is constructed by introducing penalty vectors into HCM.

3.1 Optimal Solutions of HCMP

We define the objective function of HCMP J_{HCMP} and the constraints with the quadratic penalty-vector regularization term as follows:

$$J_{\text{HCMP}}(U, V, \Delta) = \sum_{k=1}^{n} \sum_{i=1}^{c} u_{ki} \|x_k + \delta_k - v_i\|^2 + \sum_{k=1}^{n} \delta_k^T W_k \delta_k,$$
$$\sum_{i=1}^{c} u_{ki} = 1, \quad \forall k. \tag{2}$$

In a similar way to HCM, we can obtain the optimal solutions of u_{ki} and v_i as follows:

$$u_{ki} = \begin{cases} 1, & (v_i = \arg\min_l \|x_k + \delta_k - v_l\|^2) \\ 0, & (\text{otherwise}) \end{cases}$$
$$v_i = \frac{\sum_{k=1}^{n} u_{ki}(x_k + \delta_k)}{\sum_{k=1}^{n} u_{ki}}.$$

The optimal solution of δ_k can be derived as follows:
From partially differentiating J_{HCMP} by δ_k and the constraint (2), we get

$$\begin{aligned} \frac{1}{2} \frac{\partial J_{\text{HCMP}}}{\partial \delta_k} &= \left(\sum_{i=1}^{c} u_{ki} \right) \delta_k + W_k \delta_k + \sum_{i=1}^{c} u_{ki}(x_k - v_i) \\ &= (E + W_k)\delta_k + \sum_{i=1}^{c} u_{ki}(x_k - v_i). \end{aligned}$$

From $\frac{\partial J_{\text{HCMP}}}{\partial \delta_k} = 0$, we can obtain

$$\delta_k = -(E + W_k)^{-1} \cdot \sum_{i=1}^{c} u_{ki}(x_k - v_i).$$

3.2 Algorithm

In this paragraph, we construct the algorithm of HCMP using the above optimal solutions.

This algorithm needs a fixed number of clusters, c. However, it is very difficult to determine the most suitable cluster number when we classify the dataset whose cluster number is unknown like Polaris dataset in Section 6. Thus, we

Algorithm 2. HCMP

Give X.
$c \leftarrow$ a fixed value
$V, \Delta \leftarrow$ initial values
while The stop criterion does not satisfy **do**
  for all k such that $1 \leq k \leq n$ **do**
    for all i such that $1 \leq i \leq c$ **do**
      $u_{ki} \leftarrow \begin{cases} 1 & (v_i = \arg\min_l \|x_k - v_l\|^2) \\ 0 & (\text{otherwise}) \end{cases}$
    end for
  end for
  for all i such that $1 \leq i \leq c$ **do**
    $v_i \leftarrow \frac{\sum_{k=1}^{n} u_{ki} x_k}{\sum_{k=1}^{n} u_{ki}}$
  end for
  for all k such that $1 \leq k \leq n$ **do**
    $\delta_k \leftarrow -(E + W_k)^{-1} \cdot \sum_{i=1}^{c} u_{ki}(x_k - v_i)$
  end for
end while

consider sequential extraction methods of clustering to automatically determine the suitable cluster number in the next section.

4 Sequential Extraction Hard c-Means Using Quadratic Penalty-Vector Regularization

In this section, we construct a sequential extraction method for uncertain data clustering based on HCM. The reason to consider the sequential extraction is that the method doesn't need to determine the cluster number. As above mentioned, it is very difficult to determine the most suitable cluster number in many cases.

Dave et al. proposed noise clustering in Refs. [11, 12]. In the method, the given dataset is classified into some clusters and one noise cluster. Miyamoto et. al proposed sequential extraction hard c-means in Ref [9] using the noise clustering. The whole dataset are classified into one cluster and one noise dataset, and data of the cluster are removed. The sequential extraction HCM classifies the dataset by repeating the above procedure.

We propose sequential extraction hard c-means using quadratic penalty-vector regularization (SHCMP) which is constructed by introducing the concept of penalty vectors into sequential extraction hard c-means (SHCM). SHCM has not proposed by anyone but I believe the reason is that the procedure of SHCM is very simple and trivial.

4.1 Sequential Extraction Hard c-Means

The objective function of HCM J_{SHCM} and the constraints are defined as follows:

$$J_{\mathrm{SHCM}}(U,V) = \sum_{k=1}^{n} u_{k1}\|x_k - v\|^2 + \sum_{k=1}^{n} u_{k0} D^2,$$

$$\sum_{i=0}^{1} u_{ki} = 1,$$

here D is a noise parameter. The optimal solutions u_{ki} and v which minimize J_{SHCM} are as follows:

$$u_{ki} = \begin{cases} i, & (\|x_k - v\|^2 \le D^2) \\ 1-i, & (\text{otherwise}) \end{cases} \quad (i = 0, 1)$$

$$v = \frac{u_{k1}(x_k + \delta_k)}{u_{k1}}.$$

The algorithm of SHCM is as follows:

Algorithm 3. SHCM

Give X.
$v \leftarrow$ an initial value
while $X \neq \phi$ **do**
  while The stop criterion does not satisfy **do**
    for all k such that $1 \le k \le |X|$ **do**
      for all i such that $0 \le i \le 1$ **do**
        $u_{ki} \leftarrow \begin{cases} i, & (\|x_k - v\|^2 \le D^2) \\ 1-i. & (\text{otherwise}) \end{cases}$
      end for
    end for
    $v \leftarrow \frac{u_{k1}(x_k+\delta_k)}{u_{k1}}$
  end while
  $X \leftarrow X \setminus \{x_k \mid u_{k1} = 1\}$
  Renumber each datum in X.
end while

---

5 Sequential Extraction Hard c-Means Using Quadratic Penalty-Vector Regularization

In this paragraph, we propose sequential extraction hard c-means using quadratic penalty-vector regularization (SHCMP) by introducing the concept of penalty vectors into sequential extraction hard c-means (SHCM).

The objective function of HCM J_{SHCMP} and the constraints are defined as follows:

$$J_{\mathrm{SHCMP}}(U,V,\Delta) = \sum_{k=1}^{n} u_{k1}\|x_k+\delta_k-v\|^2 + \sum_{k=1}^{n} \delta_k^T W_k \delta_k + \sum_{k=1}^{n} u_{k0} D^2,$$

$$\sum_{i=0}^{1} u_{ki} = 1,$$

here D is a noise parameter. The optimal solutions u_{ki} and v which minimize J_{SHCMP} are as follows:

$$u_{ki} = \begin{cases} i, & (\|x_k+\delta_k-v\|^2 \le D^2) \\ 1-i, & (\text{otherwise}) \end{cases} \quad (i=0,1)$$

$$v = \frac{u_{k1}(x_k+\delta_k)}{u_{k1}}.$$

The optimal solution of δ_k can be derived as follows:
From partially differentiating J_{SHCMP} by δ_k, we get

$$\begin{aligned}\frac{1}{2}\frac{\partial J}{\partial \delta_k} &= u_{k1}(x_k+\delta_k-v) + W_k\delta_k \\ &= (u_{k1}E + W_k)\delta_k + u_{k1}(x_k - v).\end{aligned}$$

From $\frac{\partial J_{\mathrm{SHCMP}}}{\partial \delta_k} = 0$, we can obtain

$$\delta_k = -(u_{k1}E+W_k)^{-1} \cdot u_{k1}(x_k - v_i).$$

We construct the algorithm of SHCMP using the above optimal solutions as follows:

6 Numerical Examples

In this section, we verify our proposed algorithms through some numerical examples. We consider the following three datasets. Fig. 1, Fig. 2 and Fig. 3 show an artificial dataset, another one and Polaris dataset consisting of five data, 99 data and 51 data, respectively.

6.1 Results

Results of HCM. First, we show the results of HCM for Fig. 1, Fig. 2, and Fig. 3 in Fig. 4 with $c=2$, in Fig. 5 with $c=5$, and Fig. 6 with $c=3$.

Results of HCMP. Second, we show the results of HCMP for Artificial dataset 1 (Fig. 1) and Polaris dataset (Fig. 3). We consider three cases of $W_k = 0.1E$, $W_k = 1.0E$, and $W_k = 10E$ in the form of (1). Similarly to the above paragraph, we set $c=2$ and $c=3$ for Fig. 1 and Fig. 3. The stop criterion is that $\|v^{(L+1)} - v^{(L)}\|^2 < 0.001^2$ where L means the iteration time.

Algorithm 4. SHCMP

Give X.
$v, \Delta \leftarrow$ initial values
while $X \neq \phi$ **do**
  while The stop criterion does not satisfy **do**
    for all k such that $1 \leq k \leq |X|$ **do**
      for all i such that $0 \leq i \leq 1$ **do**

$$u_{ki} \leftarrow \begin{cases} i, & (\|x_k + \delta_k - v\|^2 \leq D^2) \\ 1 - i, & (\text{otherwise}) \end{cases}$$

      end for
    end for

$$v \leftarrow \frac{u_{k1}(x_k + \delta_k)}{u_{k1}}$$

    for all k such that $1 \leq k \leq |X|$ **do**

$$\delta_k \leftarrow -(u_{k1}E + W_k)^{-1} \cdot u_{k1}(x_k - v_i)$$

    end for
  end while
  $X \leftarrow X \setminus \{x_k \mid u_{k1} = 1\}$
  Renumber each datum in X.
end while

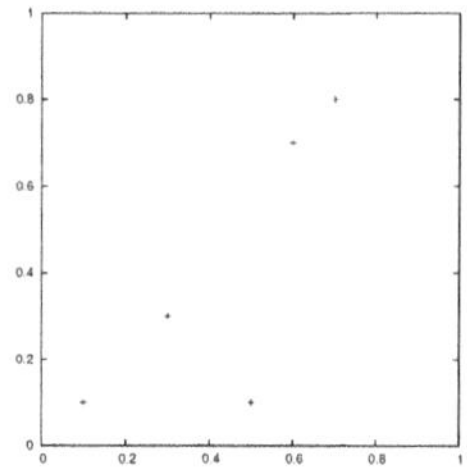

Fig. 1. Artificial dataset 1

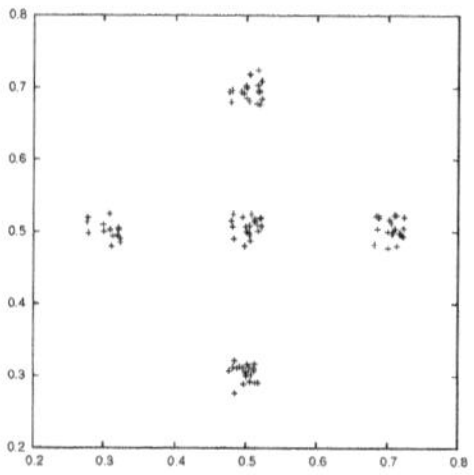

Fig. 2. Artificial dataset 2

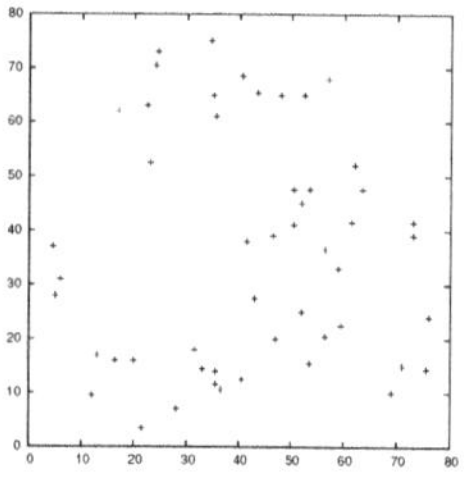

Fig. 3. Polaris dataset

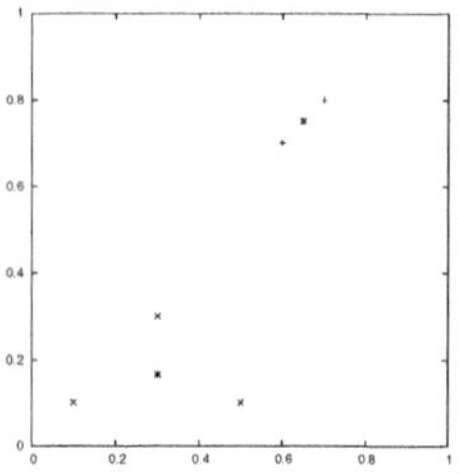

Fig. 4. Result of HCM for artificial dataset 1 ($c = 2$)

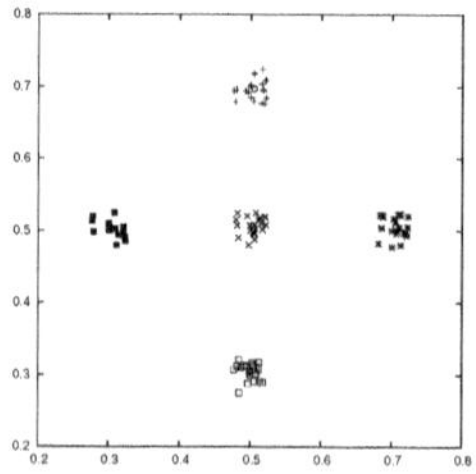

Fig. 5. Result of HCM for artificial dataset 2 ($c = 5$)

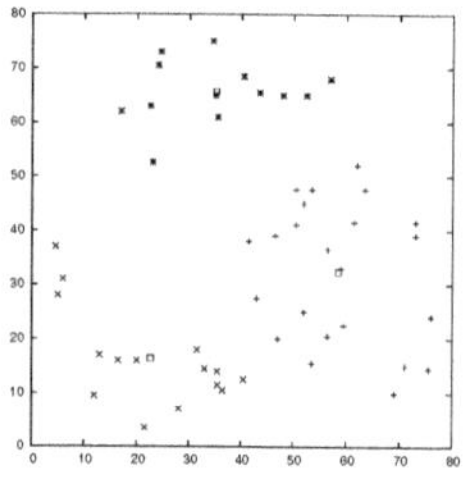

Fig. 6. Result of HCM for Polaris dataset ($c = 3$)

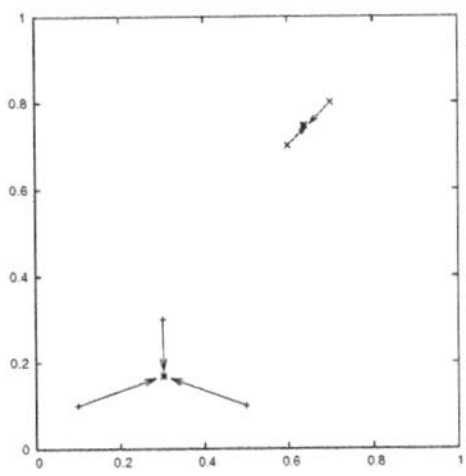

Fig. 7. Result of HCMP for Artificial dataset 1 ($c = 2$, $W_k = 0.1E$)

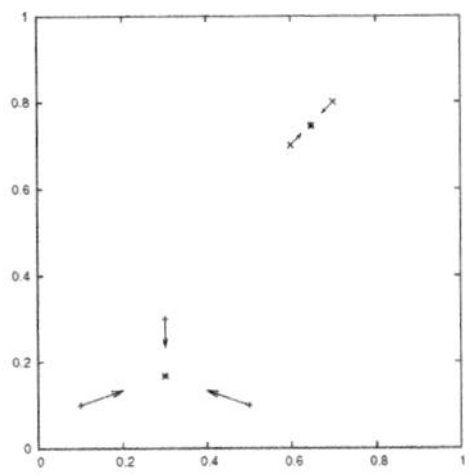

Fig. 8. Result of HCMP for Artificial dataset 1 ($c = 2$, $W_k = 1.0E$)

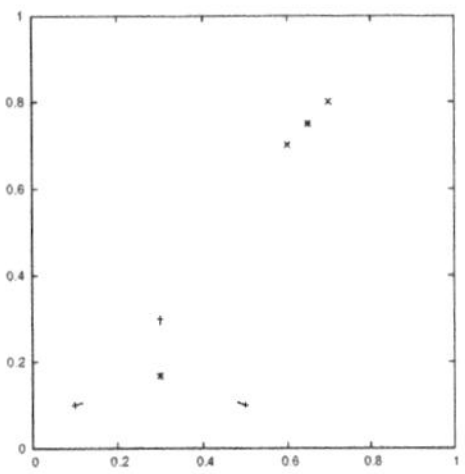

Fig. 9. Result of HCMP for Artificial dataset 1 ($c = 2$, $W_k = 10E$)

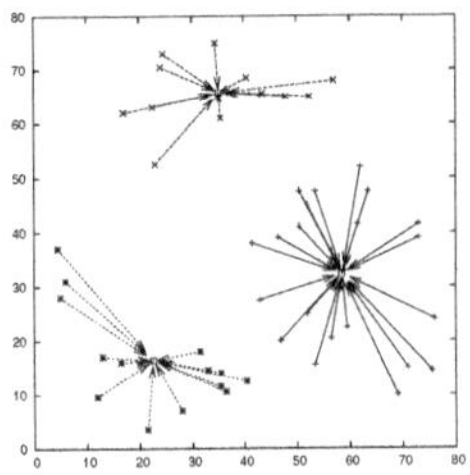

Fig. 10. Result of HCMP for Polaris dataset ($c = 3$, $W_k = 0.1E$)

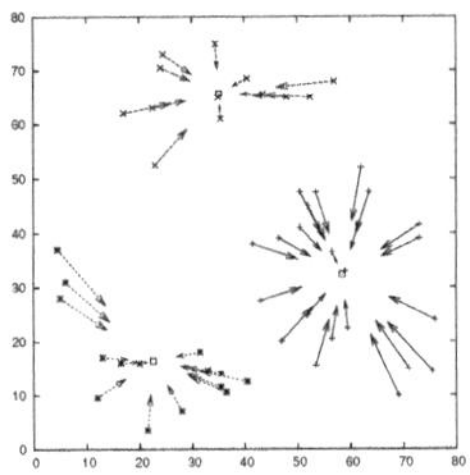

Fig. 11. Result of HCMP for Polaris dataset ($c = 3$, $W_k = 1.0E$)

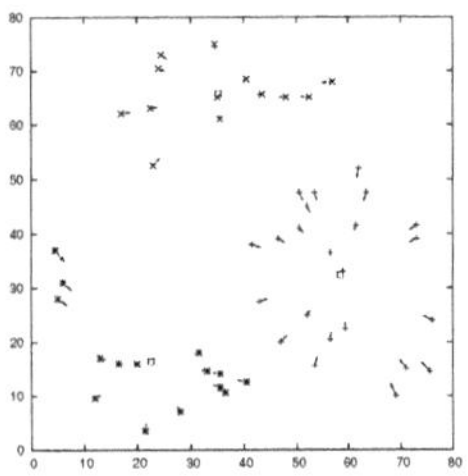

Fig. 12. Result of HCMP for Polaris dataset ($c = 3$, $W_k = 10E$)

Results of SHCM. For Artificial dataset 2 in Fig. 2, we show the process that clusters are sequentially extracted from Fig. 13 to Fig. 17, where $D = 0.1$. First, it puts the cluster extracted in Fig. 13 to be the 1st cluster and remove those from the dataset. Next, it puts the cluster extracted in Fig. 14 to be the 2st cluster and remove those from the dataset. The procedure is repeated until the dataset is empty. In case of Artificial dataset 2 in Fig. 2, the dataset becomes empty in five iteration times and we finally obtain the result in Fig. 18. The symbols of white box, times, asterisk, plus and black box mean from the 1st to the 5th clusters, respectively.

In a similar way to the above, we obtain the results of SHCM for Polaris dataset in Fig. 19 and Fig. 20, where $D = 10$ and $D = 20$. The symbols of plus, times, asterisk, white box, black box, white circle, black circle, white up triangle, black up triangle, white down triangle, and black down triangle mean from the 1st to the 11th clusters, respectively.

Results of SHCMP. We show the result of SHCMP for Polaris dataset in Fig. 3. Similar to SHCM in the above paragraph, we omit to show the process. We show four results in Fig. 21 with $D = 10$ and $W_k = 1.0E$, Fig. 22 with $D = 10$ and $W_k = 10E$, Fig. 23 with $D = 20$ and $W_k = 1.0E$, and Fig. 24 with $D = 20$ and $W_k = 10E$. The symbols of plus, times, asterisk, white box, black

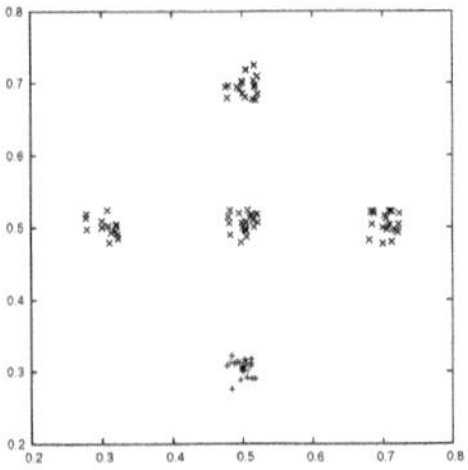

Fig. 13. The 1st cluster for Artificial dataset 2 ($D = 0.1$)

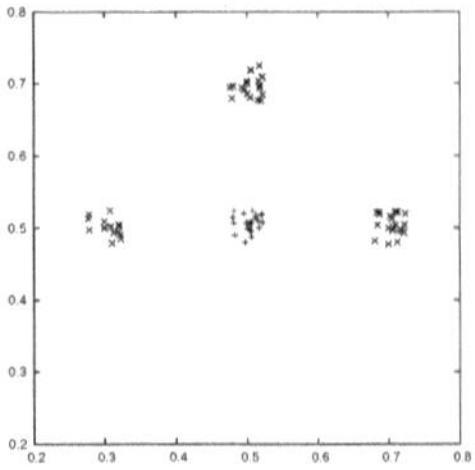

Fig. 14. The 2nd cluster for Artificial dataset 2 ($D = 0.1$)

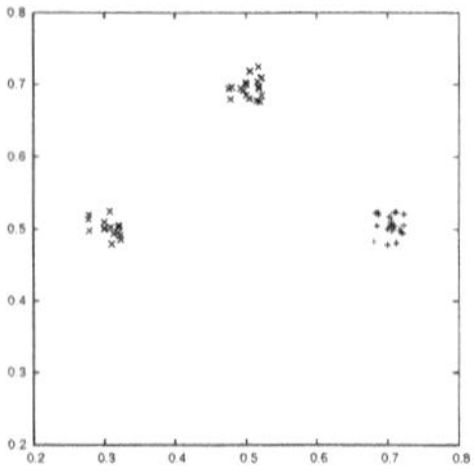

Fig. 15. The 3rd cluster for Artificial dataset 2 ($D = 0.1$)

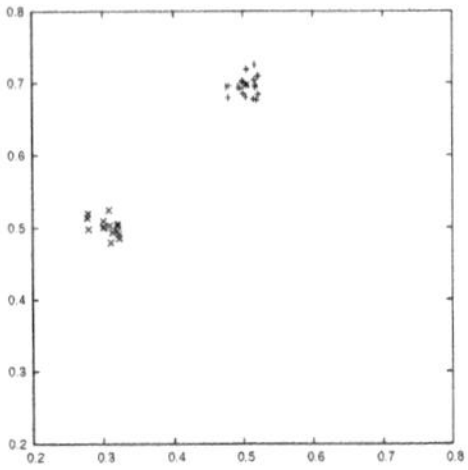

Fig. 16. The 4th cluster for Artificial dataset 2 ($D = 0.1$)

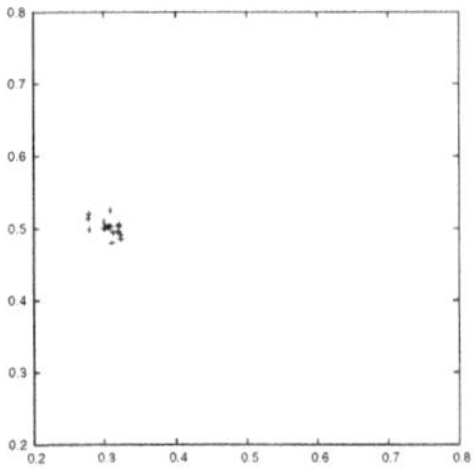

Fig. 17. The 5th cluster for Artificial dataset 2 ($D = 0.1$)

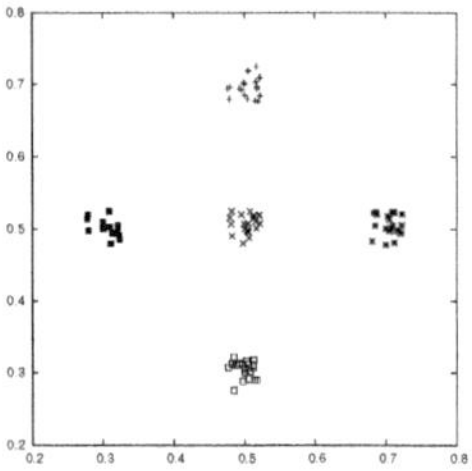

Fig. 18. Final result of SHCM for Artificial dataset 2 ($D = 0.1$)

box, white circle, black circle, white up triangle, black up triangle, and white down triangle mean from the 1st to the 10th clusters, respectively.

6.2 Consideration

First, we consider HCMP. We know that the norm of δ_k is small as the value of W_k is large. The reason is that, δ_k has great effect on the objective function as the value of W_k is large so that it makes δ_k small in order to minimize the function in this case. We expected that the values of W_k affect the belongingness of data to clusters, but we didn't show the effects in these examples.

Next, we consider SHCMP. In comparison with SHCM on the same D, the cluster number is smaller and each cluster is more massive in any examples. These points seem advantages of SHCMP over SHCM.

On the other side, the values of D and W_k have great effect on the results. For example, the range in which $delta_k$ is allowed to be given becomes too large when the value of W_k is small, and finally, clusters can not be determined frequently. Therefore, we need to estimate suitable W_k and D in advance.

Moreover, there is sometimes a datum such that the value of dissimilarity between the datum and the cluster center to which the datum belongs are larger than between the datum and other centers.

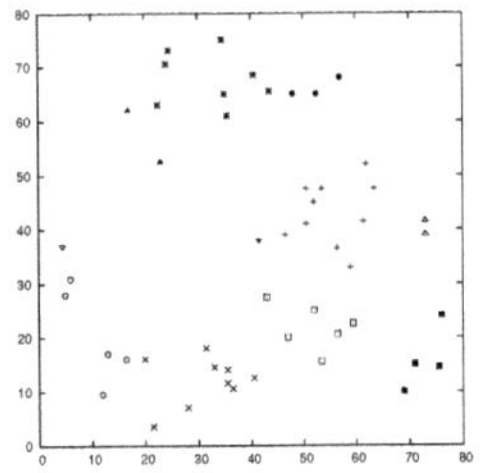

Fig. 19. Result of SHCM for Polaris dataset ($D = 10$)

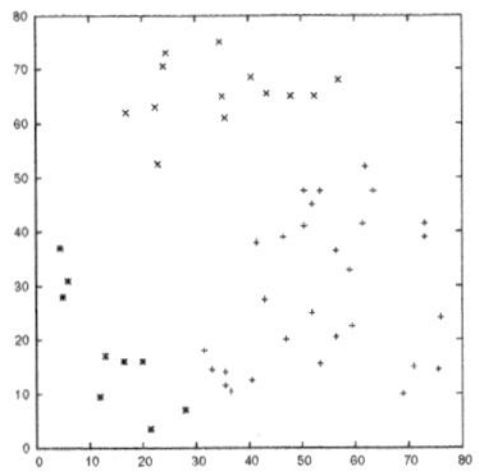

Fig. 20. Result of SHCM for Polaris dataset ($D = 20$)

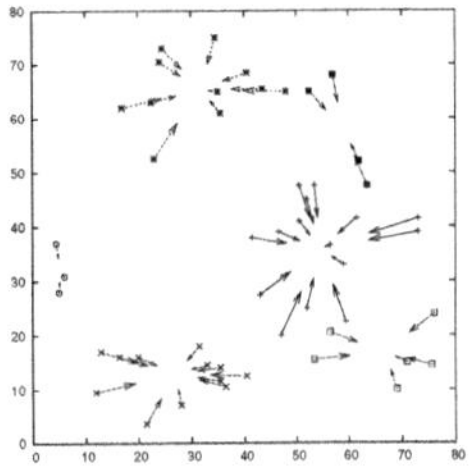

Fig. 21. Final result of SHCMP for Polaris dataset ($D = 10$, $W_k = 1.0E$)

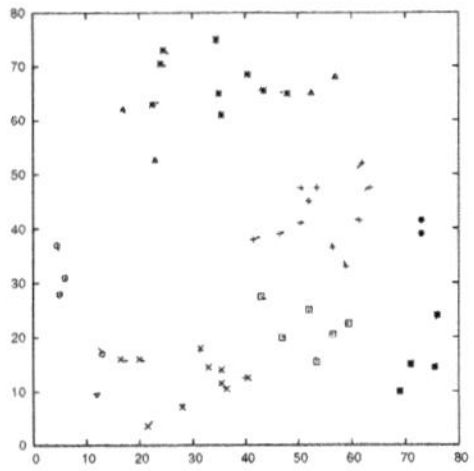

Fig. 22. Final result of SHCMP for Polaris dataset ($D = 10$, $W_k = 10E$)

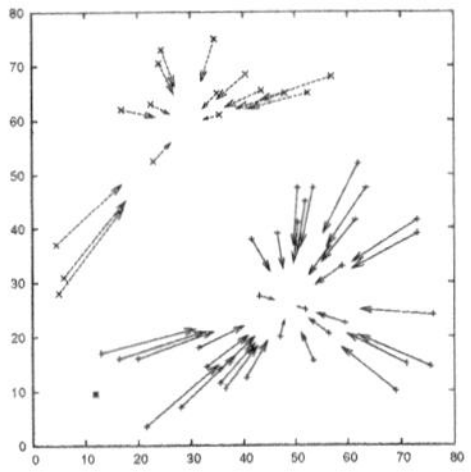

Fig. 23. Final result of SHCMP for Polaris dataset ($D = 20$, $W_k = 1.0E$)

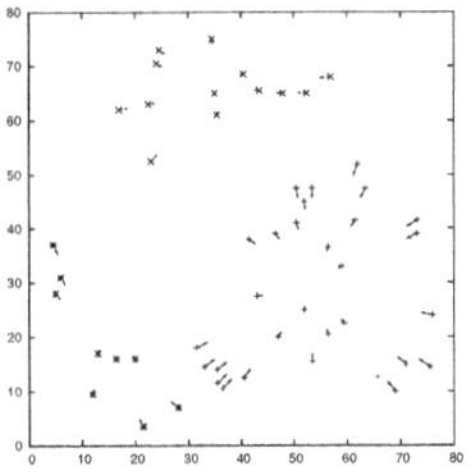

Fig. 24. Final result of SHCMP for Polaris dataset ($D = 20$, $W_k = 10E$)

For example, six data in the lower part of the center belongs to different clusters in comparison with Fig. 6 and Fig. 20. The reason is that the extracted data are removed from the dataset and they aren't considered in the process since that. Only the low-left datum in Fig. 23 belongs to the other cluster from the same reason.

7 Conclusion

In this paper, we proposed two new clustering algorithms for uncertain data, one is constructed by introducing the concept of penalty-vector regularization into hard c-means, and the other into sequential hard c-means. The latter can handle datasets whose cluster number is unknown and doesn't need the cluster number in advance.

As mentioned above, the algorithms needs the suitable W_k and D. Moreover, there is sometimes a datum such that the value of dissimilarity between the datum and the cluster center to which the datum belongs are larger than between the datum and other centers. We'll discuss the problems in the forthcoming paper.

Acknowledgment

We would like to thank gratefully and sincerely Associate Professor KANZAWA Yuchi of Shibaura Institute of Technology for his advice. This study is partly supported by the Grant-in-Aid for Scientific Research (C) (Project No.21500212) from the Ministry of Education, Culture, Sports, Science and Technology, Japan.

References

1. Miyamoto, S.: Introduction to Cluster Analysis. Morikita-Shuppan, Tokyo (1999) (in Japanese)
2. Endo, Y., Hasegawa, Y., Hamasuna, Y., Kanzawa, Y.: Fuzzy c-Means Clustering for uncertain Data using Quadratic Regularization of Penalty Vectors. Journal of Advance Computational Intelligence and Intelligent Informatics 15(1), 76–82 (2011)
3. Endo, Y., Murata, R., Haruyama, H., Miyamoto, S.: Fuzzy c-Means for Data with Tolerance. In: Proc. 2005 International Symposium on Nonlinear Theory and Its Applications, pp. 345–348 (2005)
4. Murata, R., Endo, Y., Haruyama, H., Miyamoto, S.: On Fuzzy c-Means for Data with Tolerance. Journal of Advance Computational Intelligence and Intelligent Informatics 10(5), 673–681 (2006)
5. Kanzawa, Y., Endo, Y., Miyamoto, S.: Fuzzy c-Means Algorithms for Data with Tolerance based on Opposite Criterions. IEICE Trans. Fundamentals E90-A(10), 2194–2202 (2007)
6. Endo, Y., Hasegawa, Y., Hamasuna, Y., Miyamoto, S.: Fuzzy c-Means for Data with Rectangular Maximum Tolerance Range. Journal of Advanced Computational Intelligence and Intelligent Informatics 12(5), 461–466 (2008)
7. Kanzawa, Y., Endo, Y., Miyamoto, S.: Fuzzy c-Means Algorithms for Data with Tolerance using Kernel Functions. IEICE Trans. Fundamentals E91-A(9), 2520–2534 (2008)
8. Hasegawa, Y., Endo, Y., Hamasuna, Y.: On Fuzzy c-Means for Data with Uncertainty using Spring Modulus. In: Proc. of SCIS & ISIS 2008 (2008)
9. Miyamoto, S., Arai, K.: Different Sequential Clustering Algorithms and Sequential Regression Models. In: Proc. of FUZZ-IEEE 2009 (2009)
10. MacQueen, J.B.: Some Methods of Classification and Analysis of Multivariate Observations. In: Proc. of 5th Berkeley Symposium on Math. Stat. and Prob., pp. 281–297 (1967)
11. Dave, R.N.: Characterization and Detection of Noise in Clustering. Pattern Recognition Letters 12, 657–664 (1991)
12. Dave, R.N., Krishnapuram, R.: Robust Clustering Methods: a Unified View. IEEE Trans. on Fuzzy Systems 5(2), 270–293 (1997)

Grey Synthetic Clustering Method for DoS Attack Effectiveness Evaluation

Zimei Peng, Wentao Zhao, and Jun Long

National University of Defense Technology, Changsha, Hunan 410073, China
pengzimei@126.com

Abstract. Effectiveness evaluation of DoS attack is a complex problem, in which the information is incomplete and vague. The grey theory, which deals with the "less data uncertainty" matter, is a powerful tool to solve the problem. We propose a grey synthetic clustering method for DoS attack effectiveness evaluation in this paper. Firstly, we calculate grey clustering coefficient with general grey clustering method. Secondly, if there is no significant difference about grey clustering coefficient, we calculate the synthetic clustering coefficient. Finally, clustering objects can be clustered accurately with the synthetic clustering coefficient. The experimental results show that the approach is feasible and correct.

Keywords: DoS attack effectiveness, grey synthetic clustering, effectiveness evaluation.

1 Introduction

Denial of Service attack (DoS) is a widespread means of attack on the network. This attack is easy to carry out and difficult to prevent, which poses a great threat to the normal operation of Internet network system. The aim of DoS attack is to affect the normal service of attacked target system or make the functions of system be lost partially or completely. Such attack usually aims at some weakness of TCP/IP protocol or holes in computer systems. Using improper connection method, it will make the attacked target system be flooded with a lot of useless information in a short time so as to consume network bandwidth or system resources. The result is that the attacked target system is overwhelmed and could not provide normal services to legitimate users[1]. The evaluation of DoS attack effectiveness is one of the important parts of integrated assessment and safety evaluation for network attack, and it is an important and urgent task of the computer attack and defense research. There are a variety of methods for DoS attack effectiveness evaluation, such as index-based analysis[2], BP (Back-Propagation) neural network based analysis[3], and network performance based analysis[4].

Grey theory is raised for the uncertain problem lack of experience and data, in other words, the "less data uncertainty" issues[5]. The evaluation of DoS attack effectiveness is a complex problem, in which the information is incomplete and uncertain. The grey theory, which deals with the "less data uncertainty" matter, is a powerful

V. Torra et al. (Eds.): MDAI 2011, LNAI 6820, pp. 139–149, 2011.

tool to solve the problem. Evaluation methods such as the variable weight clustering method pioneered by Professor Deng Julong, fixed weight grey clustering evaluation analysis and grey clustering evaluation based on triangle whitening weight function proposed by Professor Liu Sifeng, have been widely used in part of the project areas, such as knowledge management capability evaluation[6], military information network evaluation[7]. Wang Huimei has proposed the application of grey theory in network attack effectiveness evaluation, as well as the grey fixed weight clustering effectiveness evaluation model and evaluation algorithms of computer network attack effectiveness[8]. However, the method of grey clustering evaluation proposed in paper [8] determines which grey class the clustering object belongs to with the method that compares the size of the grey clustering coefficient vector, but in practice, it is common that there is no significant difference about grey clustering coefficient. In such case, the evaluation objects can't be determined accurately with this method. After study on grey theory, we have found that grey synthetic clustering method can solve the problem. Therefore, this paper proposes a grey synthetic clustering method for DoS attack effectiveness evaluation.

The structure of this paper is as follows: Section 1 is the introduction. Section 2 describes the index system of DoS attack effectiveness evaluation. Section 3 gives the details of the grey synthetic clustering method for DoS attack effectiveness evaluation. Section 4 presents an experiment and analysis of the result. Finally, Section 5 gives the conclusion.

2 Index System of Effectiveness Evaluation of DoS Attack

The DoS attack effectiveness evaluation mainly focuses on the impact of attack on the attacked target system[9]. The purpose of DoS attack is mainly to destroy the availability and reliability of attacked target, consume up its resources, and thus make the target unable to provide normal service to legitimate users. Therefore, we can choose the following evaluation indexes.

Network bandwidth utilization rate. When DoS attack occurs, the attacker intends to make a lot of useless information to occupy the limited network resources, which results in block of network bandwidth and realization of the attacker's intent, so the network bandwidth utilization rate will change significantly.

Server's CPU and memory usage. In other words, it is the server's CPU utilization and memory utilization before and after attack. When DoS attack occurs, the attacked target will receive a large number of packets requesting for service, which will consume a large amount of CPU and memory, so the usage of CPU and memory will change greatly.

Service response delay. It is the time that the attacked target requires from receiving service request signal to providing the service. It is an important index of DoS attack effectiveness evaluation. The difference of service response delay before and after attack can reflect DoS attack effectiveness directly.

Packet loss rate. After DoS attack, the service ability of attacked target will be reduced, and it will not be able to provide normal network service to the legitimate users. At the same time, the network bandwidth will be occupied by a large number of attack packets that created by attackers deliberately, resulting in serious packet loss.

Recovery time. After DoS attack, the attacked target needs some time to recover in order to provide normal service to legitimate users. The length of recovery time can reflect the strength of DoS attack effect.

Attack mechanism. It means the way that an attack influences the attacked target. DoS attack mechanism can be divided into three types: resource consuming, service crashing and system crashing. Resource consuming means that the attacker tries to consume the legitimate resources of target, such as network bandwidth, memory, disk space, CPU, and so on. Service crashing means that the attacker makes the service of target crashed or suspended by using some weakness of service. System crashing means that the attacker makes the system crashed by using some defects of the system. It can be concluded that the attack effectiveness of these three types of attack mechanism increases step by step. In other words, system crashing attack is the most devastating, which can make the target system unaccessible; service crashing attack only makes a particular service of target unaccessible; resource consuming only consumes resources of target in order to make the target system respond more slowly, and the attacker must send packets to target system continuously to keep the attacking going on, since the system will become normal if the attacker stops sending packets.

3 Grey Synthetic Clustering Evaluation Model

This paper presents the grey synthetic clustering algorithm of DoS attack effectiveness evaluation. Based on the effectiveness evaluation index of DoS attack and in accordance of the whitening weight function of grey number, it summarizes the attack effectiveness that need to be evaluated according to grey classes, in order to determine the grey class that the effectiveness of each attack belongs to.

Definition 1[10, 11]. Assume there are n clustering objects, m clustering indexes, s grey classes, the quantitative evaluation value of clustering object i on clustering index j is $d_{ij}(i=1,2,...,n; j=1,2,...,m)$, then, $f_j^k(*)(j=1,2,...,m; k=1,2,...,s)$ is called the whitening weight function of clustering index j on grey class k. If the clustering weight of clustering index j on grey class k is independent of k, that $w_j(j=1,2,...,m)$ is the clustering weight of cluster index j, and $\sum_{j=1}^{m} w_j = 1$, then call

$$\sigma_i^k = \sum_{j=1}^{m} f_j^k(d_{ij}) w_j \tag{1}$$

the clustering coefficient of clustering object i on grey class k. Call

$$\sigma = \begin{pmatrix} \sigma_1^1 & \sigma_1^2 & \cdots & \sigma_1^s \\ \sigma_2^1 & \sigma_2^2 & \cdots & \sigma_2^s \\ \vdots & \vdots & \ddots & \vdots \\ \sigma_n^1 & \sigma_n^2 & \cdots & \sigma_n^s \end{pmatrix} \tag{2}$$

the fixed weight clustering coefficient matrix.

Definition 2[10, 11]. Set

$$\delta_i^k = \frac{\sigma_i^k}{\sum_{k=1}^{s} \sigma_i^k}, \tag{3}$$

call δ_i^k the normalized clustering coefficient of clustering object i on grey class k.

Call $\delta_i = (\delta_i^1, \delta_i^2, \ldots, \delta_i^s)(i = 1, 2, \ldots, n)$ the normalized clustering coefficient vector of clustering object i. Call

$$\Pi = (\delta_i^k) = \begin{pmatrix} \delta_1^1 & \delta_1^2 & \cdots & \delta_1^s \\ \delta_2^1 & \delta_2^2 & \cdots & \delta_2^s \\ \vdots & \vdots & \ddots & \vdots \\ \delta_n^1 & \delta_n^2 & \cdots & \delta_n^s \end{pmatrix} \tag{4}$$

the normalized clustering coefficient matrix.

Definition 3[12]. Assume there are n clustering objects, s grey classes, let $\eta = (1, 2, \ldots, s-1, s)^T$, then call

$$\omega_i = \delta_i \cdot \eta = \sum_{k=1}^{s} k \cdot \delta_i^k \ (i = 1, 2, \ldots, n) \tag{5}$$

synthetic clustering coefficient of clustering object i. Where call $\eta = (1, 2, \ldots, s-1, s)^T$ synthetic clustering coefficient weight vector. It can be proved that $1 \le \omega_i \le s, i = 1, 2, \ldots, n$.

Definition 4[12, 13]. When there is no significant difference about grey clustering coefficient of clustering object i, if synthetic clustering coefficient of object i $\omega_i \in [1 + (k-1)(s-1)/s, 1 + k(s-1)/s]$, we call that object i belongs to grey class k.

The grey synthetic clustering evaluation algorithm of DoS attack effectiveness is as follows.

Step 1: Determine the evaluation index system. According to the index system of DoS attack effectiveness evaluation as discussed in section 2, we can identify the grey synthetic clustering evaluation index set $I = \{I_1, I_2, \ldots, I_m\}$.

Step 2: Determine the weight of each index. There are many means to determine the index weight, such as AHP (Analytic Hierarchy Process) method and Rough Set method. Through Rough Set method, the weight of each index can be identified: $W = \{w_1, w_2, \ldots, w_m\}$. The method is described in detail as follows.

We use the knowledge representation system $S = (U, A)$ in rough set theory to represent the attack samples, where U is a finite nonempty set of objects, called domain of discourse; A is a finite nonempty set of indexes including the condition indexes set I and the decision indexes set J, and $I \cup J = A$, $I \cap J = \Phi$.

The dependence degree of the decision indexes set J on the condition indexes set I is defined as:

$$\gamma_I(J) = card(pos_I(J)) / card(U) \ . \tag{6}$$

The dependence degree of the decision indexes set J on the condition indexes set $I-\{a\}$ ($a \in I$) is defined as:

$$\gamma_{I-\{a\}}(J) = card(pos_{I-\{a\}}(J)) / card(U) \tag{7}$$

where $card(\bullet)$ is the radix of set, $pos_I(J)$ is the I positive domain of J, that is,

$$pos_I(J) = \bigcup_{X \in U/J} \underline{I}X \ . \tag{8}$$

The I positive domain of J is the object set which can be precisely partitioned to the equivalence class of J according to the information of U/I .

The importance degree of condition index a is defined as:

$$\sigma_J(a) = 1 - \frac{\gamma_{I-\{a\}}(J)}{\gamma_I(J)} \ . \tag{9}$$

We can calculate the importance degree of each condition index according to the method described above. Let

$$w_i = \sigma_J(i) \Big/ \sum_{a \in I} \sigma_J(a) \ , \tag{10}$$

and we can get the weight vector of evaluation indexes as $W = \{w_1, w_2, ..., w_m\}$.

Using the definition of attribute importance in rough set theory, it needn't any prior information beyond the research data set in data processing to get the weight of each index, so the subjectivity brought by experts in subjective weighting methods such as AHP method can be effectively overcame, and the weights obtained can be more objective.

Step 3: Determine the sample matrix. Assuming the value of attack i on index j is $d_{ij}(i = 1, 2, ..., n; j = 1, 2, ..., m)$, then construct the sample matrix D according to the data sample of DoS attacks as follows:

$$D = \begin{pmatrix} d_{11} & d_{12} & \cdots & d_{1m} \\ d_{21} & d_{22} & \cdots & d_{2m} \\ \vdots & \vdots & \ddots & \vdots \\ d_{n1} & d_{n2} & \cdots & d_{nm} \end{pmatrix} \ . \tag{11}$$

Step 4: Determine the evaluation grey class and the whitening functions. Determine the grade of grey types and grey number in accordance with the evaluation requirement. Assuming there are s grey classes, we can give the whitening function of index j on grey class k $f_j^k(*)(j = 1, 2, ..., m; k = 1, 2, ..., s)$.

(1)Upper bound level, grey number $\otimes \in [0, \infty]$, the corresponding whitening function is as follows:

$$f_j^k(d_{ij}) = \begin{cases} \dfrac{d_{ij}}{c_j^k}, d_{ij} \in [0, c_j^k] \\ 1, d_{ij} \in (c_j^k, \infty) \\ 0, d_{ij} \notin [0, \infty) \end{cases} . \tag{12}$$

(2) Middle level, grey number $\otimes \in [0, c_j^k, 2c_j^k]$, the corresponding whitening function is:

$$f_j^k(d_{ij}) = \begin{cases} \dfrac{d_{ij}}{c_j^k}, d_{ij} \in [0, c_j^k] \\ \dfrac{d_{ij} - 2c_j^k}{-c_j^k}, d_{ij} \in (c_j^k, 2c_j^k] \\ 0, d_{ij} \notin [0, 2c_j^k] \end{cases} . \tag{13}$$

(3) Low bound level, grey number $\otimes \in [0, c_j^k, 2c_j^k]$, the corresponding whitening function is:

$$f_j^k(d_{ij}) = \begin{cases} 1, d_{ij} \in [0, c_j^k] \\ \dfrac{d_{ij} - 2c_j^k}{-c_j^k}, d_{ij} \in (c_j^k, 2c_j^k] \\ 0, d_{ij} \notin [0, 2c_j^k] \end{cases} . \tag{14}$$

Step 5: Calculate clustering coefficient. According to the whitening function $f_j^k(*)(j = 1, 2, ..., m; k = 1, 2, ..., s)$, the index weight $w_j(j = 1, 2, ..., m)$ and the sample value of attack i on index j $d_{ij}(i = 1, 2, ..., n; j = 1, 2, ..., m)$, we can calculate the clustering coefficient of attack i on grey class k σ_i^k using Eq. (1).

Step 6: Calculate the normalized clustering coefficient. Calculate the normalized clustering coefficient of attack i on grey class k δ_i^k using Eq. (3).

Step 7: Construct the normalized clustering coefficient vector. Construct the normalized clustering coefficient vector of attack i as follows: $\delta_i = (\delta_i^1, \delta_i^2 ... \delta_i^s); (i = 1, 2, ..., n)$.

Step 8: Calculate the synthetic clustering coefficient. According to the normalized clustering coefficient vector δ_i and weight vector of clustering coefficient $\eta = (1, 2, ..., s-1, s)^T$, we can calculate the synthetic clustering coefficient of attack i ω_i using Eq. (5).

Step 9: Determine the grey class, and give out the evaluation results. The range of synthetic clustering coefficient is divided into s disjoint intervals of equal length, which is $\left[1,1+\frac{s-1}{s}\right],\left[1+\frac{s-1}{s},1+\frac{2(s-1)}{s}\right],\ldots,\left[s-\frac{s-1}{s},s\right]$. When synthetic clustering coefficient $\omega_i \in \left[1+(k-1)(s-1)/s, 1+k(s-1)/s\right]$, we can determine that the attack i belongs to grey class k.

4 Example and Verification

DoS attack effectiveness evaluation is conducted using the grey synthetic clustering evaluation model and algorithm of DoS attack effectiveness in order to validate the availability of the algorithm. There is no open data source found in the area of DoS attack effectiveness evaluation. Even if there is, the evaluation indexes could not be exactly as same as ours. Therefore, we need to generate the data source by ourselves. The method is to set some corresponding attack scenarios and generate sample data of various DoS attack effectiveness evaluation indexes. Table 1 are the experimental data of dozens of DoS attacks generated using simulated DoS attack scenarios, where: a1, said the network bandwidth utilization, a2, said the change in CPU utilization, a3, said the change in memory utilization, a4, said the response delay, a5, said packet loss rate, a6, said the recovery time, a7 that the mechanism of attack, the sample space is {wr, srvc, sysc}, where wr that resources consuming, srvc said the service crashing, sysc that system crashing. Results of the evaluation is represented with K = {1,2,3,4} 4 grey types, where k=1 means "very good", k=2 means "good", k=3 means "general" and k=4 said "poor". We can use the rough set method described in paper [8] to determine the index weight, and we get the weights of DoS attack effectiveness evaluation index:

$$w = \{0.122, 0.213, 0.162, 0.113, 0.089, 0.136, 0.165\} . \tag{15}$$

Table 1. Data of DoS attacks experiment

attack	a1	a2	a3	a4	a5	a6	a7
1	81%	92%	78%	9.5s	34%	2.5m	srvc
2	30%	10%	61%	1.1s	14%	7.5m	wr
3	30%	28%	40%	4.5s	16%	0.9m	wr
4	94%	71%	81%	6.5s	35%	9m	sysc
5	22%	20%	10%	0.1s	1%	0.15m	wr
6	68%	94%	79%	3.2s	40%	2.7m	sysc
7	58%	74%	71%	1.8s	10%	3.25m	srvc
8	20%	13%	10%	7.0s	20%	5.35m	wr
9	48%	80%	40%	3.1s	11%	4.25m	wr
10	20%	20%	17%	4.4s	11%	7m	wr

According to the data of Table 1, we can get the sample matrix after quantifying a7:

$$D=\begin{vmatrix} 0.81 & 0.92 & 0.78 & 9.5 & 0.34 & 2.5 & 2 \\ 0.30 & 0.10 & 0.61 & 1.1 & 0.14 & 7.5 & 1 \\ 0.30 & 0.28 & 0.40 & 4.5 & 0.16 & 0.9 & 1 \\ 0.94 & 0.71 & 0.81 & 6.5 & 0.35 & 9 & 3 \\ 0.22 & 0.20 & 0.10 & 0.1 & 0.01 & 0.15 & 1 \\ 0.68 & 0.94 & 0.79 & 3.2 & 0.4 & 2.7 & 3 \\ 0.58 & 0.74 & 0.71 & 1.8 & 0.1 & 3.25 & 2 \\ 0.20 & 0.13 & 0.10 & 7.0 & 0.2 & 5.35 & 1 \\ 0.48 & 0.80 & 0.40 & 3.1 & 0.11 & 4.25 & 1 \\ 0.20 & 0.20 & 0.17 & 4.4 & 0.11 & 7 & 1 \end{vmatrix}. \tag{16}$$

According to the steps of grey synthetic clustering evaluation algorithm, we firstly give out the whitening weight functions of each index on each grey class according to sample set, and then summarize the attack effectiveness that need to be evaluated according to whitening weight functions, and at last determine the grey class that effectiveness of each attack belongs to. According to the training sample set, whitening weight function is given in Table 2.

Table 2. Whitening weight function of each index on each grey class

Grey class 1	Grey class 2	Grey class 3	Grey class 4
$f_1^1(c_1^1,\infty)=f_1^1(0.8,\infty)$	$f_1^2(-,c_1^2,+)=f_1^2(-,0.6,+)$	$f_1^3(-,c_1^3,+)=f_1^3(-,0.4,+)$	$f_1^4(0,c_1^4)=f_1^4(0,0.2)$
$f_2^1(c_2^1,\infty)=f_2^1(0.9,\infty)$	$f_2^2(-,c_2^2,+)=f_2^2(-,0.7,+)$	$f_2^3(-,c_2^3,+)=f_2^3(-,0.5,+)$	$f_2^4(0,c_2^4)=f_2^4(0,0.3)$
$f_3^1(c_3^1,\infty)=f_3^1(0.8,\infty)$	$f_3^2(-,c_3^2,+)=f_3^2(-,0.65,+)$	$f_3^3(-,c_3^3,+)=f_3^3(-,0.4,+)$	$f_3^4(0,c_3^4)=f_3^4(0,0.1)$
$f_4^1(c_4^1,\infty)=f_4^1(8,\infty)$	$f_4^2(-,c_4^2,+)=f_4^2(-,6,+)$	$f_4^3(-,c_4^3,+)=f_4^3(-,1,+)$	$f_4^4(0,c_4^4)=f_4^4(0,0.1)$
$f_5^1(c_5^1,\infty)=f_5^1(0.4,\infty)$	$f_5^2(-,c_5^2,+)=f_5^2(-,0.2,+)$	$f_5^3(-,c_5^3,+)=f_5^3(-,0.1,+)$	$f_5^4(0,c_5^4)=f_5^4(0,0.01)$
$f_6^1(c_6^1,\infty)=f_6^1(8,\infty)$	$f_6^2(-,c_6^2,+)=f_6^2(-,6,+)$	$f_6^3(-,c_6^3,+)=f_6^3(-,2,+)$	$f_6^4(0,c_6^4)=f_6^4(0,0.25)$
$f_7^1(c_7^1,\infty)=f_7^1(3,\infty)$	$f_7^2(-,c_7^2,+)=f_7^2(-,2,+)$	$f_7^3(-,c_7^3,+)=f_7^3(-,2,+)$	$f_7^4(0,c_7^4)=f_7^4(0,1)$

The mathematical expressions of four whitening functions of index 1 are

$$f_1^1(d_{ij})=\begin{cases} \frac{1}{0.8}d_{ij}, & d_{ij}\in[0,0.8] \\ 1, & d_{ij}\in[0.8,\infty] \\ 0, & other \end{cases}, \tag{17}$$

$$f_1^2(d_{ij})=\begin{cases} \frac{1}{0.6}d_{ij}, & d_{ij}\in[0,0.6] \\ -\frac{1}{0.6}d_{ij}+2, & d_{ij}\in[0.6,1.2] \\ 0, & other \end{cases}, \tag{18}$$

$$f_1^3(d_{ij}) = \begin{cases} \frac{1}{0.4} d_{ij}, & d_{ij} \in [0, 0.4] \\ -\frac{1}{0.4} d_{ij} + 2, & d_{ij} \in [0.4, 0.8] \\ 0, & other \end{cases}, \tag{19}$$

$$f_1^4(d_{ij}) = \begin{cases} 1, & d_{ij} \in [0, 0.2] \\ -\frac{1}{0.2} d_{ij} + 2, & d_{ij} \in [0.2, 0.4] \\ 0, & other \end{cases}, \tag{20}$$

Similarly, we can get the mathematical expression of whitening functions of index 2, 3, 4, 5, 6, 7.

According to the evaluation index weights and the formula in step-five of grey synthetic clustering evaluation algorithm, we can get the fixed weight clustering coefficient matrix:

$$\sigma = (\sigma_i^k) = \begin{vmatrix} 0.83 & 0.65 & 0.31 & 0 \\ 0.42 & 0.51 & 0.45 & 0.44 \\ 0.36 & 0.50 & 0.55 & 0.44 \\ 0.92 & 0.66 & 0.21 & 0 \\ 0.16 & 0.22 & 0.31 & 0.99 \\ 0.82 & 0.58 & 0.24 & 0 \\ 0.62 & 0.78 & 0.54 & 0 \\ 0.37 & 0.49 & 0.24 & 0.66 \\ 0.54 & 0.67 & 0.51 & 0.16 \\ 0.37 & 0.47 & 0.38 & 0.55 \end{vmatrix}. \tag{21}$$

With further calculation, we can get the normalized clustering coefficient matrix:

$$\Pi = (\delta_i^k) = \begin{vmatrix} 0.47 & 0.36 & 0.17 & 0 \\ 0.23 & 0.28 & 0.25 & 0.24 \\ 0.19 & 0.27 & 0.30 & 0.24 \\ 0.51 & 0.37 & 0.12 & 0 \\ 0.10 & 0.13 & 0.18 & 0.59 \\ 0.50 & 0.35 & 0.15 & 0 \\ 0.32 & 0.40 & 0.28 & 0 \\ 0.21 & 0.28 & 0.14 & 0.37 \\ 0.29 & 0.35 & 0.27 & 0.09 \\ 0.21 & 0.27 & 0.21 & 0.31 \end{vmatrix}. \tag{22}$$

And the synthetic clustering coefficients of the ten attacks are: $\omega_1 = 1.7074$, $\omega_2 = 2.4968$, $\omega_3 = 2.5745$, $\omega_4 = 1.5996$, $\omega_5 = 3.2624$, $\omega_6 = 1.6425$, $\omega_7 = 1.9597$, $\omega_8 = 2.6759$, $\omega_9 = 2.1589$, $\omega_{10} = 2.6223$.

By analyzing in accordance with step-nine in the grey synthetic clustering evaluation algorithm, we can obtain the evaluation of attack effect:

$\omega_1, \omega_4, \omega_6 \in [1, 1+3/4]$

shows that the effect of attack 1, 4 and 6 is "very good";

$\omega_2, \omega_7, \omega_9 \in [1+3/4, 1+6/4]$

shows that the effect of attack 2, 7 and 9 is "good";

$\omega_3, \omega_8, \omega_{10} \in [1+6/4, 1+9/4]$

shows that the effect of attack 3, 8 and 10 against "general";

$\omega_5 \in [1+9/4, 4]$

shows that the effect of attack 5 is "poor".

Using the general fixed weight clustering evaluation algorithm proposed by paper [8], analyzing the fixed weight clustering coefficient matrix, we get the results as following: the effect of attack 1,4,6 is "very good"; attack 2,7,9 is "good"; attack 3 is "general"; attack 5,8,10 is "poor. "

By analyzing and comparing the results of the general fixed weight clustering evaluation algorithm and the synthetic clustering evaluation algorithm, we can conclude that there are some differences between results of the two methods, but it is in accordance with the conclusion in paper [12], that is, when the significant difference of clustering coefficient of clustering objects satisfies $\theta \geq 1 - 2/s$, the evaluation results of the two methods is of the same.

This evaluation method, which uses the definition of attribute importance in rough set theory, determines the weight of each index according to the samples of discourse domain. Compared with other evaluation methods such as AHP-based method and index-based analysis method, it needs no experts' marking and thus decreases subjective influence and increases the objectivity of the evaluation results. Different with the traditional evaluation methods such as fuzzy comprehensive evaluation method and regression analysis method, this evaluation method not only needs very little information and does not require the sample subject to any distribution, but also greatly simplifies the complex horizontal comparison of indexes which makes calculating simple and convenient. With this method we can not only evaluate the effectiveness of a single attack, but also do some sorting work on effectiveness of different attacks of the same attack type. By analyzing the experimental results, we can conclude that the evaluation results accord with reality.

5 Conclusion

DoS attacks are widespread network attacks, they are diverse in means and very destructive. How to evaluate the effectiveness of DoS attack is an important and urgent task of the computer attack and defense research. This paper presents a grey synthetic clustering method for DoS attack effectiveness evaluation. It solves the problem that the evaluation model of network attack effectiveness, which is based on general grey clustering, can not accurately evaluate the object when there is no significant difference between the clustering coefficients. Finally, we compare the results of the two methods through an experiment, and it is proved that the grey synthetic clustering evaluation model is correct and feasible.

References

1. Qi, J., Zhou, X.: Simulation and evaluating efficiency of DoS attacks. Journal of Information Engineering University 8(3), 360–363 (2007) (in Chinese)
2. Wang, Y., Xian, M., Wang, G., Xiao, S.: Study on effectiveness evaluation of computer network attacks. Computer Engineering and Design 26(11), 2868–2870 (2005) (in Chinese)
3. Cheng, W., Lu, Y., Xia, Y., Yang, G.: Research on the vulnerability evaluation of computer network. Journal of Anhui University Natural Science Edition 31(4), 29–32 (2007) (in Chinese)
4. Zhang, W., Zhao, R., Zhang, Z., Shan, Z.: Simulation and effect evaluation of DoS attacks. Computer Engineering and Design 30(3), 544–546 (2009) (in Chinese)
5. Deng, J.: Elements on Grey Theory. Huazhong University of Science and Technology Press, Wuhan (2002) (in Chinese)
6. Zheng, W., Hu, Y.: Grey evaluation method of knowledge management capability. In: Second International Workshop on Knowledge Discovery and Data Ming, pp. 256–260. IEEE Computer Society Press, Washington (2009)
7. Tang, H., Zhang, J., Su, K.: On evaluation model of military information network based on multilevel grey evaluation method. In: Proceedings of the 27th Chinese Control Conference, pp. 113–116. IEEE Press, New York (2008)
8. Wang, H., Jiang, L., Xian, M., Wang, G.: Grey evaluation model and algorithm of network attack effectiveness. Journal on Communications 30(11A), 17–22 (2009) (in Chinese)
9. Zhang, L., Cao, Y., Wang, Q.: A DoS attack effect evaluation method based on multisource data fusion. In: 2010 International Conference on Communications and Mobile Computing, pp. 91–96. IEEE Computer Society Press, Washington (2010)
10. Liu, S., Guo, T., Dang, Y.: Grey System Theory and Application. The Science Press, Beijing (1999) (in Chinese)
11. Liu, S., Lin, Y.: An Introduction to Grey Systems: Foundations, Methodology and Applications. IIGSS Academic Publisher, Slippry Rock (1998)
12. Dang, Y., Liu, S., Liu, B., Zhai, Z.: Research on the grey synthetic clustering method in clustering coefficient of no significant difference. Chinese Journal of Management Science 13(4), 69–73 (2005) (in Chinese)
13. Dang, Y., Liu, S., Liu, B., Tang, X.: Study on Grey Synthetic Clusters Appraisals Model. In: 2004 IEEE International Conference on Systems, Man and Cybernetics, pp. 2398–2402. IEEE Press, New York (2004)

Fuzzy-Possibilistic Product Partition: A Novel Robust Approach to c-Means Clustering

László Szilágyi

Sapientia - Hungarian Science University of Transylvania,
Faculty of Technical and Human Science, Tîrgu-Mureş, Romania
lalo@ms.sapientia.ro

Abstract. One of the main challenges in the field of c-means clustering models is creating an algorithm that is both accurate and robust. In the absence of outlier data, the conventional probabilistic fuzzy c-means (FCM) algorithm, or the latest possibilistic-fuzzy mixture model (PFCM), provide highly accurate partitions. However, during the 30-year history of FCM, the researcher community of the field failed to produce an algorithm that is accurate and insensitive to outliers at the same time. This paper introduces a novel mixture clustering model built upon probabilistic and possibilistic fuzzy partitions, where the two components are connected to each other in a qualitatively different way than they were in earlier mixtures. The fuzzy-possibilistic product partition c-means ($\mathrm{FP^3CM}$) clustering algorithm seems to fulfil the initial requirements, namely it successfully suppresses the effect of outliers situated at any finite distance and provides partitions of high quality.

Keywords: fuzzy c-means algorithm, probabilistic partition, possibilistic partition, robust clustering.

1 Introduction

Robustness in c-means clustering refers to the stability or reproducibility of the achieved partition, and insensitivity to several kinds of noise including severely outlier data. The fuzzy c-means (FCM) clustering introduced by Bezdek [3] is a very popular clustering model due to the fine partitions it makes and its easy comprehensible alternating optimization (AO) scheme. However, the probabilistic constraints involved in FCM makes it sensitive to outlier data. To combat this problem, several solution have been proposed that produce a relaxation of this probabilistic constraint.

An early solution was given by Davé [4], who introduced an extra, specially treated noisy class to attract feature vectors situated far from all normal cluster prototypes. This theory was later improved by Menard et al [7]. Alternately, Krishnapuram and Keller came up with the possibilistic c-means algorithm (PCM) [6], which distributes the partition matrix elements based on statistical rules. This approach seemed to have solved the sensitivity to outliers, but it cannot be called a robust algorithm due to the coincident clusters it frequently

V. Torra et al. (Eds.): MDAI 2011, LNAI 6820, pp. 150–161, 2011.

produces [2]. Timm et al [10] set up a repulsive force between all couples of cluster prototypes of PCM, the strength of which decreased with distance. Their method succeeded in avoiding coincident clusters, but failed to correctly treat cases when two clusters are really close to each other. Two versions of fuzzy-possibilistic partition mixtures were proposed by Pal et al [8,9], out of which the second one appears to be a reliable clustering model. Recently, Xie et al introduced a novel possibilistic c-means clustering [12] algorithm that produces a gap between fuzzy memberships with respect to winner and non-winner clusters, similarly to the symmetrical margin between classes provided by support vector machines [11] in supervised classification problems.

All the endeavors during the last three decades failed to create a clustering model that would suppress the effect of outliers similarly to gravity systems. If we add a distant object to any working gravity system, the strength of its effect will be in reversed proportion with distance. A very distant object would hardly be observable, it would hardly influence anything within the system. On the other hand, in all existing clustering models, if we increase the distance between an outlier input vector and normal input vectors, at a certain threshold distance the partitioning will fail. It would be an excellent achievement to create a clustering model that behaves similarly to gravity systems, and would totally suppress the effect of distant outliers, while keeping or even improving the accuracy in the absence of outliers. The total suppression of the outliers' effect would mean that the further the outlier stands, the less effect it has on the normal clusters.

In this paper we introduce the novel fuzzy-possibilitistic product partition c-means clustering model (FP^{3}CM), in which the degrees of membership are given as the product of a probabilistic and a possibilistic term. This new approach can eliminate all adverse effects of distant outliers, while producing high quality partitions. The algorithm uses a reduced number of parameters, making it easily adjustable to various scenarios.

The rest of this paper is structured as follows. Section 2 summarizes the background works and counter candidates of our approach. Section 3 introduces the novel FP^{3}CM clustering model. Section 4 produces a numerical analysis of the proposed and earlier methods. Conclusions are given in the last section.

2 Preliminaries

2.1 Fuzzy c-Means Clustering

The conventional FCM partitions a set of object data into a number of c clusters based on the minimization of a quadratic objective function, formulated as:

$$J_{\text{FCM}} = \sum_{i=1}^{c}\sum_{k=1}^{n} u_{ik}^{m}||\boldsymbol{x}_k - \boldsymbol{v}_i||^2 = \sum_{i=1}^{c}\sum_{k=1}^{n} u_{ik}^{m} d_{ik}^2 \ , \tag{1}$$

constrained by the probabilistic condition $\sum_{i=1}^{c} u_{ik} = 1 \ \forall k = 1 \dots n$ where $\boldsymbol{x}_k$ represents the input data ($k = 1 \dots n$), $\boldsymbol{v}_i$ represents the prototype or centroid value or representative element of cluster i ($i = 1 \dots c$), $u_{ik} \in [0, 1]$ is the fuzzy membership function showing the degree to which input vector $\boldsymbol{x}_k$ belongs to cluster i, $m > 1$ is the fuzzyfication parameter, and $d_{ik} = ||\boldsymbol{x}_k - \boldsymbol{v}_i||$.

The minimization of the objective function is reached by alternately applying the optimization of J_{FCM} over $\{u_{ik}\}$ with $\boldsymbol{v}_i$ fixed, and the optimization of J_{FCM} over $\{\boldsymbol{v}_i\}$ with u_{ik} fixed, [3]. During each cycle, the optimal values are computed from the zero gradient conditions, and obtained as follows:

$$u_{ik}^{\star} = \frac{d_{ik}^{-2/(m-1)}}{\sum_{j=1}^{c} d_{jk}^{-2/(m-1)}} \quad \forall i = 1\ldots c, \forall k = 1\ldots n \ , \tag{2}$$

$$\boldsymbol{v}_i^{\star} = \frac{\sum_{k=1}^{n} u_{ik}^{m} \boldsymbol{x}_k}{\sum_{k=1}^{n} u_{ik}^{m}} \quad \forall i = 1\ldots c \ . \tag{3}$$

According to the alternating optimization scheme, Eqs. (2) and (3) are alternately applied, until cluster prototypes stabilize.

2.2 Possibilistic c-Means Clustering

In order to avoid the sensibility of the probabilistic partition to outlier data, Krishnapuram and Keller [6] introduced the possibilistic c-means algorithm. The elements of the possibilistic partition are denoted by t_{ik}, $i = 1\ldots c$, $k = 1\ldots n$. The value of t_{ik} characterizes the compatibility of data vector $\boldsymbol{x}_k$ with the cluster represented by prototype v_i.

The objective function of the PCM algorithm is

$$J_{\text{PCM}} = \sum_{i=1}^{c}\sum_{k=1}^{n} \left[t_{ik}^{p} d_{ik}^{2} + (1 - t_{ik})^{p} \eta_i \right] \ . \tag{4}$$

constrained by $0 \leq t_{ik} \leq 1$ $\forall i = 1\ldots c$, $\forall k = 1\ldots n$, and $0 < \sum_{i=1}^{c} t_{ik} < c$ $\forall k = 1\ldots n$, where $p > 1$ represents the possibilistic exponent, and parameters η_i are the penalty terms that control the variance of the clusters.

The iterative AO algorithm, that results from zero gradient conditions of the objective function, repeatedly applies the following formulas until convergence is reached:

$$t_{ik}^{\star} = \left[1 + \left(\frac{d_{ik}^{2}}{\eta_i} \right)^{1/(p-1)} \right]^{-1} \quad \forall i = 1\ldots c, \forall k = 1\ldots n \ , \tag{5}$$

$$\boldsymbol{v}_i^{\star} = \frac{\sum_{k=1}^{n} t_{ik}^{p} \boldsymbol{x}_k}{\sum_{k=1}^{n} t_{ik}^{p}} \quad \forall i = 1\ldots c \ . \tag{6}$$

In the probabilistic fuzzy partition, the degrees of membership assigned to an input vector $\boldsymbol{x}_k$ with respect to cluster i depends on the distances of the given vector to all cluster prototypes: d_{1k}, d_{2k}, ..., d_{ck}. On the other hand, in the possibilistic partition, the typicality value assigned to input vector $\boldsymbol{x}_k$ with respect to any cluster i depends on only one distance: d_{ik}.

PCM efficiently suppresses the effects of outlier data, at the price of frequently producing coincident cluster prototypes. This latter is the result of the highly independent clusters [2].

2.3 Existing Fuzzy-Possibilistic Mixture Partitions

In order to avoid the pitfalls of independently computed possibilistic partitions, several solutions have been proposed. The most remarkable ones are the possibilistic-fuzzy mixture clustering models proposed by Pal et al in [8] and [9].

The so-called fuzzy-possibilistic c-means (FPCM) algorithm, introduced by Pal et al [8], minimizes the following objective function

$$J_{\mathrm{FPCM}} = \sum_{i=1}^{c} \sum_{k=1}^{n} [u_{ik}^m + t_{ik}^p] d_{ik}^2 \ , \tag{7}$$

constrained by two probabilistic conditions

$$\sum_{i=1}^{c} u_{ik} = 1 \quad \forall k = 1 \dots n \quad \text{and} \quad \sum_{k=1}^{n} t_{ik} = 1 \quad \forall i = 1 \dots c \ . \tag{8}$$

Using the zero gradient conditions of the above cost function, we obtain the following optimization formulas for the iterative AO scheme of the algorithm:

$$u_{ik}^\star = \frac{d_{ik}^{-2/(m-1)}}{\sum_{j=1}^{c} d_{jk}^{-2/(m-1)}} \quad \forall i = 1 \dots c, \forall k = 1 \dots n \ , \tag{9}$$

$$t_{ik}^\star = \frac{d_{ik}^{-2/(p-1)}}{\sum_{l=1}^{n} d_{il}^{-2/(p-1)}} \quad \forall i = 1 \dots c, \forall k = 1 \dots n \ , \tag{10}$$

$$\boldsymbol{v}_i^\star = \frac{\sum_{k=1}^{n} [u_{ik}^m + t_{ik}^p] \boldsymbol{x}_k}{\sum_{k=1}^{n} [u_{ik}^m + t_{ik}^p]} \quad \forall i = 1 \dots c \ . \tag{11}$$

FPCM has the main advantage of not using the penalty terms η_i, thus making the parameter adjustment easier. However, Eq. (10) suggests that the possibilistic effect of the algorithm loses its strength as the number of input vectors grows. In case of thousands of vectors, FPCM practically reduces to FCM, regardless of the value of the exponent p.

Later, Pal et al [9] proposed another mixture clustering model, called possibilistic-fuzzy c-means (PFCM) clustering, which minimizes the objective function

$$J_{\mathrm{PFCM}} = \sum_{i=1}^{c} \sum_{k=1}^{n} [a u_{ik}^m + b t_{ik}^p] d_{ik}^2 + \sum_{i=1}^{c} \eta_i \sum_{k=1}^{n} (1 - t_{ik})^p \ , \tag{12}$$

constrained by the conventional probabilistic and possibilistic conditions of FCM and PCM, respectively.

Here a and b are two tradeoff parameters that control the strength of the possibilistic and probabilistic term in the mixed partition. All other parameters are the same as in FCM and PCM.

The minimization formulas include Eq. (2) for updating the probabilistic fuzzy partition, further on

$$t_{ik}^\star = \left[1 + \left(\frac{b d_{ik}^2}{\eta_i} \right)^{1/(p-1)} \right]^{-1} \quad \forall i = 1 \dots c, \forall k = 1 \dots n \ , \tag{13}$$

is the update formula for typicality values, while cluster prototypes are computed as:

$$\boldsymbol{v}_i^\star = \frac{\sum_{k=1}^n [a u_{ik}^m + b t_{ik}^p] \boldsymbol{x}_k}{\sum_{k=1}^n [a u_{ik}^m + b t_{ik}^p]} \quad \forall i = 1 \dots c \ . \tag{14}$$

This latter algorithm was found accurate and robust, but as we will see in later sections, it is still sensitive to outlier data.

3 Methods

3.1 Intuition

In a probabilistic fuzzy partition, any outlier input vector $\boldsymbol{x}_{\text{out}}$ receives high membership values with respect to all clusters, that is, $u_{i,\text{out}} \approx 1/c$, which strongly influence all cluster prototypes.

On the other hand, in a possibilistic approach, outlier input vectors receive very low typicality values with respect to all clusters.

In our opinion, it would be a robust solution to have an objective function whose zero gradient conditions give the following cluster prototype update formula:

$$\boldsymbol{v}_i^\star = \frac{\sum_{k=1}^n \mu_{ik}^m \tau_{ik}^p \boldsymbol{x}_k}{\sum_{k=1}^n \mu_{ik}^m \tau_{ik}^p} \quad \forall i = 1 \dots c \ . \tag{15}$$

where $\mu_{ik}, i = 1 \dots c, k = 1 \dots n$ describe a probabilistic fuzzy partition that is not necessarily equivalent with the FCM's one, and $\tau_{ik}, i = 1 \dots c, k = 1 \dots n$ stand for the elements of a possibilistic partition matrix. We will attempt to propose such an objective function in the next subsection.

3.2 The Proposed Clustering Model

Now let us introduce the fuzzy-possibilistic product partition c-means clustering model, which minimizes

$$J_{\text{FP}^3\text{CM}} = \sum_{i=1}^c \sum_{k=1}^n \mu_{ik}^m \left[\tau_{ik}^p d_{ik}^2 + (1 - \tau_{ik})^p \eta_i \right] \ , \tag{16}$$

constrained by the conventional probabilistic condition written as $\sum_{i=1}^c \mu_{ik} = 1$ $\forall k = 1 \dots n$, and the conventional possibilistic conditions $0 \leq \tau_{ik} \leq 1$ $\forall i = 1 \dots c$, $\forall k = 1 \dots n$, and $0 < \sum_{i=1}^c \tau_{ik} < c$ $\forall k = 1 \dots n$. The only parameters of FP^{3}CM are the fuzzy exponent $m > 1$, the possibilistic exponent $p > 1$, and the conventional penalty terms of the possibilistic partition denoted by η_i, $i = 1 \dots n$.

The minimization formulas are obtained using zero gradient conditions, aided by Lagrange multipliers in case of the probabilistic term. We will compute the partial derivatives of the functional:

$$\mathcal{L} = J_{\text{FP}^3\text{CM}} + \sum_{k=1}^n \lambda_k \left(1 - \sum_{i=1}^c \mu_{ik} \right) \ , \tag{17}$$

where λ_k stands for the Lagrange multipliers. The zero crossing of the partial derivatives with respect to τ_{ik}, $\forall i = 1 \dots c$, $\forall k = 1 \dots n$, leads to:

$$\frac{\partial \mathcal{L}}{\partial \tau_{ik}} = 0 \quad \Rightarrow \quad \mu_{ik}^m \left[p\tau_{ik}^{p-1} d_{ik}^2 - \eta_i p (1 - \tau_{ik})^{p-1} \right] = 0.$$

If $\mu_{ik} = 0$, the value of τ_{ik} does not make a difference. Otherwise we get

$$\left(\frac{1 - \tau_{ik}}{\tau_{ik}} \right)^{p-1} = \frac{d_{ik}^2}{\eta_i} \quad \Rightarrow \quad \frac{1}{\tau_{ik}} - 1 = \left(\frac{d_{ik}^2}{\eta_i} \right)^{1/(p-1)},$$

which finally leads to a formula that is identical with Eq. (5):

$$\tau_{ik}^\star = \left[1 + \left(\frac{d_{ik}^2}{\eta_i} \right)^{1/(p-1)} \right]^{-1} \qquad \forall i = 1 \dots c, \forall k = 1 \dots n \ . \tag{18}$$

Further on, let us examine the zero crossing of partial derivatives with respect to μ_{ik}. For any $i = 1 \dots c$ and any $k = 1 \dots n$ we get

$$\frac{\partial \mathcal{L}}{\partial \mu_{ik}} = 0 \quad \Rightarrow \quad m\mu_{ik}^{m-1} \left[\tau_{ik}^p d_{ik}^2 + \eta_i (1 - \tau_{ik})^p \right] = \lambda_k,$$

which implies

$$\mu_{ik} = \left(\frac{\lambda_k}{m} \right)^{1/(m-1)} \times \left[\tau_{ik}^p d_{ik}^2 + \eta_i (1 - \tau_{ik})^p \right]^{-1/(m-1)} \ . \tag{19}$$

The probabilistic condition says $\sum_{j=1}^{c} \mu_{jk} = 1$, which by the means of Eq. (19) becomes

$$1 = \left(\frac{\lambda_k}{m} \right)^{1/(m-1)} \times \sum_{j=1}^{c} \left[\tau_{jk}^p d_{jk}^2 + \eta_j (1 - \tau_{jk})^p \right]^{-1/(m-1)} \ . \tag{20}$$

Dividing Eq. (19) by Eq. (20) term by term, leads to

$$\mu_{ik}^\star = \frac{[\tau_{ik}^p d_{ik}^2 + \eta_i (1 - \tau_{ik})^p]^{-1/(m-1)}}{\sum_{j=1}^{c} [\tau_{jk}^p d_{jk}^2 + \eta_j (1 - \tau_{jk})^p]^{-1/(m-1)}} , \tag{21}$$

which holds for any $i = 1 \dots c$, and any $k = 1 \dots n$. Finally, let us investigate the zero crossings of the partial derivatives with respect to $\boldsymbol{v}_i$, $i = 1 \dots n$:

$$\frac{\partial \mathcal{L}}{\partial \boldsymbol{v}_i} = 0 \quad \Rightarrow \quad -2 \sum_{k=1}^{n} \mu_{ik}^m \tau_{ik}^p (\boldsymbol{x}_k - \boldsymbol{v}_i) = 0 \ ,$$

which implies

$$\boldsymbol{v}_i \sum_{k=1}^{n} \mu_{ik}^m \tau_{ik}^p = \sum_{k=1}^{n} \mu_{ik}^m \tau_{ik}^p \boldsymbol{x}_k \quad \Rightarrow \quad \boldsymbol{v}_i^\star = \frac{\sum_{k=1}^{n} \mu_{ik}^m \tau_{ik}^p \boldsymbol{x}_k}{\sum_{k=1}^{n} \mu_{ik}^m \tau_{ik}^p} , \tag{22}$$

valid for any $i = 1 \dots c$, exactly as we wished in Eq. (15). Let us remark the followings:

1. The possibilistic memberships τ_{ik} are established exactly the same way, as in the PCM algorithm. This is why the penalty terms η_i can be set as recommended by Krishnapuram and Keller [6].
2. The probabilistic partition is somewhat similar to FCM's partition, but distances are distorted, and the partition is influenced by the possibilitic penalty terms η_i.
3. Outlier input vectors $\boldsymbol{x}_k$ are indicated by the algorithm with a low value of $\max\{\sqrt[m+p]{\mu_{ik}^m \tau_{ik}^p}, i = 1 \ldots c\}$.
4. The defuzzyfication of the final partition should be performed according to the following rule: $\boldsymbol{x}_k$ is assigned to cluster with index w_k, where

$$w_k = \arg\max_j \left(\mu_{jk}^m \tau_{jk}^p | j = 1 \ldots c \right) . \tag{23}$$

In case of equal η_i values, for any $i = 1 \ldots c$, the rule becomes more simple: $w_k = \arg\max_j (\mu_{jk} | j = 1 \ldots c)$.

3.3 The Alternative Optimization Algorithm of FP³CM

Let us summarize the optimization algorithm of the proposed clustering model:

1. Set fuzzy exponent m and possibilistic exponent p, both greater than 1.
2. Set possibilistic penalty terms η_i, $i = 1 \ldots c$, as recommended by Krishnapuram and Keller in [6].
3. Update possibilistic membership values using Eq. (18).
4. Update probabilistic membership values using Eq. (21).
5. Update cluster prototypes using Eq. (22).
6. Repeat steps 3-5 until cluster prototypes converge.
7. If it is necessary for the application, perform defuzzyfication of the obtained product partition as indicated in Eq. (23).

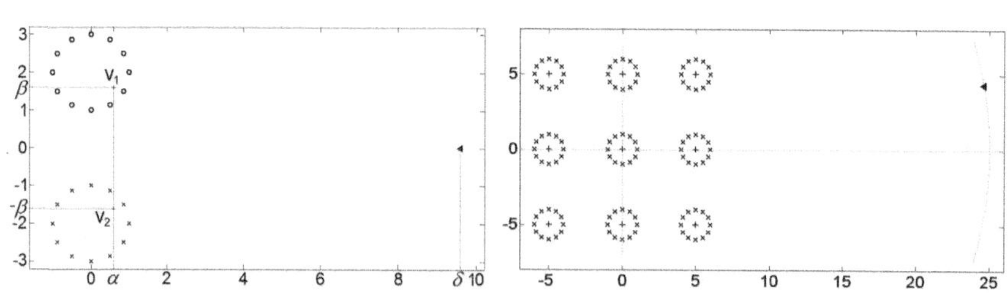

Fig. 1. Two scenarios for the numerical test of robustness: (left) two clusters and an outlier, (right) nine clusters and an outlier. We investigate the positions of cluster prototypes and the resulting partition accuracy, versus the outlier's position.

4 Results and Discussion

In the followings, we will perform some numerical tests to evaluate the robustness and accuracy of the proposed algorithm. We will compare its performances with counter candidates like FCM, FPCM, and PFCM. The pure possibilistic PCM algorithm is excluded from these tests due to its frequently coincident cluster prototypes.

4.1 Two Clusters and One Outlier Input Vector

Let us consider two sets of ν data points each, uniformly distributed along unit-radius circles: $\boldsymbol{x}_k = (\cos \frac{2k\pi}{\nu}, 2 + \sin \frac{2k\pi}{\nu})^T$ and $\boldsymbol{x}_{\nu+k} = (\cos \frac{2k\pi}{\nu}, -2 + \sin \frac{2k\pi}{\nu})^T$, $\forall k = 1 \dots \nu$.

The input data set also includes an outlier, situated at $\boldsymbol{x}_{2\nu+1} = (\delta, 0)^T$. We will attempt to classify these $n = 2\nu + 1$ vectors into $c = 2$ clusters, setting the initial cluster prototypes in the middle of the two circles: $\boldsymbol{v}_1 = (0, 2)^T$ and $\boldsymbol{v}_2 = (0, -2)^T$.

During the iterative optimization of all tested algorithms, the cluster prototypes will be attracted by the outlier vector. As long as the outlier cannot tear off any of the two prototypes, $\boldsymbol{v}_1$ and $\boldsymbol{v}_2$ will behave symmetrically, having their coordinates $\boldsymbol{v}_1 = (\alpha, \beta)^T$ and $\boldsymbol{v}_2 = (\alpha, -\beta)^T$. A graphical representation of the problem is shown in Fig. 1(left).

The question is, how α and β will depend on the outlier's position δ in case of all tested algorithms, and how far the outlier vector can go without tearing off one of the cluster prototypes.

Figure 2 presents the outcome of numerical simulations performed on all mentioned algorithms in various circumstances. The α coordinate of the symmetrical cluster prototypes is shown in two different plots in Fig. 2(b) and (c). In case of all existing algorithms, the further the outlier goes, the stronger it attracts the centroids, and at a certain boundary, one of the prototypes is torn out by the outlier.

On the other hand, FP^{3}CM behaves like a gravity system: the further the outlier is situated, the weaker its effect is upon the cluster centroids. No matter how far the outlier is, the obtained partition is correct. The outlier receives such a low membership value to both clusters that it can be easily assigned to the noisy class at defuzzyfication. Figure 2(d) shows the behavior of FP^{3}CM in case of various values of possibilistic exponent p, at a constant value of fuzzy exponent $m = 2$. The plots reveal that stronger possibilistic component or lower values of p lead to more efficient rejection of the outlier effect. However, when the outlier is not too far, lower exponent values also cause stronger deviation of the cluster centroids.

4.2 Accuracy Test with Nine Regular Clusters and an Outlier

As it is shown in Fig. 1(right), the input data in this second test consists of 9 sets of vectors uniformly distributed along unit radius circles, situated in the neighborhood of the origin. Initially, the cluster prototypes are placed in the middle

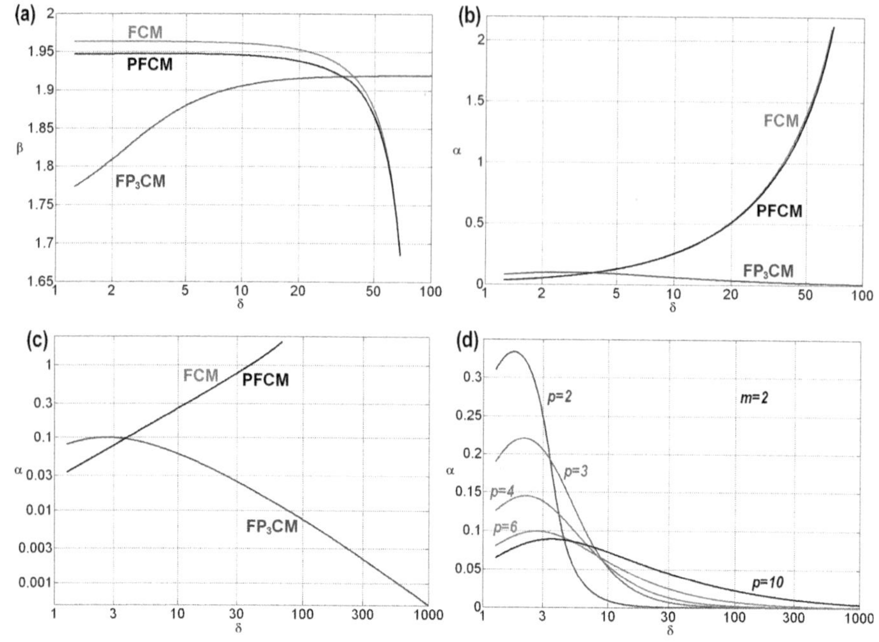

Fig. 2. (a)-(c) Position of the two symmetrical cluster prototypes at $m = 2$ and $p = 6$: (a) the β coordinate plotted against the position of the outlier δ, (b) the α coordinate plotted against the position of the outlier δ, (c) logarithmic plot of α coordinate against the distance of the outlier. Some of these graphs end at the threshold value of δ where the algorithms fail. In case of FP^{3}CM, the further the outlier wanders, the less influence it has upon cluster prototypes; (d) The α coordinate produced by the proposed algorithm FP^{3}CM, at $m = 2$ and various values of p. The algorithm manages to suppress the effect of departed outliers.

Table 1. The limit distance δ, in case of various algorithms and circumstances, where the tested algorithm fails to produce nine accurate clusters

Algo-rithm	Circumstances					Limit distance	Algo-rithm	Circumstances					Limit distance
	m	p	$\sqrt{\eta_i}$	a	b			m	p	$\sqrt{\eta_i}$	a	b	
FCM	2					361	PFCM	2	3	1.0	1	5	437
FPCM	2	5				361	PFCM	2	3	1.5	1	5	521
FPCM	2	2				367	PFCM	2	3	2.0	1	5	593
FPCM	2	1.2				401	PFCM	2	3	2.5	1	5	546
PFCM	2	2	1.0	2	3	410	PFCM	2	2	1.0	1	5	459
PFCM	2	2	1.5	2	3	479	PFCM	2	2	1.5	1	5	602
PFCM	2	2	2.0	2	3	563	PFCM	2	2	2.0	1	5	789
PFCM	2	2	2.5	2	3	649	PFCM	2	2	2.5	1	5	1001
PFCM	2	5	1.0	1	5	394	PFCM	2	2	3.0	1	5	1220
PFCM	2	5	1.5	1	5	421	PFCM	2	2	4.0	1	5	1354
PFCM	2	5	2.0	1	5	428	PFCM	2	2	5.0	1	5	1089
PFCM	2	5	2.5	1	5	370	FP^{3}CM	wide range					$+\infty$

of the nine circles. The single outlier vector moves along the big circle of radius δ, with its center in the origin. The aim of this study is to establish, which is the boundary value for δ where tested algorithms crash in various circumstances.

The obtained boundary distances are summarized in Table 1. These values emphasize the fact that currently existing algorithms may have enhanced the robustness of FCM, they may have enabled the outlier to fall somewhat further (no more than by one order of magnitude) without making the clustering crash. The novel clustering model FP^{3}CM seems to efficiently suppress the influence of the outlier vector, leading to accurate partitions for any limited value of δ.

4.3 Numerical Tests Using IRIS Data

In the followings, we will analyze the accuracy and robustness of the investigated clustering models using the IRIS data set [1], which consist of 150 labeled feature vectors of four dimensions (sepal length and width, petal length and width), organized in three clusters ("setosa","versicolor", and "virginica") of fifty vectors each. It is a reported facts, that conventional clustering models like FCM produce 133-134 correct decisions when classifying IRIS data. PFCM produced the best reported accuracy with 140 correct decisions using $a = b = 1$, $m = p = 3$, and initializing $\boldsymbol{v}_i$ with terminal FCM prototypes [9]. Under less advantageous circumstances, PFCM reportedly produced 136-137 correct decisions.

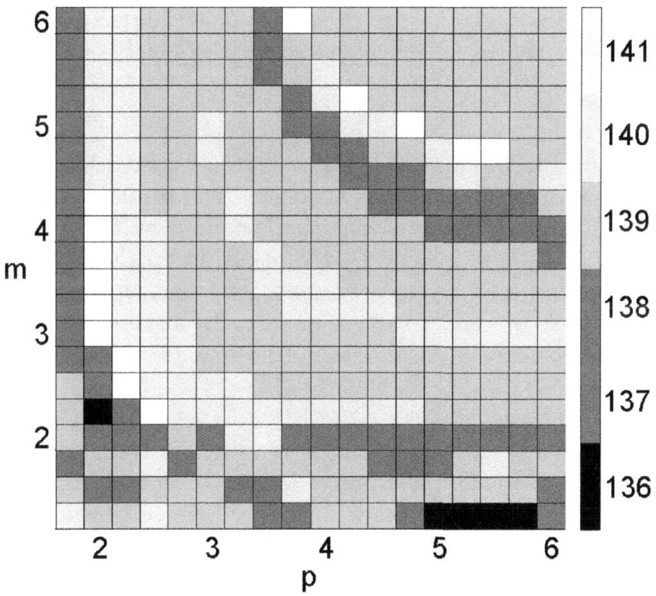

Fig. 3. Number of correct decisions (out of 150) obtained by FP^{3}CM, plotted against fuzzy exponent m and possibilistic exponent p, using $\sqrt{\eta_i} = 0.7$

Table 2. Detailed values of the final IRIS cluster prototypes: a high quality partition

Correct decisions	Centroid vector	Sepal length	Sepal width	Petal length	Petal width
141	v_1	5.0443	3.4307	1.4641	0.2337
	v_2	6.0729	2.9104	4.5693	1.4485
	v_3	6.4794	2.9876	5.2934	1.9687

Table 3. Partition accuracies and confusion matrices in various scenarios

Circumstances	IRIS type	FCM v_1	FCM v_2	FCM v_3	PFCM v_1	PFCM v_2	PFCM v_3	FP^{3}CM v_1	FP^{3}CM v_2	FP^{3}CM v_3	Correct decisions
no outlier added	Setosa	50	0	0	50	0	0	50	0	0	FCM → 136
	Versicolor	0	47	3	0	47	3	0	48	2	PFCM → 136
	Virginica	0	11	39	0	11	39	0	7	43	FP^{3}CM → 141
outlier added at 20	Setosa	50	0	0	50	0	0	50	0	0	FCM → 134
	Versicolor	0	50	0	0	50	0	0	47	3	PFCM → 135
	Virginica	0	16	34	0	15	35	0	7	43	FP^{3}CM → 140
outlier added at 30	Setosa	50	0	0	50	0	0	50	0	0	FCM → 128
	Versicolor	1	49	0	1	49	0	0	47	3	PFCM → 131
	Virginica	0	21	29	0	18	32	0	7	43	FP^{3}CM → 140
outlier added at 50 or 10^6	Setosa	50	0	0	50	0	0	50	0	0	FCM crashes
	Versicolor	3	47	0	3	47	0	0	47	3	PFCM crashes
	Virginica	0	50	0	0	50	0	0	7	43	FP^{3}CM → 140

We have tested the proposed FP^{3}CM clustering model in a wide range of both the fuzzy and the possibilistic exponents. The resulting partition quality is summarized in Fig. 3. The best partition achieved by FP^{3}CM had 141 correct decisions, which is above any reported result. Details upon the final cluster prototypes are given in Table 2. We also need to remark, that almost any parameter setting leads to good partition quality. To make sure FP^{3}CM clusters accurately, the possibilistic term should not be too strong, it is recommendable to keep parameter $p \geq 2$.

A series of numerical tests using the IRIS data targeted the clustering robustness. We artificially inserted an outlier vector into the input data set, with coordinates $\boldsymbol{x}_{151} = (\delta, \delta, \delta, \delta)^T$, and proceeded all vectors to clustering into $c = 3$ groups. Table 3 gives us an overview upon accuracy, confusion matrices, and sensibility to the outlier's position. As we can see it in the table, most existing clustering models failed somewhere between $\delta = 30$ and $\delta = 50$, while the proposed algorithm led to high quality partition even at $\delta = 10^6$, being less affected by distant outliers. All these tests were performed at $m = 2.0$, $p = 3.5$, $\sqrt{\eta_i} = 0.7$ $\forall i = 1 \ldots c$, $a = 1$, and $b = 5$.

5 Conclusions

In this paper we proposed a novel fuzzy-possibilistic mixture clustering model, in order to combat the sensitivity of existing c-means clustering models to outlier data. We performed several numerical tests on artificially created test data and the very popular IRIS data set, to evaluate the behavior of the proposed FP^{3}CM clustering model. In the presence of distant outliers, the proposed clustering model outperforms all existing c-means approaches. Further on, even in the absence of outliers, FP^{3}CM is slightly more accurate than PFCM, and outperforms conventional approaches in partition quality.

The adaptation of the proposed methodology to detect clusters of certain predefined shapes is going to be straightforward task, along the guidelines established by Davé and Bhaswan [5].

Acknowledgment. This research was funded by CNCSIS UEFISCSU, project no. PD_667, under contract no. 28/05.08.2010.

References

1. Anderson, E.: The IRISes of the Gaspe peninsula. Bull. Amer. IRIS Soc. 59, 2–5 (1935)
2. Barni, M., Capellini, V., Mecocci, A.: Comments on a possibilistic approach to clustering. IEEE Trans. Fuzzy Syst. 4, 393–396 (1996)
3. Bezdek, J.C.: Pattern recognition with fuzzy objective function algorithms. Plenum, New York (1981)
4. Davé, R.N.: Characterization and detection of noise in clustering. Patt. Recogn. Lett. 12, 657–664 (1991)
5. Davé, R.N., Bhaswan, K.: Adaptive fuzzy c-shells clustering and detection of ellipses. IEEE Trans. Neural Netw. 3(5), 643–662 (1992)
6. Krishnapuram, R., Keller, J.M.: A possibilistic approach to clustering. IEEE Trans. Fuzzy Syst. 1, 98–110 (1993)
7. Menard, M., Damko, C., Loonis, P.: The fuzzy $c + 2$ means: solving the ambiguity rejection in clustering. Patt. Recogn. 33, 1219–1237 (2000)
8. Pal, N.R., Pal, K., Bezdek, J.C.: A mixed c-means clustering model. In: Proc. IEEE Int'l Conf. Fuzzy Systems (FUZZ-IEEE), pp. 11–21 (1997)
9. Pal, N.R., Pal, K., Keller, J.M., Bezdek, J.C.: A possibilistic fuzzy c-means clustering algorithm. IEEE Trans. Fuzzy Syst. 13, 517–530 (2005)
10. Timm, H., Borgelt, C., Döring, C., Kruse, R.: An extension to possibilistic fuzzy cluster analysis. Fuzzy Sets and Systems 147, 3–16 (2004)
11. Vapnik, V.: Statistical learning theory. Wiley, New York (1998)
12. Xie, Z., Wang, S., Chung, F.L.: An enhanced possibilistic c-means clustering algorithm. Soft. Comput. 12, 593–611 (2008)

A Novel and Effective Approach to Shape Analysis: Nonparametric Representation, De-noising and Change-Point Detection, Based on Singular-Spectrum Analysis

Vasile Georgescu

Department of Mathematical Economics, University of Craiova, A.I.Cuza str. 13, 200585 Craiova, Romania
vasile.georgescu@feaa.ucv.ro

Abstract. This paper proposes new very effective methods for building nonparametric, multi-resolution models of 2D closed contours, based on Singular Spectrum Analysis (SSA). Representation, de-noising and change-point detection to automate the landmark selection are simultaneously addressed in three different settings. The basic one is to apply SSA to a shape signature encoded by sampling a real-valued time series from a radius-vector contour function. However, this is only suited for star-shaped contours. A second setting is to generalize SSA so as to apply to a complex-valued trajectory matrix in order to directly represent the contour as a time series path in the complex plan, along with detecting change-points in a complex-valued time series. A third setting is to consider the pairs (x, y) of coordinates as a co-movement of two real-valued time series and to apply SSA to a trajectory matrix defined in such a way to span both of them.

Keywords: Statistical shape analysis, Transforming planar closed contours into time series, Singular-spectrum analysis, Real- and complex-valued trajectory matrices, SSA-based change-point detection.

1 The Classical Computational Geometry Approach to Sampling Time Series from Planar Closed Contours

Statistical Shape Analysis involves methods for the geometrical study of random objects where location, rotation and scale information can be removed. By contrast, time series analysis is a widely spread technique that takes into consideration the temporal nature of data. However, despite their differences in nature, statistical shape analysis may benefit from methods commonly used in time series analysis. Indeed, there are certain ways of transforming a closed planar contour into a shape signature, represented by a contour function, and subsequently it may be possible to sample a time series from the contour function. Such functions are defined with respect to either simple or complex geometrical considerations: from metrics induced by symmetry relationships or periodicity, to the formal study of shapes based on computational differential geometry, where the quantification of differences between shapes can be achieved via a Riemannian metric on a shape manifold (namely, a

V. Torra et al. (Eds.): MDAI 2011, LNAI 6820, pp. 162–173, 2011.

finite-dimensional Riemannian manifold), and the interest naturally focuses on computing geodesic distances and geodesic paths between shapes.

Assuming that the contour has some desirable properties such as convexity or star-shapedness (i.e., given a figure A, for each point $(x, y) \in A$, the line segment connecting (x, y) with the centroid is contained in A), relatively simple contour functions, such as the radius-vector or support functions, can be introduced. Otherwise more complex contour functions should be considered, by representing the curves in a parameterized form. Their geometric properties and various quantities associated with them, such as the arc-length and the curvature, can then be expressed via derivatives and integrals using vector calculus.

The radius-vector function $r(\theta)$ is the distance from the reference point O (usually the center of gravity) to the contour in the direction of the θ-ray where $0 \le \theta \le 2\pi$. An example of a star-shaped figure and its radius-vector function is given in Figure 1. If the shape is inferred from noisy data, as it is the case with the figure below, the availability of a de-noising method becomes important. Furthermore, a change-point detection algorithm to automate the selection of salient landmarks may be of great interest.

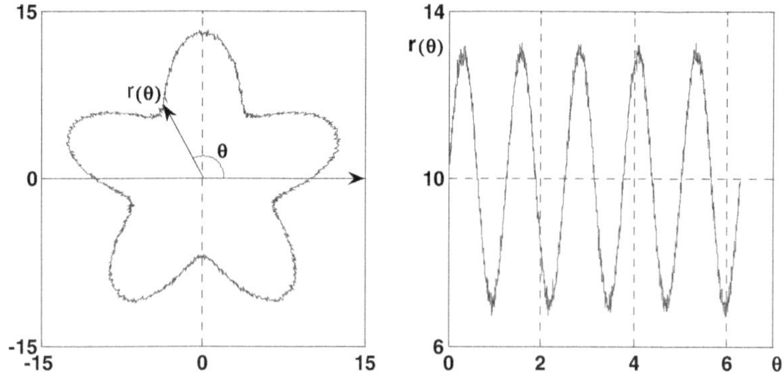

Fig. 1. A noisy star-shaped closed contour (left) and the radius-vector function $r(\theta)$ (right). The centroid has been used as the origin to generate the radius-vector function.

One can now choose a discrete sequence of (equally-spaced) values in $[0, 2\pi]$, i.e., $0 = \theta_1 < \theta_2 < \ldots < \theta_N = 2\pi$. The ordered sequence of radius-vector function values $\{r_t\}_{t=1,\ldots,N}$, with $r_t = r(\theta_t)$, can be regarded as a "time series" sampled from the contour function. The radius-vector function $r(\theta)$ is called a continuous shape signature, whereas $\{r_t\}_{t=1,\ldots,N}$ is called a discrete shape signature.

In the general case, however, description of a shape signature by the radius-vector function is not suitable for non-star-shaped contours (Figure 2).

Alternatively, there are at least two ways of representing planar curves in a parameterized form: one is using the angle (direction) function and another is using

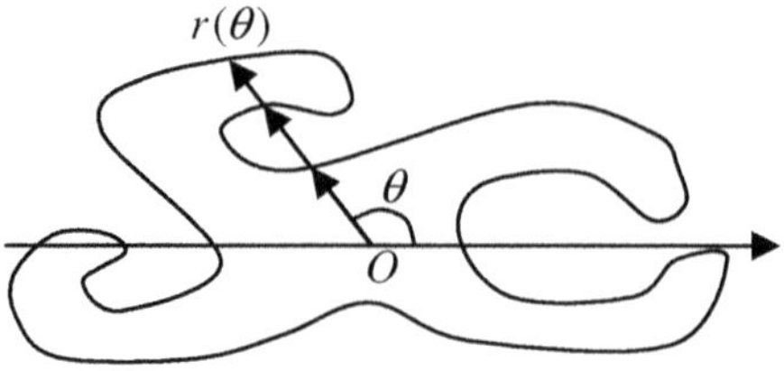

Fig. 2. Problems with the radius-vector function occur if the contour is not star-shaped

the curvature function. These involve using differential geometry, which provides a set of powerful tools for shape analysis. However, from a practical viewpoint, it is difficult to infer planar contour representations in a parameterized form.

The next section introduces a new approach to representation and de-noising of closed planar contours, along with a change-point detection algorithm to automate the selection of salient landmarks. It consists of a nonparametric, multi-resolution method, based on Singular-Spectrum Analysis.

2 A Novel and Effective Approach to Shape Analysis: Nonparametric Representation, De-noising and Change-Point Detection, Based on Singular-Spectrum Analysis

2.1 An overview of Singular-Spectrum Analysis

Singular-Spectrum Analysis (SSA) is a nonparametric method for time series structure recognition and identification. It tries to overcome the problems of finite sample length and noisiness of sampled time series not by fitting an assumed model to the available series, but by using a data-adaptive basis set.

The SSA algorithm has two basic stages: decomposition and reconstruction.

The decomposition stage is carried out in two steps:

(D1) The *Embedding* step maps the original one-dimensional time series $\{x_1, x_2, \ldots, x_N\}$ to a sequence of $K = N - M + 1$ lagged vectors of dimension M (where M is called *window length*):

$$X_i = (x_i, \ldots, x_{i+M-1})', \quad i = 1, \ldots, K, \qquad 1 < M < N. \tag{1}$$

This lagged vectors form the columns of the *trajectory matrix* X, which is actually a Hankel matrix (i.e., it has equal elements on the diagonals $i + j - 1$ =const.): $X = [X_1 : X_2 : \ldots : X_K]$.

(D2) The SVD step is the *singular value decomposition* of the trajectory matrix. Let $R = \frac{1}{N} X \cdot X'$ be an $M \times M$ matrix, called *lag-covariance matrix*. Denote by $\lambda_1, \ldots, \lambda_M$ the eigenvalues of R taken in the decreasing order of magnitude

$(\lambda_1 \geq \ldots \geq \lambda_M \geq 0)$ and by $U_1, \ldots, U_M$ the orthonormal system of the eigenvectors of the matrix R corresponding to these eigenvalues. Let $d = \max\{ i \text{ , such that } \lambda_i > 0 \}$. If we denote $V_i = X'U_i / \sqrt{\lambda_i}$ ($i = 1, \ldots, d$), then the SVD of the trajectory matrix X can be written as $X = X_1 + \ldots + X_d$, where $X_i = \sqrt{\lambda_i}\, U_i \otimes V_i'$ and $\otimes$ is the outer product. The matrices X_i are elementary matrices (have rank one).

The reconstruction stage is also carried out in two steps:

(R1) The *grouping* step consists of partitioning the set of indices $\{1, \ldots, d\}$ into m disjoint subsets $I_1, \ldots, I_m$. The case of practical interest for our application is that of a dichotomic partitioning: split the set of indices into two groups, $\{1, \ldots, d\} = I + \bar{I}$, where $I = \{i_1, \ldots, i_\ell\}$ and $\bar{I} = \{1, \ldots, d\} \setminus I$, and sum the matrices X_i within each group:

$$X = X_I + X_{\bar{I}} \tag{2}$$

where $X_I = \sum_{i \in I} X_i$ and $X_{\bar{I}} = \sum_{i \in \bar{I}} X_i$.

The choice of the ℓ most contributing eigenvalues λ_i, $i \in I$, and thus of the corresponding ℓ eigenvectors is an appropriate way to control and reduce the distance between the M-dimensional vectors that form the columns of trajectory matrix and the ℓ-dimensional hyperplane determined by the ℓ eigenvectors. For example, we can choose the index set I such that $\sum_{j=1}^{\ell} \lambda_{i_j} \Big/ \sum_{j=1}^{d} \lambda_j > 0.95$ corresponding to the set of eigenvalues whose cumulated contribution exceeds 95%.

(R2) The last step transforms each matrix of the grouped decomposition (2) into a new series of length N, by *diagonal averaging*. It consists of averaging over the diagonals $i + j - 1 = \text{const.}$ ($i = 1, \ldots, M$, $j = 1, \ldots, K$) of the matrices X_I and $X_{\bar{I}}$. Applying then twice the one-to-one correspondence between the series of length N and the Henkel matrices of size $M \times K$ (with $K = N - M + 1$), we obtain the SSA decomposition of the original series $\{x_t\}$ into a sum of two series: $x_t = z_t + \varepsilon_t$, $t = 1, \ldots, N$. In this context, the series z_t (obtained from the diagonal averaging of X_I) can often be associated with signal and the residual series ε_t with noise.

2.2 The First Setting: Applying SSA to a Shape Signature Encoded by Sampling a Real-Valued Time Series from a Radius-Vector Contour Function

This setting is well suited for star-shaped planar closed contours and starts with sampling a "time series" $\{r_t\}_{t=1,\ldots,N}$ from the radius-vector contour function, i.e., $r_t = r(\theta_t)$, $0 = \theta_1 < \theta_2 < \ldots < \theta_N = 2\pi$. The trajectory matrix X is then constructed by mapping the time series $\{r_t\}_{t=1,\ldots,N}$ to a sequence of $K = N - M + 1$ lagged vectors of dimension M :

$$X = \begin{pmatrix} r_1 & r_2 & \cdots & r_K \\ r_2 & r_3 & \cdots & r_{K+1} \\ \vdots & \vdots & \vdots & \vdots \\ r_M & r_{M+1} & \cdots & r_N \end{pmatrix} \tag{3}$$

Noise reduction is attained by reducing the rank of the trajectory matrix. In the absence of noise one should be able to recover the data trajectory matrix with the first L singular vectors. Thus, the SVD reconstruction of the trajectory matrix X can be truncated to obtain an estimate of the noise-reduced trajectory matrix, i.e., a reduced-rank form:

$$\hat{X} = \sum_{i=1}^{L} X_i, \qquad X_i = \sqrt{\lambda_i}\, U_i \otimes V_i', \qquad L < M \tag{4}$$

where $\otimes$ is the outer product and the matrices X_i are rank one matrices.

It is important to stress that in the absence of noise, $\hat{X} = X$. Thus, in the presence of noise the L strongest singular values and their associated singular vectors span the noise-free signal. It is clear that we are not interested in recovering the trajectory matrix but the signal itself. For this purpose we average the elements of the filtered trajectory matrix along the anti-diagonals of $\hat{X}$ to obtain an estimate of the enhanced signal, denoted by $\{\hat{r}_t\}_{t=1,\ldots,N}$.

Let us consider the time series $\{r_t\}_{t=1,\ldots,N}$ corresponding to the noisy star-shaped closed contour depicted in Fig.1, where $N = 1441$, $M = 54$, $K = N - M + 1 = 1388$. Figure 3 shows the contribution of each of the 54 singular values.

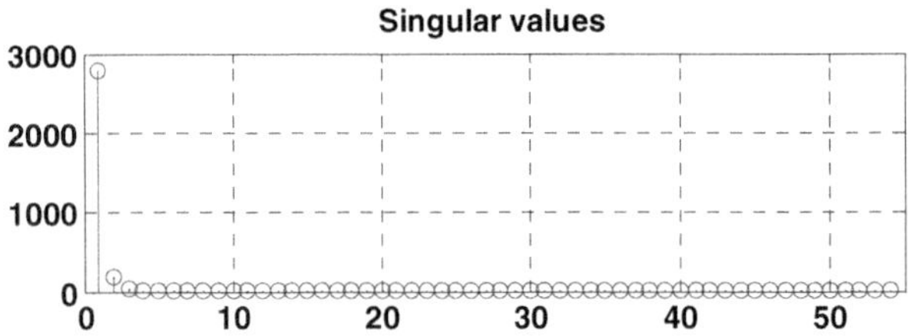

Fig. 3. The contribution of each of the 54 singular values

One can easily see that only the first two singular values have a significant contribution in recovering the smooth part of the signal (noise reduction). However, Figures 4 and 5 show that the signal can not be consistently recovered using only the first largest singular value. The second largest singular value is also needed.

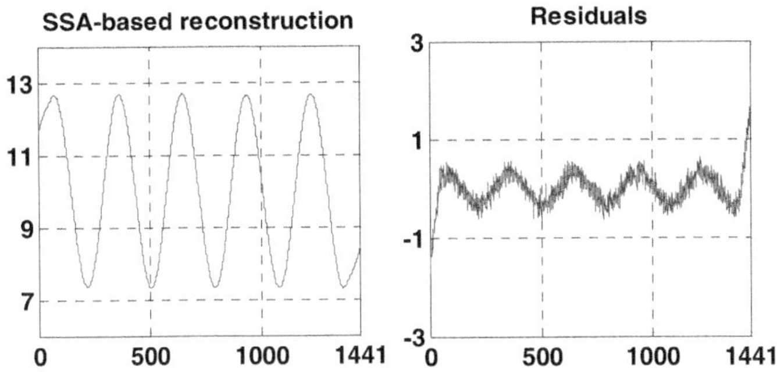

Fig. 4. SSA-based reconstruction of time series $\{\hat{r}_t\}_{t=1,\dots,N}$ using only the first largest singular value (left) and the corresponding residuals (right)

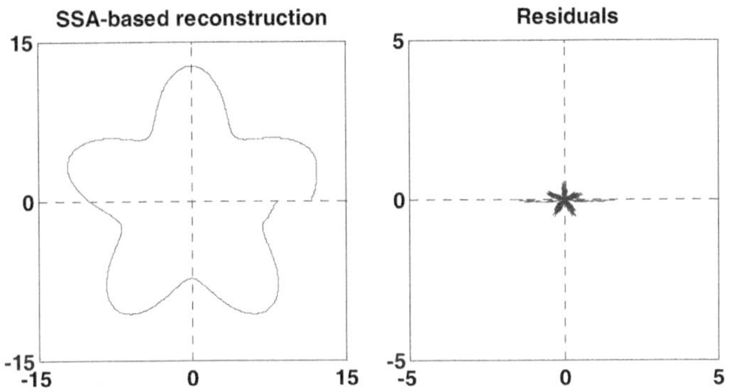

Fig. 5. SSA-based reconstruction of the contour $(\hat{x}_t, \hat{y}_t)_{t=1,\dots,N}$ using only the first largest singular value (left) and the corresponding residuals, at a scale magnified 3 times (right)

Figures 6 and 7 show that the best choice for de-noising both the time series $\{\hat{r}_t\}_{t=1,\dots,N}$ and the planar contour $(\hat{x}_t, \hat{y}_t)_{t=1,\dots,N}$ is to recover the data trajectory matrix with the first 2 singular vectors. This results in a smooth reconstruction.

The exact reconstruction of the initial noisy closed contour can be also performed if all the singular vectors corresponding to non-zero singular values are used when recovering the trajectory matrix.

In the final part of this section, a SSA-based change-point detection algorithm is presented, with the aim of automating the landmark selection.

SSA-Based Change-Point Detection. A frequentist, non-parametric algorithm for multiple change-point detection in time series based on sequential application of the Singular Spectrum Analysis was developed in [5]. The idea behind the algorithm is to apply SSA to a windowed portion of the signal in order to pick up its structure through an ℓ-dimensional subspace spanned by the eigenvectors of the lag-covariance matrix, computed in a sequence of moving time intervals $[n+1;\ n+m]$ of a given length m, where $n = 0, 1, \ldots$ is the iteration number. If at a certain time moment τ the mechanism generating the time series x_t has changed then an increase in the distance between the ℓ-dimensional hyperplane and the M-lagged vectors $(x_{\tau+1}, \ldots, x_{\tau+M})$ of trajectory matrix is to be expected. This increase will indicate the change. However, if the generating mechanism does not change further along the signal, then the corresponding lagged vectors will stay close to this hyperplane.

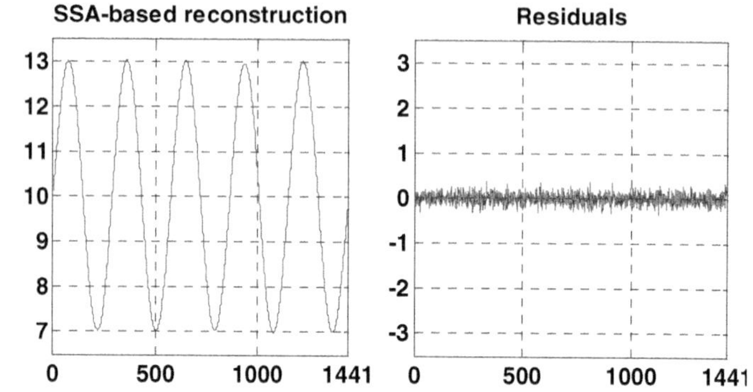

Fig. 6. SSA-based reconstruction of time series $\{\hat{r}_t\}_{t=1,\ldots,N}$ using the first two largest singular values (left) and the corresponding residuals (right)

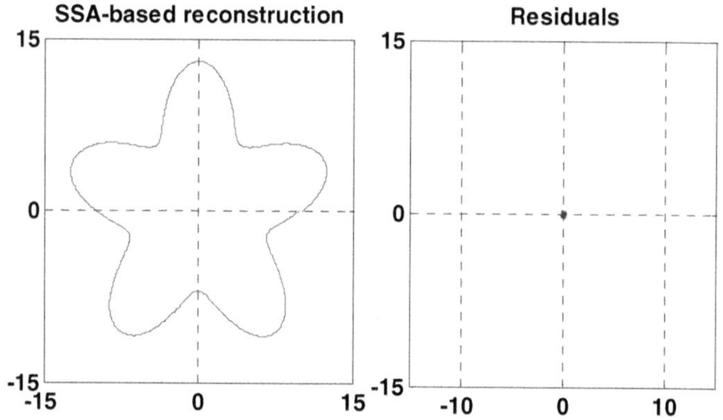

Fig. 7. SSA-based reconstruction of the contour $(\hat{x}_t, \hat{y}_t)_{t=1,\ldots,N}$ using the first two largest singular values (left) and the corresponding residuals (right)

Let $\{x_1, x_2, \ldots, x_N\}$ be a time series, where N is large enough. Two parameters have to be chosen: the window width m ($m < N$), and the lag parameter M ($M \leq m/2$). Define also $K = m - M + 1$.

For each $n = 0, 1, \ldots, N - m$, a three-stage procedure is executed:

Stage 1. Perform the SSA algorithm for the time interval $[n+1, n+m]$.

1. Construct the trajectory matrix $X^{(n)}$ (here called *base matrix*), whose columns are the vectors $X_j^{(n)}$:

$$X^{(n)} = \left(x_{n+i+j-1}\right)_{i=1:M;\ j=1:K} = \begin{pmatrix} x_{n+1} & x_{n+2} & \cdots & x_{n+K} \\ x_{n+2} & x_{n+3} & \cdots & x_{n+K+1} \\ \vdots & \vdots & \vdots & \vdots \\ x_{n+M} & x_{n+M+1} & \cdots & x_{n+m} \end{pmatrix} \tag{5}$$

$$X_j^{(n)} = \left(x_{n+j}, \ldots, x_{n+M+j-1}\right)', \quad j = -n+1, -n+2, \ldots, N-n-M+1$$

2. Perform the SDV of the lag-covariance matrix $R_n = 1/K \cdot X^{(n)}\left(X^{(n)}\right)'$. This gives us a collection of M eigenvectors.

3. Select a particular group I of $\ell < M$ of these eigenvectors; this determines an ℓ-dimensional subspace $S_{n,\ell}$ in the M-dimensional space of vectors $X_j^{(n)}$. Denote the ℓ eigenvectors that determine the subspace $S_{n,\ell}$ by $U_{i_1}, \ldots, U_{i_\ell}$.

Stage 2. Construction of the test matrix.

Denote $Q = q - p$ (thus $q = p + Q$) and construct the following $M \times Q$ trajectory matrix (called *test matrix*):

$$X_{test}^{(n)} = \left(x_{n+p+i+j-1}\right)_{i=1:M;\ j=1:Q} = \begin{pmatrix} x_{n+p+1} & x_{n+p+2} & \cdots & x_{n+q} \\ x_{n+p+2} & x_{n+p+3} & \cdots & x_{n+q+1} \\ \vdots & \vdots & \vdots & \vdots \\ x_{n+p+M} & x_{n+p+M+1} & \cdots & x_{n+q+M-1} \end{pmatrix} \tag{6}$$

The part of sample $x_{n+1}, \ldots, x_{n+m}$ that is used to construct the (base) trajectory matrix $X^{(n)}$ will be called 'base sample', and another part, $x_{n+p+1}, \ldots, x_{n+q+M-1}$, which is used to construct the vectors $X_j^{(n)}$ $(j = p+1, \ldots, q)$ and thus to compute the sum of squared distances $\mathcal{D}_{n,I,p,q}$ will be called "test sample".

Stage 3. Computation of the detection statistics

The detection statistics are:

$\mathcal{D}_{n,I,p,q}$, the sum of squares of the (Euclidean) distances between the vectors $X_j^{(n)}$ $(j = p+1, \ldots, q)$ and the ℓ-dimensional subspace $S_{n,\ell}$. Since the eigenvectors are orthogonal, the square of the Euclidean distance between an M–vector $ZY = X_j^{(n)}$ and the subspace $S_{n,\ell}$ spanned by the ℓ eigenvectors $P_1, \ldots, P_\ell$, is just $\|Z\|^2 - \|P'Z\|^2 = Z'Z - Z'PP'Z$, where $\|\cdot\|$ is the Euclidean norm and P is the $M \times \ell$-matrix with columns $P_1, \ldots, P_\ell$. Therefore

$$\mathcal{D}_{n,I,p,q} = \sum_{j=p+1}^{q} \left(X_j^{(n)}\right)' X_j^{(n)} - \left(X_j^{(n)}\right)' P P' X_j^{(n)} \tag{7}$$

The normalized sum of squared distances

$$\mathcal{D}_{n,\ell,p,q} / \mu_{n,\ell,p,q} \geq h \tag{8}$$

$S_n = \mathcal{D}_{n,I,p,q} / v_n$. Here v_j is an estimate of the sum of squared distances $\mathcal{D}_{n,I,p,q}$ at the time intervals $[\,j+1,\ \ j+r\,]$ where the hypothesis of no change can be accepted. Actually, $v_n = \dfrac{1}{n-m/2} \sum_{i=0}^{n-m/2-1} \mathcal{D}_{n,I,p,q}$ or $v_n = \mathcal{D}_{r,I,0,K}$ can be two alternative choices for v_n, where r is the largest value of $r \leq n$ so that the hypothesis of no change is accepted.

The decision rule in the algorithm, denoted by $A(M, m, \ell, p, q, h)$, is to announce that a change in the mechanism generating x_t occurs at a certain point τ, if for a certain n

$$\mathcal{D}_{n,I,p,q} / v_n \geq h \tag{9}$$

where h is a fixed threshold. Then we would expect than the vectors $X_j = X_{j-n}^{(n)}$ with $j > \tau$ lie further away from the ℓ-dimensional subspace $S_{n,l}$ than the vectors X_j with $j \leq \tau$. This means that the sequence $D(n) = \mathcal{D}_{n,I,p,q}$, considered as a function of n, is expected to start growing somewhere around $\hat{n}$, such that $\hat{n} + q + M - 1 = \tau$. The value $\hat{n} = \tau - q - M + 1$ is the first value of n such that the test sample $x_{n+p+1}, \ldots, x_{n+q+M-1}$ contain the change point.

In other words, $q + M - 1$ should be interpreted as a latency of test statistic in detecting change-points. Therefore, a corresponding backshift of the starting point on the contour with respect to the first position should be considered. Since closed contours are periodic in nature, such a task is easy to be done. For the time series $\{\hat{r}_t\}_{t=1,\ldots,N}$ encoding the shape signature of our star-shaped contour, a backshift of $q + M - 1 = 161$ positions is required. The test statistic is depicted in Figure 8. Actually, the location of change-points is in the local minima of test statistic function.

Figure 9 shows the landmark positions, automatically selected through change-point detection. Here, the detection statistics have been computed as normalized sum of squares of the distances between the vectors $X_j^{(n)}$ and the ℓ-dimensional subspace $S_{n,\ell}$, assuming $\ell = 1$. Increasing ℓ results in an increasing number of change-points.

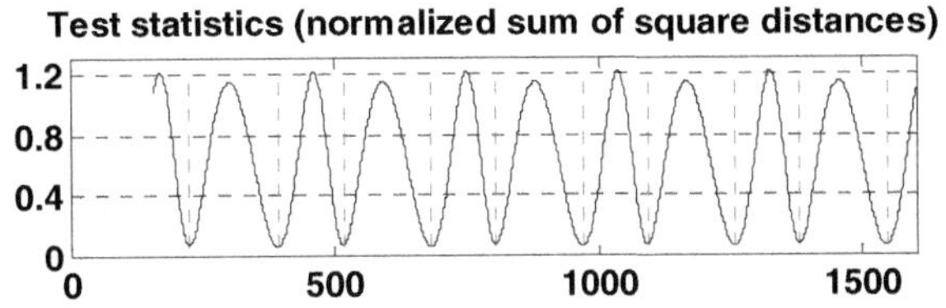

Fig. 8. Detecting change points from Distance detection statistic. The location of change points is in the local minima of test statistic function.

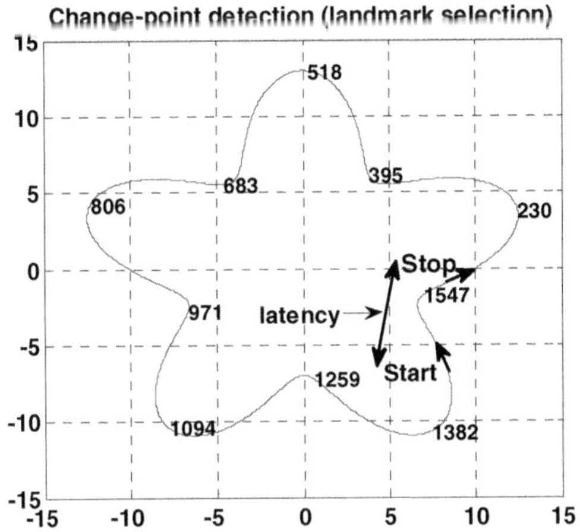

Fig. 9. Automatic selection of landmark positions through change-point detection

2.3 The Second Setting: Applying SSA to a Shape Signature Encoded by Sampling a Complex-Valued Time Series from the Contour Itself, Represented in the Complex Plane

For closed contours that are not star-shaped, a shape signature encoded by means of radius-vector function is inappropriate. In the general case, a complex-valued time series can be sampled from the contour itself, whose trace can be encoded as a sequence of complex numbers: $\{z_t\}_{t=1,\dots,N}$, where $z_t = x_t + i\, y_t \in C$. Now, the trajectory matrix has complex elements, too:

$$X = \begin{pmatrix} z_1 & z_2 & \cdots & z_K \\ z_2 & z_3 & \cdots & z_{K+1} \\ \vdots & \vdots & \vdots & \vdots \\ z_M & z_{M+1} & \cdots & z_N \end{pmatrix}, \quad z_t \in C \tag{10}$$

The generalization of SSA for this setting is founded on the ACM Algorithm 358 for Singular Value Decomposition of a complex matrix. The decomposition theorem (Businger and Golub, [1]) can be stated as follows: each and every $M \times K$ complex-valued matrix X can be reduced to diagonal form by unitary transformations U and V, $X = U\, diag[\sigma_1, \dots, \sigma_K] V^H$, where $\sigma_1 \geq \dots \geq \sigma_K \geq 0$ are real-valued scalars, called the singular values of X. Here U is an $M \times K$ column orthogonal matrix, V an $K \times K$ unitary matrix and V^H is a Hermitian transpose of V. The columns of U and V are called the left and right singular vectors of X, respectively.

As concerning the SSA-based change point detection algorithm, the sum of squared distances in equation (7) changes accordingly:

$$\mathcal{D}_{n,I,p,q} = \sum_{j=p+1}^{q} abs\left[\left(X_j^{(n)}\right)' X_j^{(n)} - \left(X_j^{(n)}\right)' P P' X_j^{(n)}\right] \tag{11}$$

Figures 10 and 11 show that the SSA-based algorithm I proposed for complex-valued trajectory matrices can be successfully applied to non star-shaped contours.

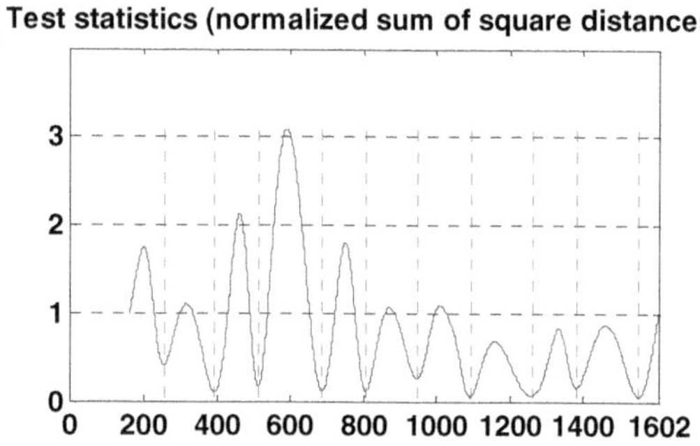

Fig. 10. Detecting change points from Distance detection statistic. The location of change points is in the local minima of test statistic function.

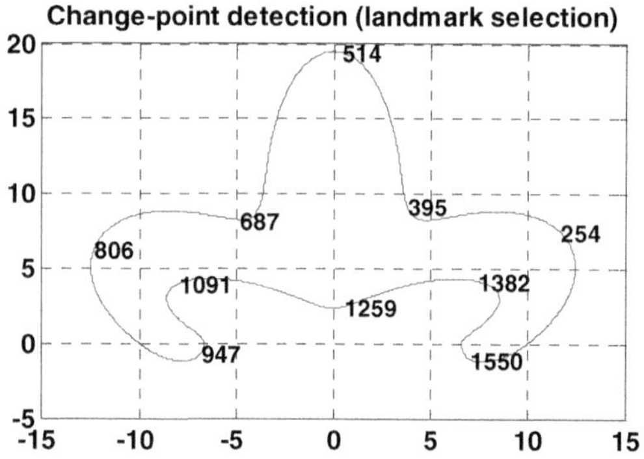

Fig. 11. Automatic selection of landmark positions through change-point detection

2.4 The Third Setting: Applying SSA to a Shape Signature Encoded by Sampling Two Real-Valued Time Series from the *x* and *y* Coordinates

Given the coordinate pairs $(x_t, y_t)_{t=1,\ldots,N}$, the trajectory matrix can now be written as follows:

$$X = \begin{pmatrix} x_1 & x_2 & \cdots & x_K \\ y_1 & y_2 & \cdots & y_K \\ x_2 & x_3 & \cdots & x_{K+1} \\ y_2 & y_3 & \cdots & y_{K+1} \\ \cdots & \cdots & \cdots & \cdots \\ x_M & x_{M+1} & \cdots & x_N \\ y_M & y_{M+1} & \cdots & y_N \end{pmatrix}, \quad x_t, y_t \in \Re \tag{12}$$

The generalization is straightforward; however, in my experiments, this version of the SSA-based change-point detection algorithm underperforms when comparing with that presented in section 2.3. A possible explanation is provided next.

3 Conclusion

For star-shaped contours the first and the second settings to my approach give identical and very accurate results, whereas the third setting produces comparable results when using for de-noising, but less accurate results when using for change-point detection.

Both the second and third settings are suitable for general-purpose applications. However, the second setting proved to be more reliable in practical cases, presumably because the number of rows in the trajectory matrix is twice as less as in the third case, when its reconstruction have to collect contributions from higher dimensional subspaces.

References

1. Businger, P.A., Golub, G.H.: Algorithm 358: Singular value decomposition of a complex matrix. Comm. ACM 12, 564–565 (1969)
2. Dryden, I.L., Mardia, K.V.: Statistical shape analysis. John Wiley and Sons, Chichester (1998)
3. Georgescu, V.: Clustering of Fuzzy Shapes by Integrating Procrustean Metrics and Full Mean Shape Estimation into K-Means Algorithm. In: 13th IFSA World Congress and 6th Conference of EUSFLAT, Lisbon, Portugal, pp. 1791–1796 (2009)
4. Goljadina, N., Nekrutkin, V., Zhigljavky, A.: Analysis of Time Series Structure: SSA and related techniques. Chapman and Holl, London (2001)
5. Moskvina, V.: Applications of the singular-spectrum analysis for change point detection in time series. Ph.D. thesis, School of Mathematics, Cardiff University, Cardiff (2001)

A SSA-Based New Framework Allowing for Smoothing and Automatic Change-Points Detection in the Fuzzy Closed Contours of 2D Fuzzy Objects

Vasile Georgescu

Department of Mathematical Economics, University of Craiova, 13 A.I.Cuza str., 200585 Craiova, Romania
vasile.georgescu@feaa.ucv.ro

Abstract. The aim of this paper is to propose a new framework, based on Singular-Spectrum Analysis, allowing for smoothing and automatic change-point detection in the fuzzy closed contours of 2D fuzzy objects. The representation of fuzzy objects is first addressed, by distinguishing between fuzzy regions and fuzzy closed curves. Fuzzy shape signatures are derived in special cases, from which fuzzy time series can be subsequently sampled. Geodesic and Euclidean fuzzy paths and distances between two points in a fuzzy region are next contrasted. Finally, a novel approach to decomposing and reconstructing a fuzzy shape and to automatic change-point detection is proposed, based on a generalization of Singular-Spectrum Analysis so as to deal with complex-valued trajectory matrices. The coordinates themselves, represented as complex numbers are used as a shape signature. This approach is suitable for non-convex and non-star-shaped fuzzy contours.

Keywords: 2D fuzzy regions and fuzzy closed curves, Fuzzy geodesic paths and distances, Singular-spectrum analysis, Complex-valued trajectory matrices encoding fuzzy closed contours, SSA-based change-point detection.

1 Representations of Fuzzy Objects: Fuzzy Regions vs. Fuzzy Closed Curves

Shapes and textures are extremely important features in human as well as machine vision systems. Shape analysis is concerned with two main classes of methods: boundary-based (when only the shape boundary points are used for the description) and region-based (when the whole interior of a shape is considered for description).

In fuzzy shape analysis, however, boundary points are neither strictly delimited, nor independent from texture information, but have assigned to them a fuzzy membership value according to the extent of their belongingness to the object; there is a progressive transition of the membership values from the support outline to the core outline.

Continuous fuzzy shapes can be described as fuzzy geometric objects. A continuous fuzzy geometric object S in $\Re^n$ is defined as a set of pairs $\{(x, \mu_S(x)) \mid x \in \Re^n\}$ where

V. Torra et al. (Eds.): MDAI 2011, LNAI 6820, pp. 174–185, 2011.

$\mu_S : \Re^n \to [0, 1]$ is the membership function of S in $\Re^n$. An alternative representation of fuzzy geometric objects is given by a set of $\alpha-$cuts (also called $\alpha-$supports). For any value $\alpha \in [0, 1]$, the $\alpha-$*support* of S, denoted by $S_\alpha = Supp_\alpha(S)$, is the hard subset $\{x \mid x \in \Re^n$ and $\mu_S(x) \geq \alpha\}$ of $\Re^n$. The 0-support will often be referred to as *support* and be denoted by $Supp\ (S)$, while the 1-support will be referred to as *core*. A fuzzy subset with a bounded support is called *bounded*. S is said to be *convex* if, for every three collinear points x, y, and z in $\Re^n$ such that y lies between x and z, $\mu_S(y) \geq \min[\mu_S(x), \mu_S(z)]$. A fuzzy subset is called *smooth* if its membership function is differentiable at every location $x \in \Re^n$.

We have to take into account two distinct classes of fuzzy geometric objects: fuzzy regions and fuzzy closed curves. The major difference between them is the shape of the fuzzy boundary.

The membership function of a fuzzy region is non-increasing away from the interior of the object. This means that the $\alpha-$supports of a fuzzy region are nested, i.e., for membership values $1 = \alpha_1 > \ldots > \alpha_{n+1} = 0$, one has $S_{\alpha_1} \subseteq \ldots \subseteq S_{\alpha_{n+1}}$.

By contrary, the membership function of a fuzzy closed curve has values greater than zero only on the fuzzy boundary and is typically LR-shaped (i.e., it is first increasing from the interior to the modal point of the frontier and then is decreasing to the exterior).

Let's start with the first case. Figure 1 shows a star-shaped fuzzy region. The centroid of the fuzzy shape will be denoted by $C_S = (x_c, y_c)$, where

$$x_c = \frac{\iint x \cdot \mu(x, y)\, dxdy}{\iint \mu(x, y)\, dxdy}; \qquad y_c = \frac{\iint y \cdot \mu(x, y)\, dxdy}{\iint \mu(x, y)\, dxdy} \tag{1}$$

The straight path from the centroid along a radial direction defined with respect to a given angle θ can be parameterized with respect to a parameter $t \in [0,1]$ as follows:

$$\pi^\theta(t) = \{(x^\theta(t), y^\theta(t)) \mid x^\theta(t) = x_c + \rho^\theta(t) \cdot \cos\theta,\ y^\theta(t) = y_c + \rho^\theta(t) \cdot \sin\theta\} \tag{2}$$

where

$$\rho^\theta(t) = \left\| x^\theta(t) - x_c,\ y^\theta(t) - y_c \right\| = \begin{cases} 2t\rho_1^\theta & t \in [0, 1/2] \\ (3-2t)\rho_1^\theta + (2t-1)\rho_0^\theta & t \in (1/2, 1] \end{cases} \tag{3}$$

Given that a fuzzy region is non-increasing away from the interior of the object, its fuzzy boundary along the path π^θ is delimited by two points: $(x_c + \rho_1^\theta \cos\theta,\ y_c + \rho_1^\theta \sin\theta)$ from the interior and $(x_c + \rho_0^\theta \cos\theta,\ y_c + \rho_0^\theta \sin\theta)$ from the exterior, where:

$$\begin{aligned} \rho_1^\theta &= \max\left(\rho^\theta = \left\| x^\theta - x_c,\ y^\theta - y_c \right\| \mid (x^\theta, y^\theta) \in \pi^\theta,\ \mu(x^\theta, y^\theta) = 1\right) \\ \rho_0^\theta &= \min\left(\rho^\theta = \left\| x^\theta - x_c,\ y^\theta - y_c \right\| \mid (x^\theta, y^\theta) \in \pi^\theta,\ \mu(x^\theta, y^\theta) = 0\right) \end{aligned} \tag{4}$$

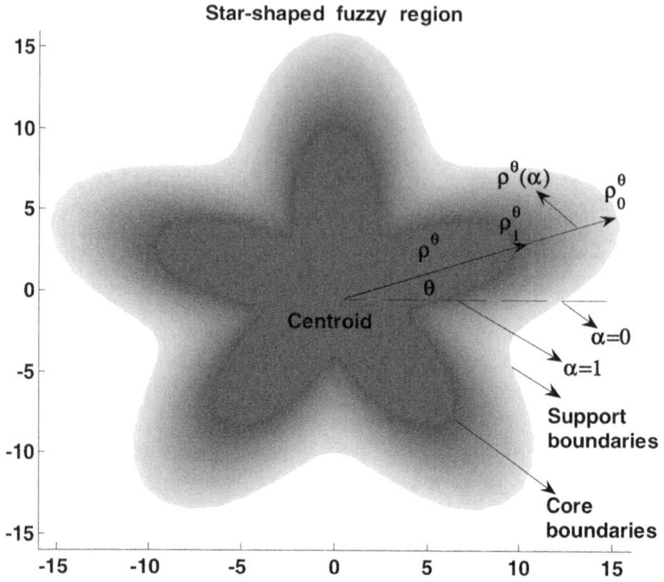

Fig. 1. A star-shaped fuzzy region; the straight path from the centroid along a radial direction

The simplest case is when the membership function along the path π^{θ} is piecewise linear:

$$\mu\left(x^{\theta}(t),\, y^{\theta}(t)\right)=\mu\left(\rho^{\theta}(t)\right)=\alpha(t)=\begin{cases}1 & t\in[0,1/2]\\ 2-2t & t\in(1/2,1]\end{cases}\tag{5}$$

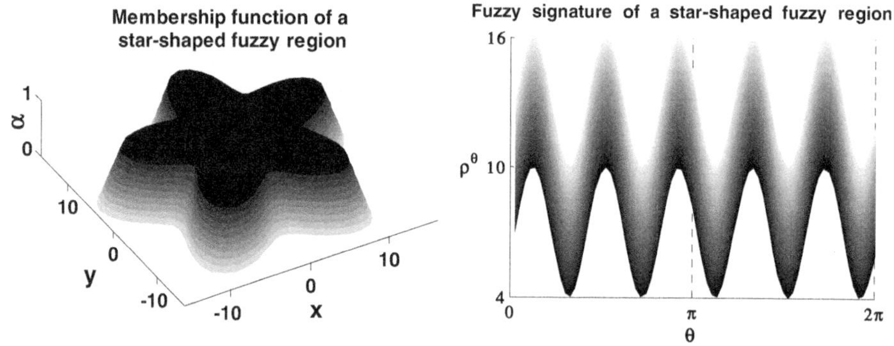

Fig. 2. 3D representation of the membership function of a star-shaped fuzzy region (left); the fuzzy signature of a star-shaped fuzzy region, based on the fuzzy radial distance (right)

Since $\alpha\rightarrow\rho_{\alpha}^{\theta}$ is an inverse of the membership function along the path $\pi^{\theta}(t)$, $t\in[1/2,1]$, we can use α to parameterize the Euclidean distance across the $\alpha-$support boundary points:

$$\rho_\alpha^\theta = \rho_1^\theta \cdot \alpha + (1-\alpha)\cdot \rho_0^\theta, \quad \alpha \in [0,1], \quad \rho_\alpha^\theta \in [\rho_1^\theta, \rho_0^\theta] \tag{6}$$

Thus, ρ^θ is defined as a fuzzy distance, with the membership function given by:

$$\mu(\rho^\theta) = \begin{cases} 1 & \rho^\theta \in [0, \rho_1^\theta] \\ \dfrac{\rho - \rho_0^\theta}{\rho_1^\theta - \rho_0^\theta} & \rho^\theta \in [\rho_1^\theta, \rho_0^\theta] \end{cases} \tag{7}$$

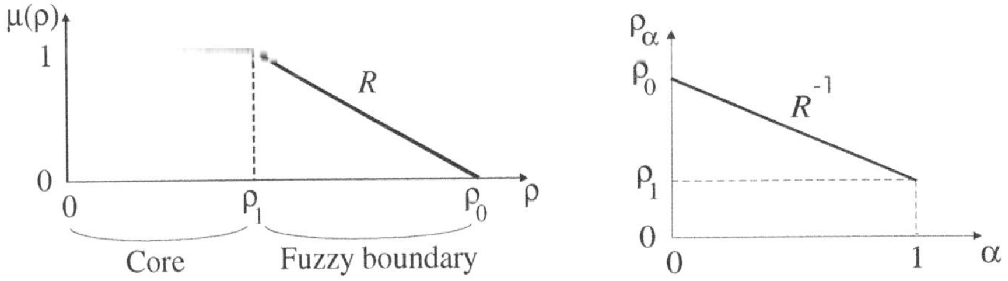

Fig. 3. The membership function of a star-shaped fuzzy region along a radial direction from the centroid

Alternatively, the linear membership function $\mu(\rho^\theta)$ can be replaced by a nonlinear function along with an appropriate parameterization of the path π^θ.

A fuzzy shape signature of a continuous star-shaped fuzzy region can be defined by ρ_α^θ as a fuzzy function of the radial angle θ, with $tg\theta$ being the slope of the straight path between the centroid and the fuzzy boundary (Figure 2, right)

In contrast with a fuzzy region, the membership function of a fuzzy closed curve is typically LR-shaped.

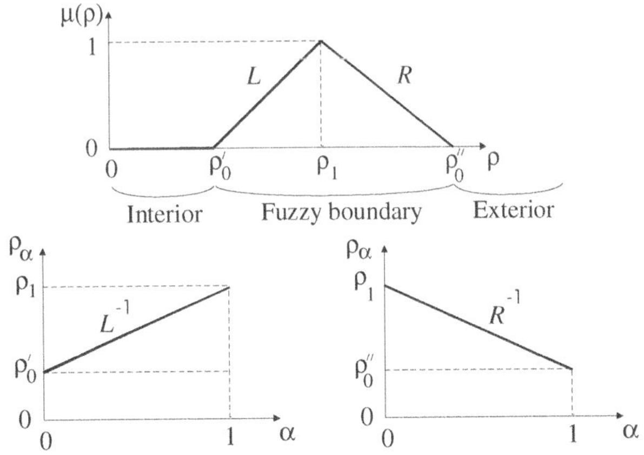

Fig. 4. The membership function of a star-shaped fuzzy closed curve along a radial direction from the centroid

A star-shaped fuzzy closed curve is depicted in Figure 5.

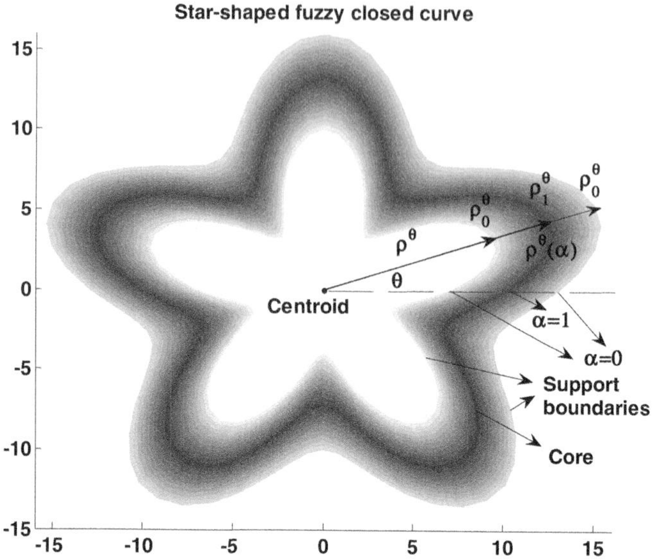

Fig. 5. A star-shaped fuzzy closed curve

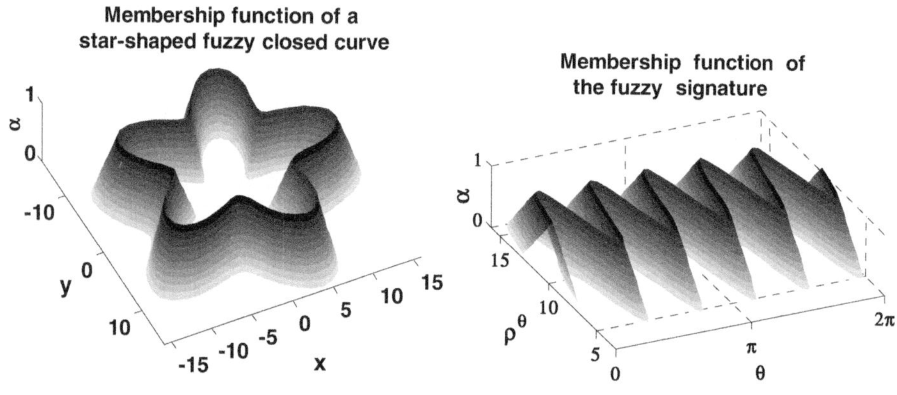

Fig. 6. 3D representation of the membership function of a star-shaped fuzzy closed curve (left); 3D representation of the membership function of the fuzzy signature (right)

The fuzzy shape signature of a star-shaped fuzzy closed curve has the membership function depicted in Figure 6 (right) and can be defined only in terms of the Euclidean notions of path and distance from the centroid to the fuzzy boundary (rather then in a geodesic sense), since the region containing the centroid does not belong to the fuzzy object itself, but to its complement (see the next section).

For a sequence of indices $\tau = \{0, 1, \ldots, n\}$ and the corresponding discrete sequence of angles $\theta_0 = 0 < \ldots < \theta_\tau < \ldots < \theta_n = 2\pi$, a fuzzy-valued time series $\{\rho_\tau\} = \{\rho(\theta_\tau)\}$, $\tau = 0, 1, \ldots, n$, can be sampled from the continuous fuzzy signature (Figure 7).

Moreover, for each α_i in a discrete sequence $\alpha_0 = 0 < \alpha_1 < \ldots < \alpha_{p-1} < 1$, a pair of real-valued time series $\left(\underline{\rho}_\tau(\alpha_i),\ \overline{\rho}_\tau(\alpha_i)\right) = \left(L_\tau^{-1}(\alpha_i), R_\tau^{-1}(\alpha_i)\right)$ can be sampled, as well as a real-valued time series from the modal values in the core ($\alpha_p = 1$).

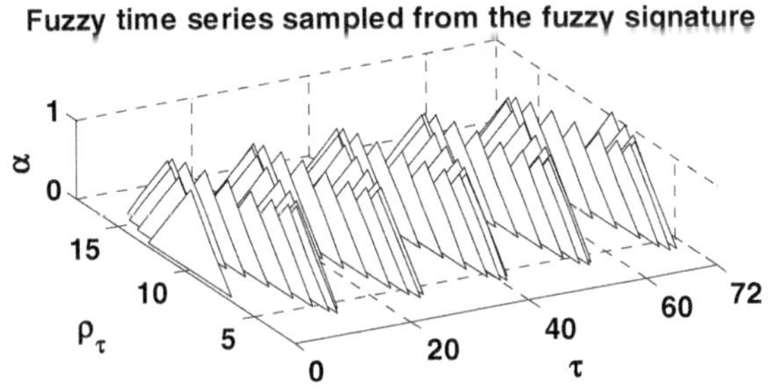

Fig. 7. Fuzzy-valued time series sampled from the continuous fuzzy signature (fuzzy radius-vector function)

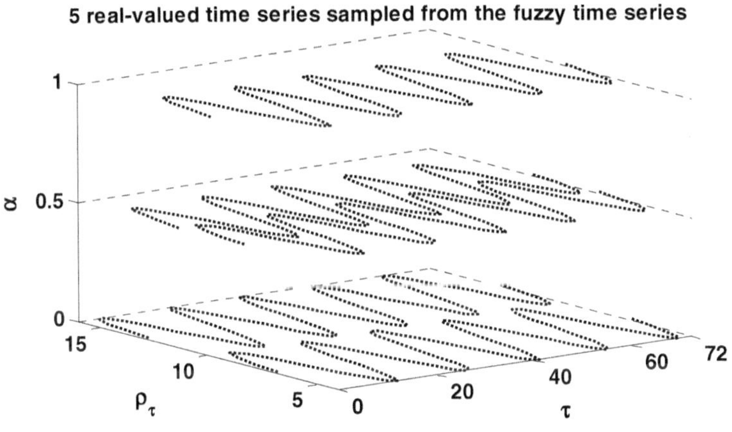

Fig. 8. Five real-valued time series sampled from the fuzzy time series, for $\alpha = 0, 0.5, 1$

2 Geodesic vs. Euclidian Fuzzy Paths and Distances

The notion of a geodesic path between two points in a fuzzy subset has been introduced with respect to different purposes and formal settings, in both the continuous and digital fuzzy geometry. For instance, this notion is the main ingredient in the definition of the fuzzy distance transform (FDT), proposed in [8] by Saha et al. (2002).

A *path* π in $\Re^n$ from a point $x \in \Re^n$ to another point (not necessarily distinct) $y \in \Re^n$ is a continuous function π : $[0, 1] \rightarrow \Re^n$ such that $\pi(0)=x$ and $\pi(1)= y$. The *length* of a path π in S, denoted by $\Pi_S(\pi)$, is the value of the following integration

$$\Pi_S(\pi) = \int_0^1 \mu_S(\pi(t)) \left| \frac{d\pi(t)}{dt} \right| dt \tag{8}$$

i.e., $\Pi_S(\pi)$ is the integral of Euclidian distances weighted by the membership values (in S) along π.

For the path defined by the equations (2), (3) and (5), we have

$$\frac{d\pi(t)}{dt} = \begin{cases} \left(-2\rho_1^\theta \sin\theta, 2\rho_1^\theta \cos\theta\right) & t \in [0, 1/2] \\ \left((2\rho_1^\theta - 2\rho_0^\theta)\sin\theta, (-2\rho_1^\theta + 2\rho_0^\theta)\cos\theta\right) & t \in (1/2, 1] \end{cases} \tag{9}$$

$$\left| \frac{d\pi(t)}{dt} \right| = \begin{cases} 2\rho_1^\theta & t \in [0, 1/2] \\ 2\left|\rho_0^\theta - \rho_1^\theta\right| & t \in (1/2, 1] \end{cases} \quad \Rightarrow \quad \Pi_S(\pi) = \rho_1^\theta + \frac{1}{2}\left|\rho_0^\theta - \rho_1^\theta\right| \tag{10}$$

which means that the length of the path across the fuzzy boundary is contracted by a factor of 1/2 with respect to the length of the equivalent Euclidean path.

When a path passes through a low density (low membership) region, its length increases slowly and the portion of the path in the complement of the support of S contributes no length. This approach is useful to measure regional object depth, object thickness distribution, etc.

Let $\zeta_S(x, y)$ denote a subset of positive real numbers defined as

$$\zeta_S(x, y) = \{\Pi_S(\pi) \mid \pi \in P(x, y)\} \tag{11}$$

i.e., $\zeta_S(x, y)$ is the set of all possible path lengths in S between x and y. The *fuzzy distance* from $x \in \Re^n$ to $y \in \Re^n$ in S, denoted as $\omega_S(x, y)$, is the infimum of $\zeta_S(x, y)$; i.e.,

$$\omega_S(x, y) = \inf \zeta_S(x, y) \tag{12}$$

Actually, the fuzzy distance ω_S is a geodesic distance, which means that the shortest paths (when they exist) in a fuzzy subset S between two points $x, y \in \Re^n$ are not necessarily a straight line segment even when S is convex.

Concepts such as connectivity and geodesic distance between pixels or voxels in a 2D or 3D digital image have been proved to play a key role in the fuzzy digital geometry, since it has been introduced in [6] by Rosenfeld (1984).

There are two main approaches in measuring distances when considering fuzzy spatial objects: the first one basically compares only the membership functions representing the concerned fuzzy object, while the other one combines spatial distance between objects and membership functions, thus taking into account both spatial information and information related to the imprecision attached to the image object.

Distances between two points in a fuzzy set are typically addressed in order to find the best path in the geodesic sense in a spatial fuzzy set.

Distances from a point to a set are used when computing distance from a point to a complement of a fuzzy set, i.e., performing distance transform.

The distances between sets are used in shape matching.

A geodesic distance between points in a fuzzy set was introduced in [1] by Bloch (2000), being defined conditionally to a reference set *X*. It naturally incorporates some concepts involved in its crisp equivalent, such as Euclidian distance, path lengths and connectivity. Thus, a geodesic distance $d_X(x, y)$ from x to y is the length of a shortest path from *x* to *y*, completely included in *X*. Let μ be a fuzzy set on the space S. The definition of the geodesic distance relies on the degree of connectivity in μ between two points *x* and *y* of *S*, as defined by Rosenfeld (1984),

$$c_\mu(x, y) = \max_{L(x,y)} \left[\min_{t \in L(x,y)} \mu(t) \right] \tag{13}$$

where $L(x, y) = t_1, \ldots, t_n$ denotes a path from $x = t_1$ to $y = t_n$, consisting of a sequence of points in *S* according to the discrete connectivity defined on *S*. Let $L^*(x, y)$ denote the shortest path between *x* and *y* on which c_μ is reached; this path is not necessarily unique and can be interpreted as a geodesic path descending as little as possible in terms of membership degrees. Let $l(L^*(x, y))$ denote its length (the number of points along the path). Then the geodesic distance in μ between *x* and *y* is defined as

$$d_\mu(x, y) = \frac{l\left(L^*(x, y)\right)}{c_\mu(x, y)} \tag{14}$$

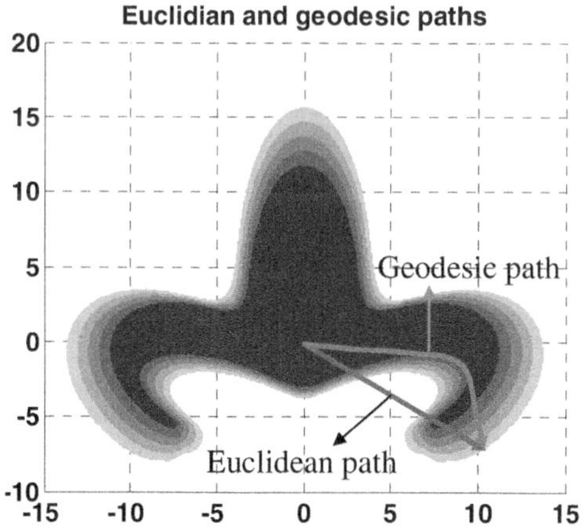

Fig. 9. Euclidean vs. geodesic paths in a non-convex, non-star-shaped fuzzy region

If $c_\mu(x, y) = 0$, then $d_\mu(x, y) = \infty$, which corresponds to the result obtained for the classical geodesic distance in the case where x and y belong to different connected components.

Figure 9 contrasts the Euclidean and geodesic paths between points in a non-convex, non-star-shaped fuzzy region.

3 A Novel Approach to Decomposing and Reconstructing a Fuzzy Shape and to Automatic Change Point-Detection, Based on SSA

In case of fuzzy shapes that are not star-shaped, abstracting a fuzzy signature based on a radial fuzzy distance is inappropriate. A fuzzy region may has the centroid belonging to its interior (or not!), while a fuzzy closed contour typically has the centroid belonging to its complement. For a connected fuzzy region with the centroid belonging to its interior, a fuzzy geodesic path (not necessarily unique) usually exists. However, geodesic paths cannot be used for the decomposition and the reconstruction of a fuzzy shape because they use weighted distances and thus may severely distort the reconstructed shape, when comparing to the original coordinates. On the other hand, a geodesic path is meaningless in case of fuzzy closed curves, when the centroid typically belongs to the complement of the fuzzy object. A Euclidean radial path is also inappropriate in case of non-star-shaped fuzzy shapes since the path may intersect the fuzzy boundary many times (see Figure 9).

The proposed novel strategy is to use the coordinates themselves as a shape signature, but represented as complex-valued numbers in the complex plane. A complex-valued fuzzy time series can then be sampled from such a continuous complex-valued fuzzy signature, allowing the machinery of time series analysis to be subsequently used. The method is general, that is, not constrained to convex, or star-shaped fuzzy regions.

Both denoising and change-point detection can be carried out by a powerful method called Singular-Spectrum Analysis (SSA). SSA is a nonparametric method for time series structure recognition and identification ([4]). The SSA algorithm has two basic stages: decomposition and reconstruction. Basically, it first builds the trajectory matrix associated to a time series, whose columns are formed by a sequence o lagged vectors extracted from the time series with a sliding window. Afterward, the *singular value decomposition* is applied to the trajectory matrix. For the reconstruction of the de-noised part of the time series the most dominant singular values (and the corresponding singular vectors) are considered, whereas the remaining singular values are used to compute the residuals (associated with noise).

Let start by assuming that the fuzzy object is represented in terms of α – supports. There are two distinct cases:

1. the case of a fuzzy region, where the membership function is non-increasing away from the interior of the object: assuming a counterclockwise rotation along the fuzzy boundary, one complex-valued α – level time series can be sampled from the continuous complex-valued α – support contour, which can be written as $z_t^\alpha = x_t^\alpha + i\, y_t^\alpha \in C$, for $t = 1, \ldots, N$. Here α is chosen from a finite non-increasing sequence $\alpha_1 = 1 < \alpha_2 < \ldots < \alpha_n = 0$.

2. the case of a fuzzy closed curve, where the membership function is typically LR-shaped: assuming a counterclockwise rotation along the fuzzy boundary, two complex-valued $\alpha-$level time series can be sampled from the continuous complex-valued $\alpha-$support contour, a left-side one and a right-side one, i.e., $(z_t^\alpha)^L = (x_t^\alpha)^L + i(y_t^\alpha)^L \in C$, $(z_t^\alpha)^R = (x_t^\alpha)^R + i(y_t^\alpha)^R \in C$, for $t = 1, \ldots, N$. Eventually, if the core reduces to a singleton, then $(z_t^1)^L = (z_t^1)^R$.

Now, a complex-valued trajectory matrix can be defined for each complex-valued time series in turn:

$$X^\alpha = \begin{pmatrix} z_1^\alpha & z_2^\alpha & \cdots & z_K^\alpha \\ z_2^\alpha & z_3^\alpha & \cdots & z_{K+1}^\alpha \\ \vdots & \vdots & \vdots & \vdots \\ z_M^\alpha & z_{M+1}^\alpha & \cdots & z_N^\alpha \end{pmatrix}, \quad z_t^\alpha \in C \tag{15}$$

The proposed generalization of SSA is founded on the ACM Algorithm 358 for Singular Value Decomposition of a complex matrix. The decomposition theorem (Businger and Golub, [2]) can be stated as follows: each and every $M \times K$ complex-valued matrix X can be reduced to diagonal form by unitary transformations U and V, $X = U\, diag[\sigma_1, \ldots, \sigma_K] V^H$, where $\sigma_1 \geq \ldots \geq \sigma_K \geq 0$ are real-valued scalars, called the singular values of X. Here U is an $M \times K$ column orthogonal matrix, V an $K \times K$ unitary matrix and V^H is a Hermitian transpose of V. The columns of U and V are called the left and right singular vectors of X, respectively.

As it is shown in Figures 10 and 11, the proposed generalization of SSA-based algorithm for complex-valued trajectory matrices can be successfully applied to decomposing and reconstructing (de-noising) non star-shaped fuzzy contours.

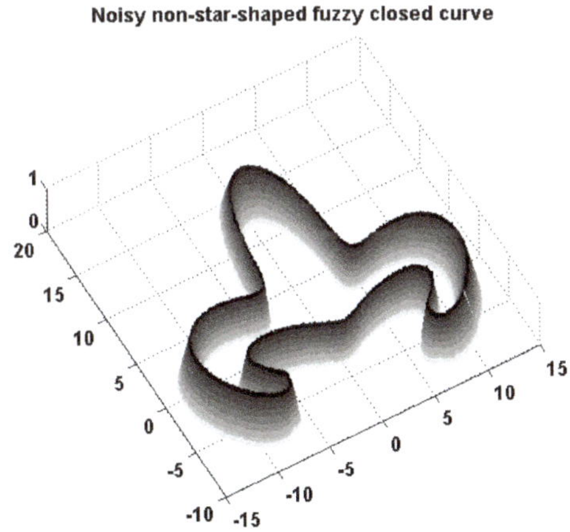

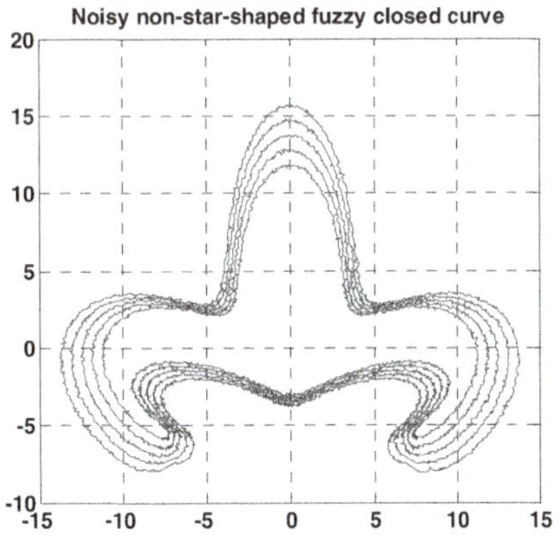

Fig. 10. The membership function of a noisy, non-star-shaped fuzzy closed curve (left); several α-cuts (right)

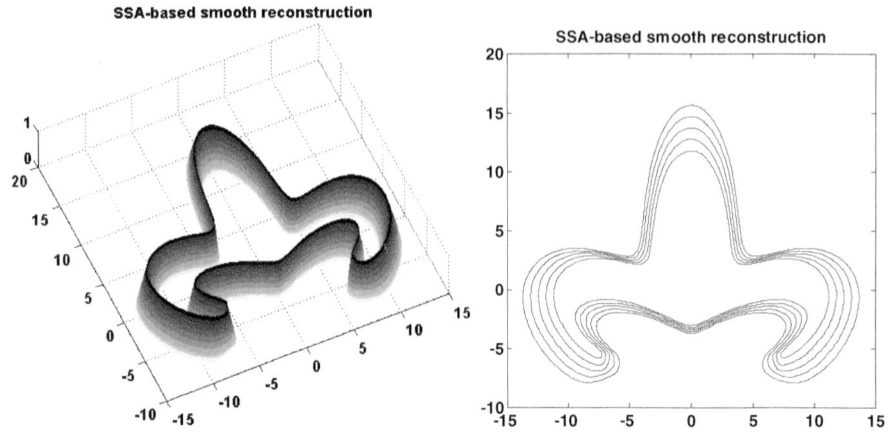

Fig. 11. The SSA-based smooth reconstruction of a non-star-shaped fuzzy closed curve (left); several α-cuts (right)

SSA is also at the core of a powerful change-point detection algorithm ([5]). The detection statistic is defined with respect to the squared distance to the subspace spanned by the eigenvectors of the lag-covariance matrix, computed in a sequence of moving time intervals. The novelty is that all calculations are adapted to be made in complex spaces (for instance, the distance between complex numbers is considered).

The detected change-points and the distance-based test statistics are shown in Figure 12. The location of change points is in the local minima of test statistic function.

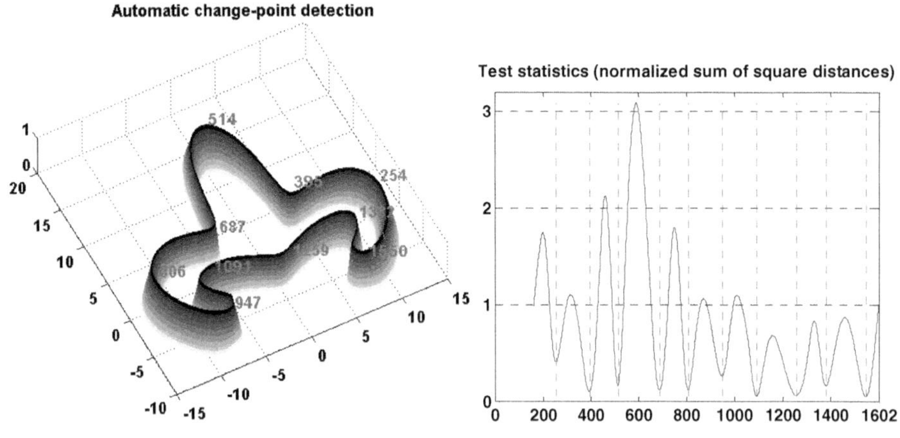

Fig. 12. Automatic selection of landmark positions through change-point detection (left); detecting change points from Distance detection statistic (right)

4 Conclusion

This paper extends my approach in [3] to a fuzzy context. SSA is generalized to deal with complex-valued trajectory matrices, encoding the coordinates themselves as complex numbers, in order to carry out the decomposition and reconstruction of a fuzzy closed contour, as well as change-point detection. This approach is suitable for any kind of fuzzy contours, including non-convex and non-star-shaped ones.

References

1. Bloch, I.: Geodesic balls in a fuzzy geodesic mathematical morphology. Pattern Recognition 33, 897–905 (2000)
2. Businger, P.A., Golub, G.H.: Algorithm 358: Singular value decomposition of a complex matrix. Comm. ACM 12, 564–565 (1969)
3. Georgescu, V.: A Novel and Effective Approach to Shape Analysis: Nonparametric Representation, De-noising and Change-Point Detection, Based on Singular-Spectrum Analysis. In: Torra, V., et al. (eds.) MDAI 2011. LNCS(LNAI), vol. 6820, pp. 162–173. Springer, Heidelberg (2011)
4. Goljadina, N., Nekrutkin, V., Zhigljavky, A.: Analysis of Time Series Structure: SSA and related techniques. Chapman and Holl, London (2001)
5. Moskvina, V.: Applications of the singular-spectrum analysis for change point detection in time series. Ph.D. thesis, School of Mathematics, Cardiff University, Cardiff (2001)
6. Rosenfeld, A.: The fuzzy geometry of image subsets. Pattern Recognition Letters 2, 311–317 (1984)
7. Rosenfeld, A.: Fuzzy geometry: An updated overview. Information Sciences 110, 127–133 (1998)
8. Saha, P.K., Wehrli, F.W., Gomberg, B.R.: Fuzzy distance transform: Theory, algorithms, and applications. Computer Vision and Image Understanding 86, 171–190 (2002)

Possibilistic Linear Programming Using General Necessity Measures Preserves the Linearity

Masahiro Inuiguchi

Graduate School of Engineering Science, Osaka University
1-3 Machikaneyama, Toyonaka, Osaka 560-8531, Japan
inuiguti@sys.es.osaka-u.ac.jp
http://www-inulab.sys.es.osaka-u.ac.jp/

Abstract. In this paper, a robust optimization approach to possibilistic linear programming problems is studied. After necessity measures and generation processes of logical connectives are reviewed, the necessity fractile optimization model of possibilistic linear programming problem is introduced as a robust optimization model. This problem is reduced to a linear semi-infinite programming problem. Assuming the convexity of the right parts of membership functions of fuzzy coefficients and the concavity of membership functions of fuzzy constraints, we investigate conditions on logical connectives for the problems to be reduced to linear programming problems. Several examples are given to demonstrate that necessity fractile optimization models are often reduced to linear programming problems.

Keywords: possibilistic linear programming, necessity measure, implication function, conjunction function.

1 Introduction

Fuzzy and possibilistic programming approaches are proposed to mathematical programming problems with ambiguity and vagueness [1,2]. By those approaches, we obtain reasonable solutions under conflicting soft constraints and goals, robust solutions under hard and soft constraints, optimistic solutions of attaining high-level goals, and so on. In possibilistic programming approaches, possibility and necessity measures are used to reduce the problems to conventional programming problems. Many results demonstrate that possibilistic linear programming problems preserve the linearity in the reduced problems when possibility and necessity measures are defined by minimum operation and Dienes implication, respectively. However, cases with the other conjunction and implication functions have not yet considerably investigated. have been proposed in calculation of linear functions with fuzzy coefficients. Inuiguchi [3] showed that the necessity fractile optimization models of possibilistic linear programming problems with soft constraints can be reduced to linear semi-infinite programming problems even when necessity measures are not defined by Dienes implication.

In this paper, we further develop the results by Inuiguchi [3]. We investigate the cases when the necessity fractile optimization models are reduced to linear

V. Torra et al. (Eds.): MDAI 2011, LNAI 6820, pp. 186–197, 2011.

programming problems. Assuming the convexity of the right parts of membership functions of fuzzy coefficients and the concavity of membership functions of fuzzy constraints, we show that the problems are reduced to linear programming problems when functions induced from implication functions defining necessity measures are convex. The results imply that the necessity fractile optimization models with many famous implication functions are reduced to linear programming problems when membership functions of fuzzy constraints are concave.

In next section, we briefly review the necessity measures and generation processes of logical connectives. Possibilistic linear programming problems with soft constraints are given and formulated as conventional programming problems through necessity fractile optimization models in Section 3. We show that the problems are reduced to linear semi-infinite programming problems. In Section 4, assuming the convexity of the right parts of membership functions of fuzzy coefficients and the concavity of membership functions of fuzzy constraints, we investigate the conditions that the problems are further reduced to linear programming problems. The results are applied to cases when implications defining necessity measures are generated from conjunction and negation functions. In Section 5, several examples are given to demonstrate that necessity fractile optimization models are often reduced to linear programming problems. In Section 6, concluding remarks are given.

2 Necessity Measures and Logical Connectives

A necessity measure [4] of a fuzzy set S under a fuzzy set V is defined by

$$N_V(S) = \inf_{u\in U} I(\mu_V(u), \mu_S(u)), \tag{1}$$

where μ_V and μ_S are membership functions of V and S. $I : [0,1] \times [0,1] \to [0,1]$ is an implication function satisfying the following properties:

(I0) I is upper semi-continuous, (semi-continuity)
(I1) $I(0,0) = I(0,1) = I(1,1) = 1$ and $I(1,0) = 0$, (boundary condition)
(I2) $I(a,b) \le I(c,d)$ if $0 \le c \le a \le 1$ and $0 \le b \le d \le 1$. (monotonicity)

$N_V(S)$ evaluates to what extent an uncertain variable u whose possible range is V surely takes a value in S. Moreover, it can also be understood as the degree of inclusion $V \subseteq S$.

As is shown in (1), a necessity measure is defined by an implication function. In the literature [5,6], implication functions are known to be generated from a conjunction function T and a strong negation n. In this paper, a conjunction function is defined as a two-place function $T : [0,1] \times [0,1] \to [0,1]$ satisfying

(T0) T is lower semi-continuous, (semi-continuity)
(T1) $T(0,0) = T(0,1) = T(1,0) = 0$ and $T(1,1) = 1$, (boundary condition)
(T2) $T(a,b) \le T(c,d)$ if $0 \le a \le c \le 1$ and $0 \le b \le d \le 1$. (monotonicity)

A conjunction function T satisfies the following properties (t1) $T(a,1) = T(1,a) = a$ for any $a \in [0,1]$, (T3) $T(a,b) = T(b,a)$ for any $a,b \in [0,1]$ (commutativity) and (T4) $T(a, T(b, c)) = T(T(a, b), c)$ for any $a, b, c \in [0, 1]$ (associativity) is known as a triangular norm. A strong negation is a continuous strictly decreasing function $n : [0, 1] \to [0, 1]$ such that (n1) $n(0) = 1$ (boundary condition) and (n2) $n(n(a)) = a$ for any $a \in [0, 1]$ (involution).

Given a conjunction function T and a strong negation n, the following three kinds of implication functions can be generated:

$$I^{\mathrm{R}}[T](a, b) = \sup\{s \in [0, 1] \mid T(a, s) \leq b\}, \tag{2}$$

$$I^{\mathrm{S}}[T](a, b) = n(T(a, n(b))), \tag{3}$$

$$I^{\mathrm{r-R}}[T](a, b) = \sup\{s \in [0, 1] \mid T(n(b), s) \leq n(a)\}. \tag{4}$$

The first one, I^{R}, is encountered in the maximum solution of a fuzzy relation equation and understood in view of modus ponens. The second one, I^{S}, is introduced in analogy to Boolean logic. The last one, $I^{\mathrm{r-R}}$ is reciprocal to the first one which is obtained by taking a contraposition of the first one. When T is a t-norm, $I^{\mathrm{R}}[T]$, $I^{\mathrm{S}}[T]$ and $I^{\mathrm{r-R}}[T]$ are called R-implication (residual implication), S-implication and reciprocal R-implication, respectively. Whereas I^{S} produces an implication function from an arbitrary conjunction function T, I^{R} and $I^{\mathrm{r-R}}$ produce an implication function from a conjunction function which satisfies

$$T(1, a) > 0 \text{ for any } a > 0. \tag{5}$$

On the other hand, a conjunction function can be generated from an implication function through a transformation,

$$T^{\mathrm{I}}[I](a, b) = n(I(a, n(b))). \tag{6}$$

This transformation is symmetrical to I^{S}. From an implication function, a conjunction function is produced through T^{I}. A conjunction function in this paper is not commutative. Thus, a new conjunction function may be generated from a conjunction function through

$$T^{\mathrm{T}}[T](a, b) = T(b, a). \tag{7}$$

For the transformations (1) to (3), (5) and (6), we have

$$T^{\mathrm{I}} \circ I^{\mathrm{S}} = \text{id.}, \quad I^{\mathrm{S}} \circ T^{\mathrm{I}} = \text{id.}, \quad T^{\mathrm{T}} \circ T^{\mathrm{T}} = \text{id.}, \quad I^{\mathrm{S}} \circ T^{\mathrm{T}} \circ T^{\mathrm{I}} \circ I^{\mathrm{R}} = I^{\mathrm{r-R}}, \tag{8}$$

where '$\circ$' denotes a composition, for example, $T^{\mathrm{I}} \circ I^{\mathrm{R}}$ is a composite transformation of I^{R} and T^{I}, i.e., $T^{\mathrm{I}} \circ I^{\mathrm{R}}[T](a, b) = T^{\mathrm{I}}[I^{\mathrm{R}}[T]](a, b)$. The notation 'id.' stands for the identical transformation.

From (T0), we have the following equalities [6]:

$$\begin{cases} I^{\mathrm{R}} \circ T^{\mathrm{I}} \circ I^{\mathrm{R}}[T] = I^{\mathrm{S}}[T], & I^{\mathrm{r-R}} \circ T^{\mathrm{I}} \circ I^{\mathrm{R}} \circ T^{\mathrm{T}}[T] = I^{\mathrm{S}}[T], \\ I^{\mathrm{R}} \circ T^{\mathrm{I}} \circ I^{\mathrm{r-R}} \circ T^{\mathrm{T}}[T] = I^{\mathrm{r-R}}[T], & I^{\mathrm{r-R}} \circ T^{\mathrm{I}} \circ I^{\mathrm{r-R}} \circ T^{\mathrm{T}}[T] = I^{R}[T]. \end{cases} \tag{9}$$

Equations (8)–(9) are summed up by Figure 1. As shown in Figure 1, the generation process from a lower semi-continuous conjunction function as well as from an upper semi-continuous implication function is closed. Note that the semi-continuity is preserved through the generation process [6].

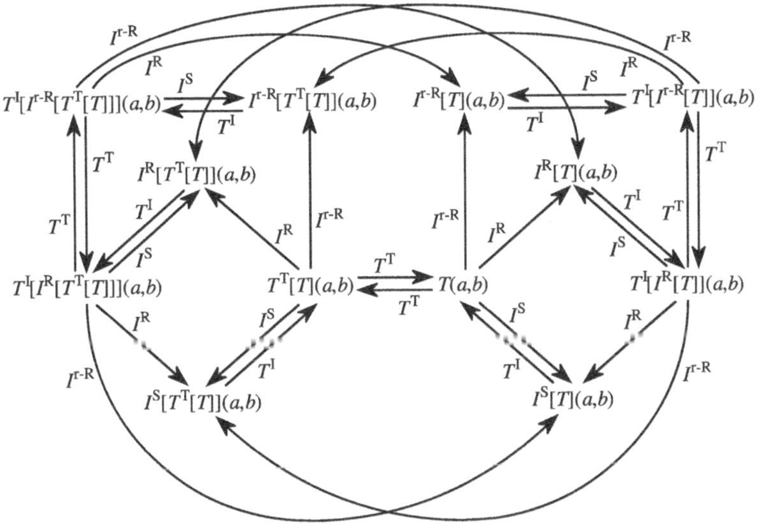

Fig. 1. Conjunction and implication generations are closed

3 Possibilistic Linear Programming Problems

We consider the following possibilistic linear programming problem,

$$\begin{array}{rl} \text{maximize} & \boldsymbol{c}^{\mathrm{T}}\boldsymbol{x}, \\ \text{subject to} & \boldsymbol{a}_i^{\mathrm{T}}\boldsymbol{x} \lesssim_i b_i,\ i = 1, 2, \ldots, m, \\ & \boldsymbol{x} \geq \boldsymbol{0}, \end{array} \tag{10}$$

where $\boldsymbol{x} = (x_1, x_2, \ldots, x_n)^{\mathrm{T}}$ is a decision vector. b_i, $i = 1, 2, \ldots, m$ are constants. c_j and a_{ij} of $\boldsymbol{c}$ and $\boldsymbol{a}_i$ are not known exactly but the possible ranges of those values are given by fuzzy numbers C_j and A_{ij}, respectively. A fuzzy number is a normal, convex and bounded fuzzy set on the real line whose membership function is upper semi-continuous. The notation $\lesssim_i$ is a fuzzified inequality so that $\lesssim_i b_i$ corresponds to a fuzzy set B_i with verbal expression 'a set of real numbers roughly smaller than b_i'. We assume that the membership function μ_{B_i} of B_i is non-increasing and upper semi-continuous and satisfies $\mu_{B_i}(b_i) = 1$.

By the extension principle, the possible ranges of $\boldsymbol{c}^{\mathrm{T}}\boldsymbol{x}$ and $\boldsymbol{a}_i^{\mathrm{T}}\boldsymbol{x}$ are obtained as fuzzy sets $\boldsymbol{C}^{\mathrm{T}}\boldsymbol{x}$ and $\boldsymbol{A}_i^{\mathrm{T}}\boldsymbol{x}$, respectively, defined by membership functions [7]:

$$\mu_{\boldsymbol{C}^{\mathrm{T}}\boldsymbol{x}}(y) = \sup_{\substack{r_1, \ldots, r_n \\ \boldsymbol{r}^{\mathrm{T}}\boldsymbol{x} = y}} \min\left(\mu_{C_1}(r_1), \ldots, \mu_{C_n}(r_n)\right), \tag{11}$$

$$\mu_{\boldsymbol{A}_i^{\mathrm{T}}\boldsymbol{x}}(y) = \sup_{\substack{r_1, \ldots, r_n \\ \boldsymbol{r}^{\mathrm{T}}\boldsymbol{x} = y}} \min\left(\mu_{A_{i1}}(r_1), \ldots, \mu_{A_{in}}(r_n)\right), \tag{12}$$

where $\boldsymbol{r} = (r_1, r_2, \ldots, r_n)^{\mathrm{T}}$.

Using a necessity measure N^i defined by an upper semi-continuous implication function I^i, in this paper, we formulate Problem (10) as a necessity fractile optimization model (see Inuiguchi and Ramík [2]):

$$\begin{aligned} \text{maximize } & q, \\ \text{subject to } & N^0_{\boldsymbol{C}^{\mathrm{T}}\boldsymbol{x}}([q,+\infty)) \geq \alpha^0, \\ & N^i_{\boldsymbol{A}_i^{\mathrm{T}}\boldsymbol{x}}(B_i) \geq \alpha^i,\ i=1,2,\ldots,m, \\ & \boldsymbol{x} \geq \boldsymbol{0}, \end{aligned} \tag{13}$$

where q is an auxiliary variable. $\alpha^0 \in (0,1]$ and $\alpha^i \in (0,1]$, $i=1,2,\ldots,m$ are certainty levels of goal achievement and constraint satisfactions specified by the decision maker. Similar to the percentile of a probability distribution, under a given $\boldsymbol{x}$, the largest value q satisfying $N^0_{\boldsymbol{C}^{\mathrm{T}}\boldsymbol{x}}([q,+\infty)) \geq \alpha^0$ is called the α_0-necessity fractile [2]. Problem (13) is a maximization problem of the α_0-necessity fractile under constraints that the necessity measure of the event that $\boldsymbol{a}_i^{\mathrm{T}}\boldsymbol{x}$ is roughly smaller than b_i is not less than α^i for $i=1,2,\ldots,m$. We note

$$N^0_{\boldsymbol{C}^{\mathrm{T}}\boldsymbol{x}}([q,+\infty)) = \inf_{r<q} I^0(\mu_{\boldsymbol{C}^{\mathrm{T}}\boldsymbol{x}}(r),0),\quad N^i_{\boldsymbol{A}_i^{\mathrm{T}}\boldsymbol{x}}(B_i) = \inf_r I^i(\mu_{\boldsymbol{A}_i^{\mathrm{T}}\boldsymbol{x}}(r),\mu_{B_i}(r)). \tag{14}$$

The selections of necessity measures and certainty levels depend on the required robustness of goal achievement/constraint satisfactions, the meanings of total goal achievement/constraint satisfactions, the estimation of fuzzy coefficients and so on. The method proposed by Inuiguchi et al. [8] would be useful for selecting suitable necessity measures.

We note also that in most of previous studies on possibilistic linear programming, a necessity measure defined by Dienes implication ($I(a,b)=\max(1-a,b)$) is used and there is almost no study on possibilistic linear programming using general necessity measures except for Inuiguchi [3].

Let $[S]_\alpha$ be an α-level set of a fuzzy set S, i.e., $[S]_\alpha = \{u \in U \mid \mu_S(u) \geq \alpha\}$. Then, because fuzzy numbers C_i and A_{ij} are bounded and have upper semi-continuous membership functions, we have (see Dubois and Prade [7])

$$[\boldsymbol{C}^{\mathrm{T}}\boldsymbol{x}]_\alpha = \sum_{j=1}^n [C_j]_\alpha x_j,\quad [\boldsymbol{A}_i^{\mathrm{T}}\boldsymbol{x}]_\alpha = \sum_{j=1}^n [A_{ij}]_\alpha x_j. \tag{15}$$

Let $c_j^{\mathrm{L}}(\alpha) = \inf[C_j]_\alpha$, $c_j^{\mathrm{R}}(\alpha) = \sup[C_j]_\alpha$, $a_{ij}^{\mathrm{L}}(\alpha) = \inf[A_{ij}]_\alpha$ and $a_{ij}^{\mathrm{R}}(\alpha) = \sup[A_{ij}]_\alpha$. Then, considering the non-negativity of $\boldsymbol{x}$, we have

$$[\boldsymbol{C}^{\mathrm{T}}\boldsymbol{x}]_\alpha = \left[\sum_{j=1}^n c_j^{\mathrm{L}}(\alpha)x_j, \sum_{j=1}^n c_j^{\mathrm{R}}(\alpha)x_j\right],\quad [\boldsymbol{A}_i^{\mathrm{T}}\boldsymbol{x}]_\alpha = \left[\sum_{j=1}^n a_{ij}^{\mathrm{L}}(\alpha)x_j, \sum_{j=1}^n a_{ij}^{\mathrm{R}}(\alpha)x_j\right]. \tag{16}$$

Inuiguchi [3] proved the following theorem.

Theorem 1. *Let N^i be a necessity measure defined by an implication function I^i. Then for any fuzzy sets V and S of a universal set U, we have*

$$N^i_V(S) \geq \alpha \Leftrightarrow \forall \beta \in [0,1];\ [V]_\beta \subseteq [S]_{f^i(\beta,\alpha)}, \tag{17}$$

where $f^i(\beta,\alpha) = T^{\mathrm{I}}[I^{\mathrm{R}}[T^{\mathrm{I}}[I^i]]](\beta,\alpha) = \inf\{s \in [0,1] \mid I(\beta,s) \geq \alpha\}$.

From the assumptions of B_i, we have $[B_i]_\beta = (-\infty, \bar{b}_i(\beta)]$, where $\bar{b}_i : [0,1] \to [0,+\infty)$ is defined by $\bar{b}_i(\beta) = \sup\{r \mid \mu_{B_i}(r) \geq \beta\}$. From (16) and Theorem 1, Problem (13) is reduced to a linear semi-infinite programming problem,

$$\begin{aligned} &\text{maximize} \ q, \\ &\text{subject to} \ \sum_{j=1}^{n} c_j^{\mathrm{L}}(\beta)x_j \geq \bar{q}(f^0(\beta,\alpha^0)), \ \forall\beta \in [0,1], \\ &\qquad\qquad \sum_{j=1}^{n} a_{ij}^{\mathrm{R}}(\beta)x_j \leq \bar{b}_i(f^i(\beta,\alpha^i)), \ \forall\beta \in [0,1], \ i = 1,2,\ldots,m, \\ &\qquad\qquad \boldsymbol{x} \geq \mathbf{0}, \end{aligned} \tag{18}$$

where we assume $-\infty \geq -\infty$ and $\infty \leq \infty$. $\bar{q} : [0,1] \to \{-\infty, q\}$ is defined by

$$\bar{q}(\beta) = \begin{cases} q \ \text{if} \ \beta > 0, \\ -\infty \ \text{if} \ \beta = 0. \end{cases} \tag{19}$$

4 Reduction to a Linear Programming Problem

In this section, we show that Problem (13) is further reduced to a linear programming problem under certain conditions. We note that the membership function $\mu_{A_{ij}}$ of A_{ij} can be decomposed to a non-decreasing function $\mu_{A_{ij}}^{\mathrm{L}} : \mathbf{R} \to [0,1]$ and a non-increasing function $\mu_{A_{ij}}^{\mathrm{R}} : \mathbf{R} \to [0,1]$ such that $\mu_{A_{ij}}(r) = \min(\mu_{A_{ij}}^{\mathrm{L}}(r), \mu_{A_{ij}}^{\mathrm{R}}(r))$, for all $r \in \mathbf{R}$. Indeed, they are obtained by

$$\mu_{A_{ij}}^{\mathrm{L}}(r) = \begin{cases} \mu_{A_{ij}}(r) & \text{if } r < a_{ij}^{\mathrm{L}}(1), \\ 1 & \text{if } r \geq a_{ij}^{\mathrm{L}}(1), \end{cases} \quad \mu_{A_{ij}}^{\mathrm{R}}(r) = \begin{cases} 1 & \text{if } r \leq a_{ij}^{\mathrm{R}}(1), \\ \mu_{A_{ij}}(r) & \text{if } r > a_{ij}^{\mathrm{R}}(1). \end{cases} \tag{20}$$

We define the following notations:

$$\hat{c}_j^{\mathrm{L}}(\alpha) = \inf_{\beta\in[0,1]} \{c_j^{\mathrm{L}}(\beta) \mid f^0(\beta,\alpha) > 0\}, \ j = 1,2,\ldots,n, \tag{21}$$

$$\hat{a}_{ij}^{\mathrm{R}}(\alpha) = \sup_{\beta\in[0,1]} \{a_{ij}^{\mathrm{R}}(\beta) \mid f^i(\beta,\alpha) > 0\}, \ i = 1,2,\ldots,m, \ j = 1,2,\ldots,n \tag{22}$$

$$\hat{b}_i(\alpha) = \sup_{\beta\in[0,1]} \{\bar{b}_i(f^i(\beta,\alpha)) \mid f^i(\beta,\alpha) > 0\}, \ i = 1,2,\ldots,m \tag{23}$$

$$\hat{\beta}^i(\alpha) = \inf\{\beta \in [0,1] \mid f^i(\beta,\alpha) = f^i(1,\alpha)\}, \ i = 0,1,\ldots,m. \tag{24}$$

The following theorem can be obtained straightforwardly.

Theorem 2. *Assume images of* $\mu_{A_{ij}}^{\mathrm{R}}$*'s and* μ_{B_i}*'s include* $(0,1]$*, i.e.,* $(0,1] \subseteq \mu_{A_{ij}}^{\mathrm{R}}(\mathbf{R})$*,* $i = 1,2,\ldots,m$*,* $j = 1,2,\ldots,n$*. and* $(0,1] \subseteq \mu_{B_i}(\mathbf{R})$*,* $i = 1,2,\ldots,m$*. If* $a_{ij}^{\mathrm{R}}(\beta)$ $(i = 1,2,\ldots,m,\ j = 1,2,\ldots,n)$ *and* $\bar{b}_i(f^i(\beta,\alpha^i))$ $(i = 1,2,\ldots,n)$ *are* *convex and concave with respect to β in the range $(0, \hat{\beta}^i(\alpha^i))$, respectively, then Problem (18) is reduced to the following linear programming problem:*

$$\begin{aligned} \text{maximize } & \sum_{j=1}^{n} \hat{c}_j^{\mathrm{L}}(\alpha^0)x_j, \\ \text{subject to } & \sum_{j=1}^{n} \hat{a}_{ij}^{\mathrm{R}}(\alpha^i)x_j \le \hat{b}_i(\alpha^i), \ i = 1, 2, \ldots, m, \\ & \sum_{j=1}^{n} a_{ij}^{\mathrm{R}}(\hat{\beta}^i(\alpha^i))x_j \le \bar{b}_i(f^i(\hat{\beta}^i(\alpha^i), \alpha^i)), \ i = 1, 2, \ldots, m, \\ & \boldsymbol{x} \ge \mathbf{0}. \end{aligned} \tag{25}$$

(**Proof**) The first constraints of Problem (18), i.e., $\sum_{j=1}^{n} c_j^{\mathrm{L}}(\beta)x_j \ge \bar{q}(f^0(\beta, \alpha^0))$, $\forall \beta \in [0, 1]$ can be reduced to $\sum_{j=1}^{n} c_j^{\mathrm{L}}(\beta)x_j \ge q, \forall \beta \in [0, 1]$ such that $f^0(\beta, \alpha^0) > 0$. Then from (19), this constraints are equivalent to

$$\sum_{j=1}^{n} \hat{c}_j^{\mathrm{L}}(\alpha^0)x_j = \sum_{j=1}^{n} \left(\inf_{\substack{\beta \in [0,1] \\ f^0(\beta, \alpha^0) > 0}} c_j^{\mathrm{L}}(\beta) \right) x_j = \inf_{\substack{\beta \in [0,1] \\ f^0(\beta, \alpha^0) > 0}} \left(\sum_{j=1}^{n} c_j^{\mathrm{L}}(\beta)x_j \right) \ge q.$$

Similarly, by the assumption of the images of $\mu_{A_{ij}}^{\mathrm{R}}$'s and μ_{B_i}'s, the convexity of $a_{ij}^{\mathrm{R}}(\beta)$ $(i = 1, 2, \ldots, m, \ j = 1, 2, \ldots, n)$ and the concavity of $\bar{b}_i(f^i(\beta, \alpha^i))$ $(i = 1, 2, \ldots, m)$, the second constraints of Problem (18), $\sum_{j=1}^{n} a_{ij}^{\mathrm{R}}(\beta)x_j \le \bar{b}_i(f^i(\beta, \alpha^i))$, $\forall \beta \in [0, 1]$ are reduced to the following two constraints:

$$\begin{aligned} \sum_{j=1}^{n} \hat{a}_{ij}^{\mathrm{R}}(\alpha^i)x_j &= \sum_{j=1}^{n} \left(\sup_{\substack{\beta \in [0,1] \\ f^i(\beta, \alpha^i) > 0}} a_{ij}^{\mathrm{R}}(\beta) \right) x_j = \sup_{\substack{\beta \in [0,1] \\ f^i(\beta, \alpha^i) > 0}} \left(\sum_{j=1}^{n} a_{ij}^{\mathrm{R}}(\beta)x_j \right) \\ &\le \sup_{\substack{\beta \in [0,1] \\ f^i(\beta, \alpha^i) > 0}} \bar{b}_i(f^i(\beta, \alpha^i)) = \hat{b}_i(\alpha^i), \end{aligned}$$

$$\sum_{j=1}^{n} a_{ij}^{\mathrm{R}}(\hat{\beta}^i(\alpha^i))x_j \le \bar{b}_i(f^i(\hat{\beta}^i(\alpha^i), \alpha^i)).$$

Then Problem (18) is reduced to Problem (25). (Q.E.D.)

By Theorem 2, we know that there are cases when Problem (13) is reduced to a linear programming problem. However, the sufficient condition shown in Theorem 2 is rather complex. Then we break down the condition into a combination of simple conditions sacrificing its generality to a small extent.

A1. $(0, 1] \subseteq \mu_{A_{ij}}^{\mathrm{R}}(\mathbf{R})$ and $\mu_{A_{ij}}^{\mathrm{R}}$ of A_{ij} $(i = 1, 2, \ldots, m, \ j = 1, 2, \ldots, n)$ is convex, i.e., $\mu_{A_{ij}}^{\mathrm{R}}(\lambda y_1 + (1 - \lambda)y_2) \le \lambda\mu_{A_{ij}}^{\mathrm{R}}(y_1) + (1 - \lambda)\mu_{A_{ij}}^{\mathrm{R}}(y_2)$ for all $\lambda \in [0, 1]$ and for any y_1, $y_2 \in \mathbf{R}$.

A2. $(0, 1] \subseteq \mu_{B_i}(\mathbf{R})$ and μ_{B_i} of B_i $(i = 1, 2, \ldots, m)$ is concave, i.e., $\mu_{B_i}(\lambda y_1 + (1 - \lambda)y_2) \ge \lambda\mu_{B_i}(y_1) + (1 - \lambda)\mu_{B_i}(y_2)$ for all $\lambda \in [0, 1]$ and for any y_1, $y_2 \in \mathbf{R}$.

Note that those assumptions are satisfied when A_{ij} and B_i have linear membership functions. Under these assumption, we have the following lemma.

Lemma 1. *Under assumptions A1 and A2, $a_{ij}^{\mathrm{R}} : [0,1] \to \mathbf{R}$ defined by $a_{ij}^{\mathrm{R}}(\alpha) = \sup[A_{ij}]_\alpha$ is convex in the domain $(0,1]$ and $\bar{b}_i : [0,1] \to \mathbf{R}$ defined by $\bar{b}_i(\beta) = \sup\{r \mid \mu_{B_i}(r) \geq \beta\}$ is concave in the domain $(0,1]$.*

(**Proof**). Both can be proved in the same way. We prove only the latter assertion. Let β_1, $\beta_2 \in (0,1]$ and $\lambda \in [0,1]$ be fixed arbitrarily. By the concavity and the upper semi-continuity of μ_{B_i}, we have

$$
\begin{aligned}
&\bar{b}_i(\lambda\beta_1 + (1-\lambda)\beta_2) = \sup\{r \mid \mu_{B_i}(r) \geq \lambda\beta_1 + (1-\lambda)\beta_2\} \\
&= \sup\{\lambda r_1 + (1-\lambda)r_2 \mid \mu_{B_i}(\lambda r_1 + (1-\lambda)r_2) \geq \lambda\beta_1 + (1-\lambda)\beta_2\} \\
&\geq \sup\{\lambda r_1 + (1-\lambda)r_2 \mid \lambda\mu_{B_i}(y_1) + (1-\lambda)\mu_{B_i}(y_2) \geq \lambda\beta_1 + (1-\lambda)\beta_2\} \\
&\geq \lambda \sup\{r_1 \mid \mu_{B_i}(r_1) \geq \beta_1\} + (1-\lambda)\sup\{r_2 \mid \mu_{B_i}(r_2) \geq \beta_2\} \\
&= \lambda\bar{b}_i(\beta_1) + (1-\lambda)\bar{b}_i(\beta_2).
\end{aligned}
$$

Then $\bar{b}_i$ is concave in the domain $(0,1]$. (Q.E.D.)

We obtain the following theorem.

Theorem 3. *In addition to assumptions A1 and A2, we assume that function $f^i(\cdot,\alpha) : [0,1] \to [0,1]$ is convex in the range $(0, \hat{\beta}^i(\alpha^i))$ for a fixed parameter $\alpha \in (0,1]$, i.e., $f^i(\lambda\beta_1 + (1-\lambda)\beta_2, \alpha) \leq \lambda f^i(\beta_1,\alpha) + (1-\lambda)f^i(\beta_2,\alpha)$ for all $\lambda \in [0,1]$ and for any β_1, $\beta_2 \in (0, \hat{\beta}^i(\alpha^i))$ such that $f^i(\beta_1,\alpha)$, $f^i(\beta_2,\alpha) \in (0,1]$. The composite function $\bar{b}_i(f^i(\cdot,\alpha)) : [0,1] \to \mathbf{R}$ is concave.*

(**Proof**). Let β_1, $\beta_2 \in (0, \hat{\beta}^i(\alpha^i))$ such that $f^i(\beta_1,\alpha)$, $f^i(\beta_2,\alpha) \in (0,1]$ be fixed arbitrarily. Let $\lambda \in [0,1]$ be fixed arbitrarily. Obviously, $\bar{b}_i$ is decreasing. Then by the convexity of $f^i(\cdot,\alpha)$ and the concavity of $\bar{b}_i$, we have

$$
\begin{aligned}
\bar{b}_i(f^i(\lambda\beta_1 + (1-\lambda)\beta_2, \alpha)) &\geq \bar{b}_i(\lambda f^i(\beta_1,\alpha) + (1-\lambda)f^i(\beta_2,\alpha)) \\
&\geq \lambda\bar{b}_i(f^i(\beta_1,\alpha)) + (1-\lambda)\bar{b}_i(f^i(\beta_2,\alpha)).
\end{aligned}
$$

Therefore, $\bar{b}_i(f^i(\cdot,\alpha))$ is concave. (Q.E.D.)

From Lemma 1 and Theorem 3, we know that the assumption of Theorem 2 is satisfied when $f^i(\cdot,\alpha) : [0,1] \to [0,1]$ is convex in the range $(0, \hat{\beta}^i(\alpha^i))$ for a fixed parameter $\alpha \in (0,1]$ under assumptions A1 and A2.

Now we investigate cases when the implication functions defining necessity measures are defined by a conjunction function T^i and a strong negation n^i. Under the assumption of the concavity of μ_{B_i} in the range $(0,1]$, we investigate the condition that $f^i(\cdot,\alpha)$ is convex in the range $(0, \hat{\beta}^i(\alpha^i))$.

When the implication function I^i is $I^{\mathrm{R}}[T^i]$ of (2), we have

$$
f^i(\beta,\alpha) = T^{\mathrm{I}}[I^{\mathrm{R}}[T^{\mathrm{I}}[I^{\mathrm{R}}[T^i]]]](\beta,\alpha) = T^i(\beta,\alpha) \tag{26}
$$

because $T^{\mathrm{I}} \circ I^{\mathrm{S}} = \mathrm{id}$ and $I^{\mathrm{R}} \circ T^{\mathrm{I}} \circ I^{\mathrm{R}}[T] = I^{\mathrm{S}}[T]$ as shown in (8) and (9). Then the condition, the convexity of $f^i(\cdot,\alpha)$ in the range $(0, \hat{\beta}^i(\alpha^i))$ is equivalent to the convexity of $T^i(\cdot,\alpha)$ in the range $(0, \hat{\beta}^i(\alpha^i))$.

When the implication function I^i is $I^{\mathrm{S}}[T^i]$ of (3), we have

$$\begin{aligned} f^i(\beta,\alpha) &= T^{\mathrm{I}}[I^{\mathrm{R}}[T^{\mathrm{I}}[I^{\mathrm{S}}[T^i]]]](\beta,\alpha) = T^{\mathrm{I}}[I^{\mathrm{R}}[T^i]](\beta,\alpha) \\ &= n^i(\sup\{s \in [0,1] \mid T^i(\beta,s) \le n^i(\alpha)\}). \end{aligned} \tag{27}$$

We define a set $BS(\alpha) \subseteq [0,1] \times [0,1]$ by

$$BS(\alpha) = \{(\beta,s) \in [0,1] \times [0,1] \mid T^i(\beta,s) \le \alpha\}. \tag{28}$$

and a function $\psi^{BS(\alpha)} : [0,1] \to [0,1]$ by

$$\psi^{BS(\alpha)}(\beta) = \sup\{s \in [0,1] \mid T^i(\beta,s) \le \alpha\}. \tag{29}$$

Then the following lemma is straightforwardly derived.

Lemma 2. *If $BS(\alpha)$ is a convex set, we know that a function $\psi^{BS(\alpha)}$ is a concave function. Moreover, if $[0,1] \times [0,1] - BS(\alpha)$ is a convex set, $\psi^{BS(\alpha)}$ is a convex function.*

As the result, we obtain the following theorem.

Theorem 4. *If T^i is quasi-concave and n^i is convex, $f^i(\cdot,\alpha)$ is also a convex function for any $\alpha \in [0,1]$.*

(**Proof**). Let $\alpha \in [0,1]$. From the quasi-convexity of T^i, $BS(n^i(\alpha))$ becomes a convex set. Then $\psi^{BS(n^i(\alpha))}$ becomes a concave function. Let β^1, $\beta^2 \in [0,1]$ and $\lambda \in [0,1]$ be fixed arbitrary. From the convexity of n^i, the convexity of $f^i(\cdot,\alpha)$ is proved by

$$\begin{aligned} f^i(\lambda\beta^1 + (1-\lambda)\beta^2,\alpha) &= n^i(\psi^{BS(n^i(\alpha))}(\lambda\beta^1 + (1-\lambda)\beta^2)) \\ &\le n^i(\lambda\psi^{BS(n^i(\alpha))}(\beta^1) + (1-\lambda)\psi^{BS(n^i(\alpha))}(\beta^2)) \\ &\le \lambda n^i(\psi^{BS(n^i(\alpha))}(\beta^1)) + (1-\lambda)n^i(\psi^{BS(n^i(\alpha))}(\beta^2)). \end{aligned}$$

(Q.E.D.)

When the implication function I^i is $I^{\mathrm{r-R}}[T^i]$ of (4), we have

$$\begin{aligned} f^i(\beta,\alpha) &= T^{\mathrm{I}}[I^{\mathrm{R}}[T^{\mathrm{I}}[I^{\mathrm{r-R}}[T^i]]]](\beta,\alpha) = T^{\mathrm{I}}[I^{\mathrm{r-R}}[T^{\mathrm{T}}[T^i]]](\beta,\alpha) \\ &= n^i(\sup\{s \in [0,1] \mid T^i(s,\alpha) \le n^i(\beta)\}). \end{aligned} \tag{30}$$

For convenience, we define

$$\varphi^{\alpha}(\beta) = \sup\{s \in [0,1] \mid T^i(s,\alpha) \le n^i(\beta)\}. \tag{31}$$

Then we have the following theorem.

Lemma 3. *If $T^i(\cdot,\alpha)$ is convex and n^i is concave, φ^α is a concave function.*
(**Proof**). Let β^1, $\beta^2 \in [0,1]$ and $\lambda \in [0,1]$ be fixed arbitrary. From the concavity of n^i and the convexity of $T^i(\cdot,\alpha)$ the concavity of φ^α is proved by

$$\begin{aligned}&\varphi^\alpha(\lambda\beta^1+(1-\lambda)\beta^2)) = \sup\{s\in[0,1] \mid T^i(s,\alpha)\le n^i(\lambda\beta^1+(1-\lambda)\beta^2)\}\\ &\ge \sup\{\lambda s_1+(1-\lambda)s_2\in[0,1] \mid \\ &\qquad\qquad T^i(\lambda s_1+(1-\lambda)s_2,\alpha)\le\lambda n^i(\beta^1)+(1-\lambda)n^i(\beta^2)\}\\ &\ge \sup\{\lambda s_1+(1-\lambda)s_2\in[0,1] \mid \\ &\qquad\qquad \lambda T^i(s_1,\alpha)+(1-\lambda)T^i(s_2,\alpha)\le\lambda n^i(\beta^1)+(1-\lambda)n^i(\beta^2)\}\\ &\ge \lambda\varphi^\alpha(\beta^1)+(1-\lambda)\varphi^\alpha(\beta^2).\end{aligned}$$

(Q.E.D.)

Theorem 5. *If $T^i(\cdot,\alpha)$ is convex and n^i is linear, $f^i(\cdot,\alpha)$ is also a convex function.*
(**Proof**). Let β^1, $\beta^2 \in [0,1]$ and $\lambda \in [0,1]$ be fixed arbitrary. From the linearity of n^i and the convexity of $T^i(\cdot,\alpha)$ the convexity of $f^i(\cdot,\alpha)$ is proved by the following inequality:

$$\begin{aligned}&f^i(\lambda\beta^1+(1-\lambda)\beta^2,\alpha) = n^i(\varphi^\alpha(\lambda\beta^1+(1-\lambda)\beta^2))\\ &\le n^i(\lambda\varphi^\alpha(\beta^1)+(1-\lambda)\varphi^\alpha(\beta^1)) = \lambda n^i(\varphi^\alpha(\beta^1))+(1-\lambda)n^i(\varphi^\alpha(\beta^2)).\end{aligned}$$

(Q.E.D.)

5 Examples of Implication Functions

In this section, we demonstrate that, for necessity measures using many famous implication functions, Problem (13) is reduced to a linear programming problem under assumptions A1 and A2.

The famous implication functions I^i's and their corresponding functions f^i's are shown in Table 1. As shown in Table 1, for each of implication functions I^i except for Reichenbach implication function has the associated function f^i to which $f^i(\cdot,\alpha)$ is convex in the range $(0,\hat\beta^i(\alpha^i))$ for any $\alpha\in(0,1)$.

We demonstrate the reduction of Problem (13) to a linear programming problem when $m=3$ and assumptions A1 and A2 are satisfied. Reichenbach implication function, $I^{\rm S}[T^1]$, $I^{\rm R}[T^2]$ and $I^{\rm r-R}[T^2]$ with the following conjunction functions T^1, T^2 and a strong negation $n^1(a)=n^3(a)=1-a$ are selected for I^0, I^1, I^2 and I^3, respectively:

$$T^1(a,b)=\begin{cases}0 & \text{if } a+b\le 1,\\ \dfrac{a+b}{2} & \text{if } a+b>1,\end{cases}\qquad T^2(a,b)=\begin{cases}0 \text{ if } a+b\le 1,\\ ab \text{ if } a+b>1,\end{cases}\tag{32}$$

$I^1=I^{\rm S}[T^1]$, $I^2=I^{\rm R}[T^2]$ and $I^3=I^{\rm r-R}[T^2]$ are obtained as follows:

$$I^1(a,b)=\begin{cases}1 & \text{if } a\le b,\\ \dfrac{1-a+b}{2} & \text{if } a>b,\end{cases}\qquad I^2(a,b)=\begin{cases}1-a & \text{if } b<a(1-a) \text{ or } a=0,\\ \dfrac{b}{a} & \text{if } b\ge a(1-a) \text{ and } a>0,\end{cases}$$

Table 1. Famous implication functions I^i and associated functions f^i

Name	$I^i(a,b)$	$f^i(\beta,\alpha)$
Dienes	$\max(1-a,b)$	$\begin{cases} 0 \text{ if } \alpha+\beta\le 1 \\ \alpha \text{ if } \alpha+\beta>1 \end{cases}$
Reichenbach	$1-a+ab$	$\begin{cases} 0 & \text{if } \alpha+\beta\le 1 \\ (\alpha+\beta-1)/\beta & \text{if } \alpha+\beta>1 \end{cases}$
Łukasiewicz	$\min(1,1-a+b)$	$\max(0,\alpha+\beta-1)$
Gödel	$\begin{cases} 1 \text{ if } a\le b \\ b \text{ if } a>b \end{cases}$	$\min(\alpha,\beta)$
reciprocal Gödel	$\begin{cases} 1 & \text{if } a\le b \\ 1-a & \text{if } a>b \end{cases}$	$\begin{cases} 0 \text{ if } \alpha+\beta\le 1 \\ \beta \text{ if } \alpha+\beta>1 \end{cases}$
Goguen	$\begin{cases} 1 & \text{if } a\le b \\ b/a & \text{if } a>b \end{cases}$	$\alpha\beta$
reciprocal Goguen	$\begin{cases} 1 & \text{if } a\le b \\ (1-a)/(1-b) & \text{if } a>b \end{cases}$	$\begin{cases} 0 & \text{if } \alpha+\beta\le 1 \\ (\alpha+\beta-1)/\alpha & \text{if } \alpha+\beta>1 \end{cases}$
Gains-Rescher	$\begin{cases} 1 \text{ if } a\le b \\ 0 \text{ if } a>b \end{cases}$	$\begin{cases} 0 \text{ if } \alpha=0 \\ \beta \text{ if } \alpha>0 \end{cases}$
Inuiguchi	$\begin{cases} 1 & \text{if } a=0 \text{ or } b=1 \\ (1-a+b)/2 & \text{otherwise} \end{cases}$	$\begin{cases} 0 & \text{if } \beta=0 \\ \min(1,\max(0,2\alpha+\beta-1)) & \text{if } \beta>0 \end{cases}$
Fodor	$\begin{cases} 1 & \text{if } a\le b \\ \min(1-a,b) & \text{if } a>b \end{cases}$	$\begin{cases} 0 & \text{if } \alpha+\beta\le 1 \\ \min(\alpha,\beta) & \text{if } \alpha+\beta>1 \end{cases}$

$$I^3(a,b) = \begin{cases} b & \text{if } 1-a < b(1-b) \text{ or } b=1, \\ \dfrac{1-a}{1-b} & \text{if } 1-a \ge b(1-b) \text{ and } b<1. \end{cases} \tag{33}$$

Because both of T^1 and n^1 are convex, from Theorem 4, $f^1(\cdot,\alpha)$ is convex for any $\alpha \in (0,1]$. $T^2(\cdot,\alpha) = f^2(\cdot,\alpha)$ is convex for any $\alpha \in (0,1]$ and n^3 is linear, from Theorem 5, $f^3(\cdot,\alpha)$ is convex for any $\alpha \in (0,1]$. Indeed, we obtain

$$f^1(\beta,\alpha) = \begin{cases} \beta & \text{if } 2\alpha > 1, \\ \max(0, 2\alpha+\beta-1) & \text{if } 2\alpha \le 1, \end{cases} \tag{34}$$

$$f^3(\beta,\alpha) = \begin{cases} \alpha & \text{if } \alpha^2 < \alpha+\beta-1 \text{ or } \alpha=0, \\ \max\left(0, \dfrac{\alpha+\beta-1}{\alpha}\right) & \text{if } \alpha^2 \ge \alpha+\beta-1 \text{ and } \alpha>0. \end{cases} \tag{35}$$

Then we can confirm the convexity of $f^1(\cdot,\alpha)$ and $f^3(\cdot,\alpha)$.

Let $\check{\beta}^i(\alpha) = \inf\{\beta \in [0,1] \mid f^i(\beta,\alpha) > 0\}$, $i = 0,1,\ldots,m$. We obtain

$$\begin{cases} \check{\beta}^0(\alpha) = 1-\alpha, \check{\beta}^1(\alpha) = \max(0, 1-2\alpha), \ \hat{\beta}^1(\alpha) = 1, \\ \check{\beta}^2(\alpha) = 1-\alpha, \ \hat{\beta}^2(\alpha) = 1, \ \check{\beta}^3(\alpha) = 1-\alpha, \ \hat{\beta}^3(\alpha) = 1-\alpha(1-\alpha). \end{cases} \tag{36}$$

By those values, the meaningful ranges of β for the second constraints of Problem (18) can be obtained under assumptions A1 and A2. Namely, $\forall \beta \in [0,1]$ is replaced with $\forall \beta \in [\check{\beta}^i(\alpha^i), \hat{\beta}^i(\alpha^i)]$. By the convexity of $f^i(\cdot,\alpha^i)$, we only consider two values near the upper and lower bounds of $[\check{\beta}^i(\alpha^i), \hat{\beta}^i(\alpha^i)]$.

Finally, the problem with $m = 3$ is reduced to a linear programming problem in the form of (25).

6 Concluding Remarks

We have shown that the necessity fractile optimization models are reduced to linear programming problems under certain conditions. The conditions on membership functions of fuzzy parameters included in the problem are satisfied by linear membership functions which are often used in the literature. On the other hand, the condition on implication functions defining necessity measures is satisfied by many of famous implication functions. Then the results of this paper are applicable to many real world problems. Moreover, the results can be applied to combinatorial programming problems with fuzzy parameters because their continuous relaxation problems are often linear.

Under other conditions, we can show that the necessity fractile optimization models can be solved rather easily by iterative applications of linear programming techniques. For the specification of suitable necessity measures, we may apply Inuiguchi's approach [8]. Those are future topics of this study.

References

1. Rommelfanger, H., Słowiński, R.: Fuzzy Linear Programming with Single or Multiple Objective Functions. In: Słowiński, R. (ed.) Fuzzy Sets in Decision Analysis, Operations Research and Statistics, pp. 179–213. Kluwer, Boston (1998)
2. Inuiguchi, M., Ramík, J.: Possibilistic Linear Programming: A Brief Review of Fuzzy Mathematical Programming and a Comparison with Stochastic Programming in Portfolio Selection Problem. Fuzzy Sets and Systems 111, 29–45 (2000)
3. Inuiguchi, M.: A Semi-infinite Programming Approach to Possibilistic Optimization under Necessity Measure Constraints. In: Proceedings of 2009 IFSA World Congress and 2009 EUSFLAT Conference, Lisbon, Portugal, pp. 873–878 (2009)
4. Dubois, D., Prade, H.: Fuzzy Sets and Systems: Theory and Applications. Academic Press, New York (1980)
5. Dubois, D., Prade, H.: A Theorem on Implication Functions Defined from Triangular Norms. Stochastica 8, 267–279 (1984)
6. Inuiguchi, M., Sakawa, M.: On the Closure of Generation Processes of Implication Functions from a Conjunction Function. In: Yamakawa, T., Matsumoto, G. (eds.) Methodologies for the Conception, Design, and Application of Intelligent Systems, vol. 1, pp. 327–330. World Scientific, Singapore (1996)
7. Dubois, D., Prade, H.: Fuzzy Numbers: An Overview. In: Bezdek, J.C. (ed.) Analysis of Fuzzy Information. Mathematics and Logic, vol. I, pp. 3–39. CRC Press, Boca Raton (1987)
8. Inuiguchi, M., Greco, S., Słowiński, R., Tanino, T.: Possibility and Necessity Measure Specification Using Modifiers for Decision Making under Fuzziness. Fuzzy Sets and Systems 137, 151–175 (2003)

An Efficient Hybrid Approach to Correcting Errors in Short Reads

Zhiheng Zhao, Jianping Yin, Yong Li, Wei Xiong, and Yubin Zhan

School of Computer, National University of Defense Technology, 410073
Changsha, China
wader_zzh@hotmail.com

Abstract. High-throughput sequencing technologies produce a large number of short reads that may contain errors. These sequencing errors constitute one of the major problems in analyzing such data. Many algorithms and software tools have been proposed to correct errors in short reads. However, the computational complexity limits their performance. In this paper, we propose a novel and efficient hybrid approach which is based on an alignment-free method combined with multiple alignments. We construct suffix arrays on all short reads to search the correct overlapping regions. For each correct overlapping region, we form multiple alignments for the substrings following the correct overlapping region to identify and correct the erroneous bases. Our approach can correct all types of errors in short reads produced by different sequencing platforms. Experiments show that our approach provides significantly higher accuracy and is comparable or even faster than previous approaches.

Keywords: High-throughput sequencing, Error correction, Suffix array, Multiple Alignments.

1 Introduction

High-throughput sequencing technologies such as Illumina's Genome Analyzer, ABI's SOLiD, and Roche's 454, e.g. [1] open up a range of new opportunities for genome research. Unlike the Sanger method, high-throughput sequencing technologies can produce a large amount of short reads in a single run. For example, the Illumina Genome Analyzer IIx can currently generate an output of up to 640 million paired-end reads in a single run with a read length between 35 and 150. This leads to many novel applications such as genome re-sequencing, de novo genome assembly and metagenomics. However, high-throughput sequencing data is more error-prone than the Sanger sequencing method. With a significant impact on the accuracy of applications such as re-sequencing and de novo genome assembly, sequencing errors have become one of the major problems in analyzing high-throughput sequencing data.

It is a difficult task to correct errors in high-throughput sequencing data, for the computational demands for large-scale short reads limit the performance of error correction algorithms and tools. The intuitive error correction method

V. Torra et al. (Eds.): MDAI 2011, LNAI 6820, pp. 198–210, 2011.

is multiple read alignments, such as MisEd [2] for Sanger reads. If the reads could be aligned correctly, we could correct the erroneous bases which are in the minority in each column. However, multiple read alignments are extremely compute-intensive for large-scale reads and do not adapt well to short reads. Hence, some algorithms were proposed based on heuristics and alignment-free methods to determine which reads align to the same genomic position. Pevzner *et al.* [3] formulated the error correction problem as a spectral alignment problem (SAP), in which k-mers are divided into solid (correct) and insolid (erroneous) according to their multiplicity and the insolid k-mers are corrected using a minimum number of edit operations to solid k-mers until all reads only contain solid k-mers. Most of previous error correction methods are based on the SAP and use heuristics to approximate the SAP. Pevzner *et al.* [3] used a simple greedy heuristics to solve the SAP. Subsequently, Chaisson *et al.* [4] proposed a dynamic programming algorithm for the SAP and implemented a heuristic algorithm based on an approximation to dynamic programming algorithm [5] in assembly tool Euler-SR similarly with Butler *et al.* [6]. Recently, some tools which optimize the k-mers classification [7, 8] or accelerate error correction using GPU [9] are also based on the SAP.

However, because all k-mers in the SAP are independent, we cannot utilize the local context of a k-mer in the sequencing reads to identify errors. The more repetitive the genome is, the greater the chance is that a sequencing error will merely change one solid k-mer to another solid k-mer, hiding the error [8]. Shrec [10] proposes a different idea for error correction which expands the local context of erroneous bases. It first searches the common correct substrings in all reads and then identifies the erroneous bases following these substrings. Shrec is based on a generalized suffix trie data structure that holds all short reads and corrects errors with a majority voting scheme. However, it requires huge memory and therefore it is difficult to be used for large-scale read data. Shrec was extended by Salmela [11] to a mixed set of reads from various sequencing technologies, with different read lengths and error characteristics. HiTEC [12] adopts the similar idea, while using the suffix array data structure instead of suffix trie. The suffix array is more memory efficient than suffix trie and HiTEC is based on a thorough statistical analysis. This makes HiTEC more accurate and efficient. However, HiTEC can only correct substitutions in short reads with identical read lengths.

In this paper, we propose an efficient hybrid approach which is based on an alignment-free method combined with multiple alignments. Our approach includes two stages. In the first stage, we construct suffix arrays on all short reads to search the correct overlapping regions as HiTEC. Each overlapping region contains a common substring shared by some reads. If the number of the reads and the length of the substring are large enough, we can consider that the common substring is from a unique genomic position and has no errors.For the reads contained in each correct overlapping region, we can consider the common substring is an anchor in the multiple aligments. Therefore, in the second stage, we only form multiple alignments for the substrings following the common substring to identify and correct the erroneous bases.

This strategy makes our approach differentiate the alignment-based methods which require O(m^3 + m^2*l^2_{ave}) time (m is the number of reads and l_{ave} is the average length of reads), for our approach does not form multiple alignments of all the entire reads. The worst case time complexity of our approach is O(ml_{ave}*log(ml_{ave})+ ml_{ave}*(l_{ave} - γ)^2) in which γ is the parameter to determine the correct overlapping region. For short reads our approach is comparable or even faster than all published approaches. Additionally, our approach is also different from Shrec and HiTEC, while they only correct errors directly following the common substring. Furthermore, the multiple alignments can be adjusted by the user-defined gap penalty and mismatch penalty. Hence, our approach can correct all types of errors in short reads produced by different sequencing platforms. Experiments show that our approach provides significantly higher accuracy than previous methods.

The rest of this paper is organized as follows. In section 2, we introduce the ideas and the methods used in our approach. We present the algorithm of our approach in section 3. In Section 4 we evaluate the performance of our approach. Finally, Section 5 concludes the paper.

2 Methods

2.1 Problem

We first give the error correction problem for our approach formally. Supposed that the reads are produced from a genome G and the length of G is N. If the sequence of G is unknown, we can use a reference genome instead of G. The genome G and the reads can be considered as strings over the alphabet {A, C, G, T, N}. Supposed the sequencing platforms produce m reads, and the length of i read is l_i, let r=c_0,c_1,...,c_{n-1} be a read of length n. $r[i,j]$ is the substring c_i, c_{i+1},..., c_j, in which $0 \leq i \leq j \leq n$. The reverse complement $\bar{r}$ of r, is obtained by first reversing s and then applying the transformation A $\leftrightarrow$ T; C $\leftrightarrow$ G. For example, if r = ACTG, then $\bar{r}$ = CAGT. if r is produced from the substring g of G that g = $G[j, k]$, we say that r maps to g. We can obtain the combined total length of the reads denoted by M, $M = \sum_{i=0}^{m} l_i$. We use *coverage* denoting the expected number of times that a position in the genome is sequenced, so *coverage* = M/N. All error correction methods rely on the coverage of the reads being moderately high so that every position of the genome is sequenced several times with high probability. Reads from low coverage regions cannot be corrected because there is insufficient data to infer the correct sequence. We use p denoting the per-base error rate, so for a read with the length l, the expected error bases in the read is l*p.

For a reads set r_0,r_1,...,r_{m-1}, supposed these reads map to the substrings of G which are g_0,g_1,...,g_{m-1}, the task of an error correction algorithm is to covert the reads r_0,r_1,...,r_{m-1} to r^*_0,r^*_1,...,r^*_{m-1} using the edit operations, making D

$= \sum_{i=0}^{m} \mid g_i - r_i^* \mid$ as small as possible. There are three types of errors: substitutions, insertions and deletions. The distribution of error types varies from one sequencing platform to another. For instance, the Roche/454 sequencing platform produces reads with insertions and deletions, due mainly to homopolymers, whereas the SOLiD and Illumina platforms are prone to substitution errors. Hence, $\mid g_i - r_i^* \mid$ can denote the Hamming distance allowing only substitutions or edit distance allowing also insertions and deletions between g_i and r_i^* according to the error characteristics of the sequencing platforms.

2.2 Solution

If the reads have been aligned correctly to the genome, we can identify and correct the errors by checking each column of the alignments. However, as the analysis in the first section, most of time we cannot obtain the correct alignments for the computational complexity and the accuracy of multiple alignments of large-scale reads. We suppose that the reads can map to a genome or a reference genome. Then, there are many substrings of the reads map to a identical segment of the genome for the high coverage of reads. We define these substrings as an overlapping region if they are identical. For a common substring S contained in an overlapping region, let *L(S)* denote the length of the common substring and *H(S)* denote the number of the reads across the overlapping region. Actually, a common substring is corresponding to a k-mer in the SAP and is similar to the *witness* in HiTEC and the path from the root to a node in the suffix trie in Shrec. Figure 1 shows an example of an overlapping region.

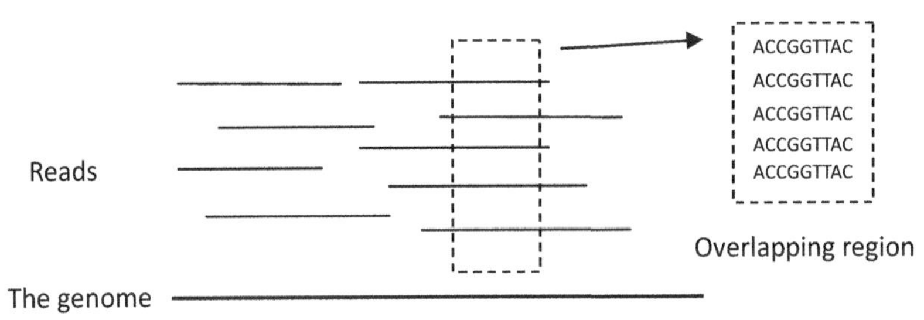

Fig. 1. An example of an overlapping region. The common substring contained in the overlapping region is S = ACCGGTTAC. *L(S)* = 9, *H(S)* = 5.

If the common substring of an overlapping region is unique , we can obtain the alignment between the reads across the overlapping region. However, the random occurrences and repeats in the genome will make a common substring identical to another one. The larger the length of the common substring is, the smaller the probability that the common substring has random occurrences or repeats. On the other hand, for a common substring S, the lager *L(S)* is, the smaller *H(S)*. If the *H(S)* is small enough, we have insufficient data to infer whether the common substring is correct. Hence, we use two parameters γ and δ. If $L(S) = \gamma$ and $H(S) > \delta$, we define the common substring S as a correct common substring and the overlapping region as a correct overlapping region.

For the reads across a correct overlapping region, we then form multiple alignments and correct errors for the strings following the common substring. We retrieve the consensus of the multiple alignments and then check each column to correct the bases which are not identical the consensus. From the figure 2, we can see that our approach can correct mixed errors in short reads. We adjust the gap penalty and the mismatch penalty of multiple alignments so that our approach can adapt to the short reads produced from different sequencing platforms or mixed short reads.

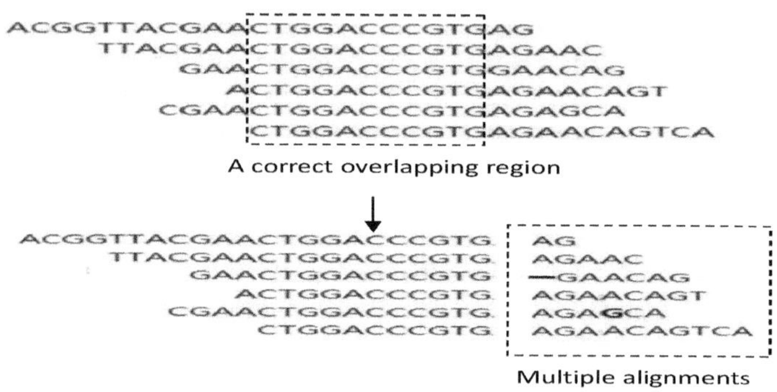

Fig. 2. Forming multiple alignments and correcting errors for the strings following the common substring in the reads across a correct overlapping region. From the multiple alignments, we can see that the third read has a deletion error and the fifth read has a substitution error.

3 Algorithm

Our approach has two stages: searching the correct overlapping regions, forming multiple alignments and correcting errors. We will present each stage detailedly as well as the full algorithm and the computational complexity.

3.1 Searching the Correct Overlapping Regions

We use the suffix array data structure to search the correct overlapping regions. We first give the definition of the suffix array. For a alphabet Σ, $|\Sigma|$ is the size of alphabet. A string is any finite sequence over Σ. Let $T{=}c_0,c_1,\ldots,c_{n-1}$ be a string of length n. $T[i,j]$ is the substring c_i, $c_{i+1},\ldots$, c_j, in which $0{\le}i{\le}j{\le}n$. We add a special character "\$" at the tail of the string T and denote the suffix of T at the position of i by $S_i{=}T[i,n]{=}c_i, c_{i+1},\ldots,c_{n-1},\$$. \$ is smaller than any character in Σ. The suffix array of T on the alphabet Σ, which we denote as SA, is an array of length n containing the positions of string T such that $S_{SA[i]}$ gives the increasing lexicographical order, i.e., $S_{SA[0]}<S_{SA[1]}< \ldots <S_{SA[n-1]}$. Besides suffix array, the longest common prefix (LCP) array is often used to compute the length of the longest common prefix between suffixes. LCP[i] is the length of the longest common prefix of $S_{SA[i]}$ and $S_{SA[i-1]}$. The suffix array data structure has been proposed by Manber and Myers [13], and the suffix array construction algorithm can be seen from [14] for a survey.Figure 3 shows the suffix array and LCP array of string ACCTATACCGTA.

i	$SA[i]$	$S_{SA}[i]$	$LCP[i]$
0	12	\$	0
1	11	A\$	0
2	6	ACCGTA\$	1
3	0	ACCTATACCGTA\$	3
4	4	ATACCGTA\$	1
5	7	CCGTA\$	0
6	1	CCTATACCGTA\$	2
7	8	CGTA\$	1
8	2	CTATACCGTA\$	1
9	9	GTA\$	0
10	10	TA\$	0
11	5	TACCGTA\$	2
12	3	TATACCGTA\$	2

Fig. 3. The suffix array and LCP array of string ACCTATACCGTA

We also use the libdivsufsort library of Yuta Mori [15] which is a fast and lightweight suffix array construction algorithm to compute the full suffix array of R as HiTEC. $R = r_1\$\overline{r_1}\$r_2\$\overline{r_2}\$\ldots\$r_m\$\ \overline{r_m}\$$. The algorithm requires O($nlogn$) time in which n is the length of R. n = 2*(M + m). If $n< 2^{32}$, it requires $5n$+O(1) bytes space for suffix array.

With the suffix array of R, we can get all the correct overlapping regions. We first computing the LCP array of R. Because each read in R has been separated by \$, we use an variation of LCP array called LCP_\$ array instead of LCP array, for we only need to compute the longest common prefix before any \$ of $S_{SA[i]}$ and $S_{SA[i-1]}$ for LCP_\$[i]. The LCP_\$ array also be computed by the algorithm of Kasai *et al.* [16] in O(n) time and space. Then, we scan the LCP_\$ array. If LCP_\$[$i + j$] $\geq \gamma$ for j = 1,2,...,k and $k \geq \delta$, $0 \leq i < n$, the γ-long prefix of $S_{SA[i]}, S_{SA[i+1]}, \ldots, S_{SA[i+k]}$ is a correct overlapping region.

Constructing the suffix array needs O($nlogn$) in time. Computing the LCP_\$[i] array and searching the correct overlapping regions both need O(n) in time. So this stage requires O($nlogn$) in time and $6n$+O(1) bytes space.

3.2 Forming Multiple Alignments and Correcting Errors

We form the multiple alignments for each correct overlapping region in the second stage. Supposed that there are k reads across the correct overlapping region which are r_{i_1}, r_{i_1},...,r_{i_k}, and for the read r_{i_j} ($0 \leq j \leq k$),the common substring ends at p_j. Hence, we need to form multiple alignments for the string set $S = \{r_{i_1}[p_1,l_{i_1}],r_{i_2}[p_2,l_{i_2}], \ldots, r_{i_k}[p_k,l_{i_k}]\}$, in which l_{i_j} is the length of the read r_{i_j}.

Next we compute the consensus $r_{consensus}$ of S using the majority voting scheme. For the i-th column of S, let occurrences(x)($x \in$ {A,C,G,T}) represents the occurrences of x. If $\exists x_i$ occurrences(x_i) $> h/2$ and occurrences(x_i) ≥ 2 (h is the number of non-null elements in the i-th column of S), $r_{consensus}[i] = x_i$; otherwise, $r_{consensus}[i]$ = N. If $r_{consensus}[j] \in$ {A,C,G,T} and the later characters of $r_{consensus}[j]$ are all N, we truncate the end of $r_{consensus}$ from the j-th character and set the length of $r_{consensus}$ to $j+1$;

Then we align each string in S to the consensus using the Needleman-Wunsch algorithm[17]. We allow free gaps both for the string and the consensus if the users choose gapped alignment for substitution errors and indel errors. Otherwise we disallow free gaps only for the substitution errors. For the gapped alignment, we do not count the gap penalty at the end of the string and the consensus for they need to be aligned from the left. If the score of the alignment exceeds the threshold, we correct the string according to the consensus. Otherwise, we consider the string is the substring of a read which is a random occurrence or a repeat of the genome and skip the string.

It seems that we only correct errors in the "right" region of each correct overlapping region. Actually, because we search the correct overlapping regions from the suffix array which is constructed from all the reads and their reverse complement. For each correct overlapping region Γ, we will find the corresponding correct overlapping region $\overline{\Gamma}$ of its reverse complement and correct errors in the "right" region of $\overline{\Gamma}$. The "right" region of $\overline{\Gamma}$ is corresponding to the "left" region of Γ, hence our approach forms multiple read alignments in fact.

There may exist inconsistency when a read traverses across two or more correct overlapping regions. To address this problem, we use $left_i$ and $right_i$ to keep the correct segment of read r_i. We set $left_i = 0$ and $right_i = 0$ initially. In the second stage, when we find r_i across correct overlapping region for the first time, we first set $right_i$ to the right end position of the common substring of r_i. Then, if the string $r_i[right_i, l_i]$ has p-long prefix aligned to the consensus in the multiple alignments, we set $right_i = right_i + p$. Similarly, when we find the reverse complement $\overline{r_i}$ of read r_i across correct overlapping region for the first time, we first set $left_i$ to the length of the right segment beyond the common substring of $\overline{r_i}$. Then, if the string $\overline{r_i}[left_i, l_i]$ has q-long prefix aligned to the consensus in the multiple alignments, we set $left_i = left_i$ - q. We consider that the substring $r_i[left_i, right_i]$ is the corrected segment of read r_i. Therefore, when we correct read r_i in another multiple alignment, if the suspicious bases are in the substring $r_i[left_i, right_i]$, we ignore them simply. Otherwise, we correct them and update $left_i$ and $right_i$.

We give the detailed algorithm of the second satge in Algorithm 1.

Algorithm 1. align_correct()

given: a correct overlapping region Γ, reads r_1,r_1,...,r_m, the length of r_i is l_i for $1 \leq i \leq m$, *left*$[m]$ and *right*$[m]$
output: the corrected bases, *left*$[m]$ and *right*$[m]$

obtain the reads r_{i_1}, r_{i_2},...,r_{i_k} across Γ
compute the end position p_j of the common substring of r_{i_j}, for $1 \leq j \leq k$
obtain the string sets $S = \{r_{i_1}[p_1,l_{i_1}],r_{i_2}[p_2,l_{i_2}], \ldots, r_{i_k}[p_k,l_{i_k}]\}$
for each column i in S **do**
    $r_{consensus}[i]$ = N
    compute occurrences(A),occurrences(C),occurrences(G),occurrences(T)respectively
    h $\leftarrow$ the number of non-null characters in the column
    if $\exists x$(occurrences(x) $> h/2$ and occurrences(x)$\geq$ 2) ($x \in$ {A,C,G,T}) **then**
        $r_{consensus}[i] = x$
    end if
end for
truncate $r_{consensus}$
for each string r_i in S **do**
    score = align(r_i,$r_{consensus}$)
    if score $< \varepsilon(\min(l_i, l_{consensus}))$($\varepsilon$ is an user-defined threshold) **then**
        continue
    else
        Get all the suspicious bases of r_i
        for each suspicious base p **do**
            if p is included in r_i[*left*$_i$, *right*$_i$] **then**
                continue
            else
                correct(p)
                update *left*$_i$ and *right*$_i$
            end if
        end for
    end if
end for

3.3 Choosing Parameters

Now we present how to choose the parameters γ and δ. The larger γ and δ, the smaller the probability that the correct overlapping region has random occurrences or repeats. However, we cannot set γ and δ too large because it will miss many errors. We use the same method as Quake [8] to determine γ. For a γ-long string, the expected occurrences in the genome G which length is N are $2N/4^\gamma$,hence, we use the following formula to determine γ.

$$\frac{2N}{4^\gamma} \approx 0.01$$

$$\gamma \approx \log_4 200N \tag{1}$$

If there are no errors, the expected occurrences of a γ-long string in the reads is *coverage*. For the p per-base error rate, the expected occurrences of a γ-long string without error is

$$occurrences(\gamma) = coverage - coverage(1 - (1-p)^{\gamma})$$

To increase the discriminative power for identifying the location of errors in the reads with low coverage, we set δ as equation 2.

$$\delta = \frac{occurrences(\gamma)}{2} = \frac{coverage - coverage(1 - (1-p)^{\gamma})}{2} \quad (2)$$

For the reads produced from an approximately 5 Mbp genome such as *E. coli*, if the coverage is 70, the per-base error rate is 1%, we will set γ to 15 and δ to 31.

3.4 The Full Algorithm and Complexity

Now we can give the full algorithm and the computational complexity of our approach. The full algorithm is presented in Algorithm 2.

Algorithm 2. The full algorithm of our approach

given: reads $r_1,r_1,\ldots,r_m$, the length of r_i is l_i for $1 \leq i \leq m$, N, p and *coverage*
output: the corrected reads

1: compute γ and δ
2: initialize the arrays *left*$[m]$ and *right*$[m]$
3: constructed R and compute SA and LCP_$ array
4: scan LCP_$ array to search the correct overlapping regions
5: **for** each correct overlapping regions Γ **do**
6:     align_correct()
7: **end for**

Let l_{ave} be the average length of the reads. Constructing the suffix array needs O($M\,logM$) time and computing LCP_$ array needs O(M) time. There are M/δ correct overlapping regions. For each correct overlapping regions, computing the consensus needs O(δ*(l_{ave} - γ)) and using the Needleman-Wunsch algorithm to align each string to the consensus needs O(δ*(l_{ave} - $\gamma)^2$), so the time complexity of align_correct function is O(δ*(l_{ave} - $\gamma)^2$). Hence, the worst case time complexity of our approach is O($M\,logM$ + M*(l_{ave} - $\gamma)^2$) which is also O(ml_{ave}*$log(ml_{ave})$+ ml_{ave}*(l_{ave} - $\gamma)^2$). Our approach is more efficient than the alignment-based methods for large-scale short reads and is comparable or even faster than the alignment-free methods. Our approach requires only $6M$+O(1) bytes space which makes it memory-efficient.

4 Evaluation

In this section, we evaluate the performance of our approach. We use the simulation data sets which are created from several bacterial genomes as previous programs. These genomes can be downloaded from GenBank under the accession numbers. We create two read data sets: S1 and S2. S1 is a read data set for evaluating our approach compared to alternative approaches (SHREC [10], HITEC [12], and Quake [8]) with only substitution errors, for SHREC and HITEC can only correct substitutions. S2 is a read data set for evaluating our approach compared to Ext-SHREC [11] and Quake with all types of errors (Ext-SHREC does the same work as SHREC when the reads have the same length and only have substitution errors). The datasets used for our performance evaluation are summarized in Table 1.

Table 1. Datasets used for performance evaluation.The reads in S1 only contain substitution errors. The reads in S2 contain all types of errors with the same probability for each type of errors and the length of reads varies from 60 bps to 120 bps.

Dataset	ID	Reference genome(GenBank)	Genome length(MB)	Error rate(%)	coverage	length of reads (bp)
S1	A1			1	70	
	A2	NC_001139	1.1	2		70
	A3			3		
	B1			1		
	B2	NC_007146	1.9	2		70
	B3			3		
	C1			1		
	C2	NC_003923	2.8	2		70
	C3			3		
	D1			1		
	D2	NC_000913	4.7	2		70
	D3			3		
S2	E1			1	70	
	E2	NC_003923	2.8	2		60∼120
	E3			3		
	F1			1		
	F2	NC_000913	4.7	2		60∼120
	F3			3		

We also use the *accuracy* to evaluate the performance of the algorithm we measured as HITEC [12]. The accuracy is defined as the ratio between the number of corrected reads and the number of initially erroneous reads. We use err_{bef} denoting the number of erroneous reads before correction and err_{aft} denoting the number of erroneous after correction. Then,

$$accuracy = \frac{err_{bef} - err_{aft}}{err_{bef}}$$

If we denote the number of erroneous reads that are corrected, correct reads that are left unchanged, correct reads that are wrongly changed, and erroneous reads that are left unchanged by *TP*, *TN*, *FP*, *FN* (true/false positive/negative) respectively, we have $err_{bef} = TP + FN$, $err_{aft} = FP + FN$ and therefore

$$accuracy = \frac{TP - FP}{TP + FN}$$

We called our approach MyHybrid for short. The tests shown in Table 2 and Table 3 were performed on a desktop computer with Intel Xeon E5420 4-core processor at 2.50GHz, 8GB RAM, running RHEL 5 x86_64 server. All algorithms to be compared use the default parameters. We also evaluate the time and memory required by the algorithm we measured in Table 4. Note that the test runs on a 64bit Linux, the memory used by the measured programs is almost two times as many as that running on 32bit operating systems.

From the tests, we can see that the accuracy of our approach is comparable to HiTEC and Quake for the data set S1 which only contains substitution errors. With various error rates, our approach performs more steadily than the other three programs. For the data set S2 which contains mixed errors, our approach is more efficient than Ext-SHREC and Quake. This makes our approach more available for the real read data which has complex error characters. Furthermore, in addition to obtaining very high accuracy, our approach has also very good time and space complexities. Our approach outperforms HiTEC and SHREC and is approximate to Quake on computational performance.

Table 2. Accuracy comparison for the data set S1

dataset	Accuracy(%)			
ID	SHREC	Quake	HiTEC	MyHybrid
A1	95.12	97.45	98.79	97.62
A2	87.04	96.38	97.60	97.04
A3	79.75	92.67	94.39	96.56
B1	92.64	98.49	99.03	98.25
B2	83.77	94.00	98.54	98.30
B3	64.30	91.78	96.27	97.16
C1	90.42	96.41	99.24	98.48
C2	73.08	95.33	97.73	98.06
C3	58.33	91.92	94.55	96.57
D1	88.05	97.17	98.62	98.40
D2	72.64	94.43	95.95	97.03
D3	57.95	92.80	92.16	94.33

Table 3. Accuracy comparison for the data set S2

dataset	Accuracy(%)		
ID	Ext-SHREC	Quake	MyHybrid
E1	85.02	94.69	93.97
E2	74.80	89.06	92.14
E3	60.75	85.33	89.36
F1	80.64	95.64	92.02
F2	68.31	90.47	90.35
F3	55.42	87.66	89.98

Table 4. Time and space comparison between SHREC, Quake, HiTEC and our approach for the data set S1

dataset	Time(s)				Memory(MB)			
ID	SHREC	Quake	HiTEC	MyHybrid	SHREC	Quake	HiTEC	MyHybrid
A1	1651	244.7	257.6	138.7	3002	448	1408	1462
A2	2540	265.9	386.4	142.2	3014	456	1408	1462
A3	3789	307.5	579.6	149.0	3528	440	1408	1462
B1	2517	278.6	478.9	205.7	3206	510	2476	2520
B2	3861	380.4	714.3	238.4	3890	518	2476	2520
B3	4960	421.9	1075.0	244.5	5742	526	2476	2520
C1	4033	346.1	785.5	291.0	3970	1048	3652	3738
C2	5580	381.0	1174.8	321.9	4864	1082	3652	3738
C3	7842	474.2	1767.3	345.6	6004	1124	3652	3738
D1	5947	429.3	1529.4	976.0	6190	1060	6014	6184
D2	8612	468.0	1911.7	1065.4	7100	1176	6014	6184
D3	14960	502.4	3441.1	1092.8	7452	1248	6014	6184

5 Conclusion

In this paper, we propose a novel and efficient hybrid approach for correcting errors in short reads. Our approach can correct all types of errors in short reads produced by different sequencing platforms. Experiments show that our approach provides significantly higher accuracy and is comparable or even faster than previous approaches.

References

1. Mardis, E.R.: The impact of next-generation sequencing technology on genetics. Trends Genet. 24, 133–141 (2008)
2. Tammi, M.T., Arner, E., Kindlund, E., Andersson, B.: Correcting errors in shotgun sequences. Nucleic Acids Res. 31, 4663–4672 (2003)
3. Pevzner, P.A., Tang, H., Waterman, M.S.: A new approach to fragment assembly in DNA sequencing. In: RECOMB 2001, pp. 256–267 (2001)
4. Chaisson, M.J., Pevzner, P.A., Tang, H.: Fragment assembly with short reads. Bioinformatics 20, 2067–2074 (2004)
5. Chaisson, M.J., Brinza, D., Pevzner, P.A.: De novo fragment assembly with short mate-paired reads: Does the read length matter? Genome Res. 19, 336–346 (2009)
6. Butler, J., MacCallum, I., Kleber, M., Shlyakhter, I.A., Belmonte, M.K., Lander, E.S., Nusbaum, C., Jaffe, D.B.: ALLPATHS: De novo assembly of whole-genome shotgun microreads. Genome Res. 18, 810–820 (2008)
7. Yang, X., Dorman, K.S., Aluru, S.: Reptile: representative tiling for short read error correction. Bioinformatics 26, 2526–2533 (2010)
8. Kelley, D., Schatz, M., Salzberg, S.: Quake: quality-aware detection and correction of sequencing errors. Genome Biology 11(11), R116 (2010)
9. Shi, H., Schmidt, B., Liu, W., Muller-Wittig, W.: A parallel algorithm for error correction in high-throughput short-read data on CUDA-enabled graphics hardware. J. Comput.Biol. 17, 603–615 (2009)

10. Schroder, J., Schroder, H., Puglisi, S.J., Sinha, R., Schmidt, B.: SHREC: a short-read error correction method. Bioinformatics 25, 2157–2163 (2009)
11. Salmela, L.: Correction of sequencing errors in a mixed set of reads. Bioinformatics. 26(10), 1284–1290 (2010)
12. Ilie, L., Fazayeli, F., Ilie, S.: HiTEC: accurate error correction in high-throughput sequencing data. Bioinformatics 27(3), 295–302 (2011)
13. Manber, U., Myers, G.: Suffix arrays: a new method for on-line search. SIAM J. Comput. 22(5), 935–948 (1993)
14. Simon, J., Puglisi, W.F., Smyth, A.: A taxonomy of suffix array construction algorithms. ACM Comput. Surv. 39(2), 1–31 (2007)
15. Mori, Y.: Short description of improved two-stage suffix sorting algorithm, `http://homepage3.nifty.com/wpage/software/itssort.txt`
16. Kasai, T., Lee, G.H., Arimura, H., Arikawa, S., Park, K.: Linear-time longest-common-prefix computation in suffix arrays and its applications. In: Amir, A., Landau, G.M. (eds.) CPM 2001. LNCS, vol. 2089, pp. 181–192. Springer, Heidelberg (2001)
17. Needleman, S.B.: A general method applicable to the search for similarities in the amino acid sequence of two proteins. Journal of Molecular Biology 48(3), 443–453 (1970)

Rule Protection for Indirect Discrimination Prevention in Data Mining

Sara Hajian, Josep Domingo-Ferrer, and Antoni Martínez-Ballesté

Universitat Rovira i Virgili
Department of Computer Engineering and Mathematics
UNESCO Chair in Data Privacy
{sara.hajian,josep.domingo,antoni.martinez}@urv.cat

Abstract. Services in the information society allow automatically and routinely collecting large amounts of data. Those data are often used to train classification rules in view of making automated decisions, like loan granting/denial, insurance premium computation, etc. If the training datasets are biased in what regards sensitive attributes like gender, race, religion, etc., discriminatory decisions may ensue. Direct discrimination occurs when decisions are made based on biased sensitive attributes. Indirect discrimination occurs when decisions are made based on non-sensitive attributes which are strongly correlated with biased sensitive attributes. This paper discusses how to clean training datasets and outsourced datasets in such a way that legitimate classification rules can still be extracted but indirectly discriminating rules cannot.

Keywords: Anti-discrimination, Indirect discrimination, Discrimination prevention, Data mining, Privacy.

1 Introduction

Automated data collection in the information society facilitates automating decision making as well. Superficially, automating decisions may give a sense of fairness: classification rules do not guide themselves by personal preferences. However, at a closer look, one realizes that classification rules are actually trained on the collected data. If those training data are biased, the learned model will be biased. For example, if the data are used to train classification rules for loan granting and most of the Brazilians in the training dataset were denied their loans, the leaned rules will also show biased behavior toward Brazilian and it is a discriminatory reason for loan denial. Unfairly treating people on the basis of their belonging to a specific group (race, ideology, gender, etc.) is known as discrimination and is legally punished in many democratic countries.

1.1 Discrimination-Aware Data Mining

The literature in law and social sciences distinguishes direct and indirect discrimination (the latter is also called systematic). Direct discrimination consists

V. Torra et al. (Eds.): MDAI 2011, LNAI 6820, pp. 211–222, 2011.

of rules or procedures that explicitly impose "disproportionate burdens" on minority or disadvantaged groups (*i.e.* discriminatory rules) based on sensitive attributes related to group membership (*i.e.* discriminatory attributes). Indirect discrimination consists of rules or procedures that, while not explicitly mentioning discriminatory attributes, impose the same disproportionate burdens, intentionally or unintentionally. This effect and its exploitation is often referred to as *redlining* and indirectly discriminating rules can be called *redlining rules* [1]. The term "redlining" was invented in the late 1960s by community activists in Chicago [2]. The authors of [1] also support this claim: even after removing the discriminatory attributes from the dataset, discrimination persists because there may be other attributes that are highly correlated with the sensitive (discriminatory) ones or there may be background knowledge from publicly available data (*e.g.* census data) allowing inference of the discriminatory knowledge (rules).

The existing literature on anti-discrimination in computer science mainly elaborates on data mining models and related techniques. Some proposals are oriented to the *discovery* and *measure* of discrimination [1,3,4,7]. Others deal with the *prevention* of discrimination. Although some methods have been proposed, discrimination prevention stays a largely unexplored research avenue. Clearly, a straightforward way to handle discrimination prevention would consist of removing discriminatory attributes from the dataset. However in terms of indirect discrimination, as stated in [1,2] there may be other attributes that are highly correlated with the sensitive ones or there may be background knowledge from publicly available data that allow for the inference of discrimination rules. Hence, one might decide to remove also those highly correlated attributes as well. Although this would solve the discrimination problem, in this process much useful information would be lost. Hence, one challenge regarding discrimination prevention is considering indirect discrimination other than direct discrimination and another challenge is to find an optimal trade-off between anti-discrimination and usefulness of the training data.

1.2 Contribution and Paper Organization

The main contributions of this paper are as follows: (1) a new preprocessing method for indirect discrimination prevention based on data transformation that can consider several discriminatory attributes and their combinations; (2) some measures for evaluating the proposed method in terms of its success in discrimination prevention and its impact on data quality. Although some methods have recently been proposed for discrimination prevention [2,5,6,10], such works only consider direct discrimination. Their approaches cannot guarantee that the transformed dataset is really discrimination-free, because it is known that discriminatory behaviors can be hidden behind non-discriminatory items. To the best of our knowledge this is the first work that proposes a discrimination prevention method for indirect discrimination.

In this paper, Section 2 elaborates on the discovery of indirect discrimination. Section 3 presents our proposed method. Evaluation measures and experimental evaluation are presented in Section 4. Conclusions are drawn in Section 5.

2 Discovering Discrimination

In this section, we present some background concepts that are used throughout the paper. Moreover, we formalize the finding of indirect discrimination.

2.1 Background

A *dataset* is a collection of records and their attributes. Let $\mathcal{DB}$ be the original dataset. An *item* is an attribute along with its value, *e.g.* `Race=black`. An *itemset* is a collection of one or more items. A *classification rule* is an expression $X \rightarrow C$, where X is an itemset, containing no class items, and C is a class item, *e.g.* `Class=bad`.

The *support* of an itemset, $supp(X)$, is the fraction of records that contain the itemset X. We say that a rule $X \rightarrow C$ *is completely supported* if both X and C appear in the record. The *confidence* of a classification rule, $conf(X \rightarrow C)$, measures how often the class item C appears in records that contain X. A *frequent classification rule* is a classification rule with a support or confidence greater than a specified lower bound. Let $\mathcal{FR}$ be the database of frequent classification rules extracted from $\mathcal{DB}$.

With the assumption that discriminatory items in $\mathcal{DB}$ are predetermined (*e.g.* `Race=black`), rules fall into one of the following two classes with respect to discriminatory and non-discriminatory items in $\mathcal{DB}$: (i) a classification rule is *potentially discriminatory* (PD) when $X = A, B$ with A a non-empty discriminatory itemset and B a non-discriminatory itemset (*e.g.* {`Race=black, City=NYC`}$\rightarrow$ `Class=bad`); (ii) a classification rule is *potentially non-discriminatory* (PND) when $X = D, B$ is a non-discriminatory itemset (*e.g.* {`Zip=10451, City=NYC`} $\rightarrow$ `Class=bad`). Let assume that the notation $X(D, B)$ means $X = D, B$. Let $\mathcal{PR}$ a database of frequent classification rules with PD and PND classification rules. The word "potentially" means that a PD rule could probably lead to discriminatory decisions, so some measures are needed to quantify the discrimination potential (direct discrimination). Also, a PND rule could lead to discriminatory decisions if combined with some background knowledge (indirect discrimination); *e.g.*, if the premise of the PND rule contains the `Zip=10451` itemset, relying on additional background knowledge one knows that zip 10451 is mostly inhabited by black people.

Pedreschi *et al.*[1,4] introduced a family of measures of the degree of discrimination of a PD rule. One of these measures is *extended lift* measure ($elift$):

$$elift(A, B \rightarrow C) = \frac{conf(A, B \rightarrow C)}{conf(B \rightarrow C)}$$

Whether the rule is to be considered discriminatory can be assessed by using a threshold: Let $\alpha \in R$ be a fixed threshold[1] and let A be a discriminatory itemset. A PD classification rule $c : A, B \rightarrow C$ is *α-protective* w.r.t. $elift$ if $elift(c) < \alpha$. Otherwise, c is *α-discriminatory*.

[1] Note that α is a fixed threshold stating an acceptable level of discrimination according to laws and regulations.

2.2 Indirect Discrimination Formalization

In terms of indirect discrimination, the purpose of discrimination discovery is identifying PND rules that are to a certain extent equivalent to α-discriminatory rules or, in other words, identifying redlining rules. To determine the redlining rules, Pedreschi *et al.* in [1] stated the theorem below which gives a lower bound for α-discrimination of PD classification rules given information available in PND rules (γ, δ) and information available from background rules (β_1, β_2). They assume that background knowledge takes the form of classification rules relating a non-discriminatory itemset D to a discriminatory itemset A within the context B.

Theorem 1 ([1]). *Let $r : X(D, B) \rightarrow C$ be a PND classification rule, and let*

$$\gamma = conf(D, B \rightarrow C) \qquad \delta = conf(B \rightarrow C) > 0.$$

Let A be a discriminatory itemset, and let β_1, β_2 such that

$$conf(r_{b1} : A, B \rightarrow D) \geq \beta_1$$

$$conf(r_{b2} : D, B \rightarrow A) \geq \beta_2 > 0.$$

Call

$$f(x) = \frac{\beta_1}{\beta_2}(\beta_2 + x - 1)$$

$$elb(x, y) = \begin{cases} f(x)/y & \text{if } f(x) > 0 \\ 0 & \text{otherwise} \end{cases}$$

It holds that, for $\alpha \geq 0$, if $elb(\gamma, \delta) \geq \alpha$, the PD classification rule $r' : A, B \rightarrow C$ is α-discriminatory.

Based on the above theorem, we propose the following formal definitions of redlining and non-redlining rules.

Definition 1. *A PND classification rule $r : X(D, B) \rightarrow C$ is a* redlining rule *if it could yield an α-discriminatory rule $r' : A, B \rightarrow C$ in combination with currently available background knowledge rules of the form $r_{b1} : A, B \rightarrow D$ and $r_{b2} : D, B \rightarrow A$, where A is a discriminatory itemset.*

Definition 2. *A PND classification rule $r : X(D, B) \rightarrow C$ is a* non-redlining rule *if it cannot yield any α-discriminatory rule $r' : A, B \rightarrow C$ in combination with currently available background knowledge rules of the form $r_{b1} : A, B \rightarrow D$ and $r_{b2} : D, B \rightarrow A$, where A is a discriminatory itemset.*

Note that the correlation between the discriminatory itemset A and the non-discriminatory itemset D with context B indicated by the background rules r_{b1} and r_{b2} holds with confidences at least β_1 and β_2, respectively; however, it is not a completely certain correlation. Let $\mathcal{RR}$ be the database of redlining rules extracted from database $\mathcal{DB}$.

3 A Proposal for Indirect Discrimination Prevention

In this section we present a new indirect discrimination prevention method. The method transforms the source data by removing indirect discriminatory biases so that no unfair decision rule can be indirectly mined from the transformed data. The proposed solution is based on the fact that the dataset of decision rules would be free of indirect discrimination if it contained no redlining rule.

For discrimination prevention using preprocessing, we should transform data by removing all evidence of discrimination in the form of α-discriminatory rules and redlining rules. In [10] and [11] we concentrated on direct discrimination and considered α-discriminatory rules. In this paper, we focus on indirect discrimination and consider redlining rules. For these rules, a suitable data transformation with minimum information loss should be applied in such a way that those redlining rules are converted to non-redlining rules.

As mentioned above, based on the definition of the indirect discriminatory measure (*i.e. elb*), to convert redlining rules into non-redlining rules, we should enforce the following inequality for each redlining rule $r : D, B \to C$ in $\mathcal{RR}$:

$$elb(\gamma, \delta) < \alpha \qquad (1)$$

By using the definitions in the statement of Theorem 1, Inequality (1) can be rewritten as

$$\frac{\frac{conf(r_{b1})}{conf(r_{b2})}(conf(r_{b2}) + conf(r : D, B \to C) - 1)}{conf(B \to C)} < \alpha \qquad (2)$$

To enforce the above inequality, there can be two situations:

- **Case 1:** Assume that discriminatory items (*i.e.* A) are removed from the original database ($\mathcal{DB}$), and the r_{b1} and r_{b2} rules are obtained from publicly available data so that their confidences are constant. Let us rewrite Inequality (2) in the following way

$$conf(r : D, B \to C) < \frac{\alpha \cdot conf(B \to C) \cdot conf(r_{b2})}{conf(r_{b1})} - (conf(r_{b2}) + 1) \quad (3)$$

  It is clear that Inequality (2) can be satisfied by decreasing the confidence of redlining rule ($r : D, B \to C$) to values less than the right-hand side of Inequality (3).
- **Case 2:** Assume that discriminatory items (*i.e.* A) are not removed from the original database ($\mathcal{DB}$), and the rules r_{b1} and r_{b2} might be obtained from $\mathcal{DB}$ so that their confidences might change by data transformation. This could be more useful to detect the non-discriminatory items that are highly correlated with the discriminatory ones and thereby discover the possibly discriminatory rules that could inferred from them. Let us rewrite Inequality (2) as Inequality (4), where the confidences of r_{b1} and r_{b2} rules are not constant.

$$conf(B \rightarrow C) > \frac{\frac{conf(r_{b1})}{conf(r_{b2})}(conf(r_{b2}) + conf(r : D, B \rightarrow C) - 1)}{\alpha} \quad (4)$$

Clearly, in this case Inequality (2) can be satisfied by increasing the confidence of the base rule ($B \rightarrow C$) of the redlining rule ($r : D, B \rightarrow C$) to values greater than the right-hand side of Inequality (4) without affecting either the confidence of the redlining rule or the confidence of the r_{b1} and r_{b2} rules.

The detailed process of our preprocessing discrimination prevention method for indirect discrimination is described by means of the following phases:

- *Phase 1.* Use Pedreschi's measure on each PND rule to discover the patterns of indirect discrimination emerged from the available data and also the background knowledge. It consists of the following steps: (i) extract frequent classification rules from $\mathcal{DB}$ using Apriori [9]; (ii) divide the rules into PD and PND, with respect to the predetermined discriminatory items in the dataset; (iii) for each PND rule, compute *elb* to determine the collection of redlining rules. Let $\mathcal{RR}$ be a database of redlining rules and their respective α-discriminatory rules ensuing from those rules through combination with background knowledge rules.
- *Phase 2.* Transform the original data to convert each redlining rule to a non-redlining rule without seriously affecting the data or other rules. Algorithms 1 and 2 show the steps of this phase.
- *Phase 3.* Evaluate the transformed dataset with the discrimination prevention and information loss measures of Section 4.1 below, to check whether they are free of discrimination and useful enough.

The second phase will be explained in detail in the following subsection.

3.1 Data Transformation Method

The data transformation method should increase or decrease some rule confidences as proposed in the previous section with minimum impact on data quality. In terms of the measures defined in Section 4.1 below, we should maximize the discrimination prevention measures and minimize the information loss measures. It is worth mentioning that data transformation methods were previously used for knowledge hiding [8] in privacy-preserving data mining (PPDM). Here we propose a data transformation method for hiding discriminatory and redlining rules.

Algorithms 1 and 2 detail our proposed data transformation method for each of the aforementioned cases. Without loss of generality, we assume that the class attribute C is binary (any non-binary class attribute can be expressed as the Cartesian product of binary class attributes).

1. **No discriminatory attributes in the dataset.** For each redlining rule in this case, Inequality (3) should be enforced. Note that $conf(r_{b2} : D, B \rightarrow A)$

and $conf(r_{b1} : A, B \to D)$ are constant. The values of both sides of Inequality (3) are not independent; hence, a transformation is required that decreases the left-hand side of the inequality without any impact on the right-hand side. A possible solution for decreasing

$$conf(r : D, B \to C) = \frac{supp(D, B, C)}{supp(D, B)} \tag{5}$$

In inequality (3) to the target value is to perturb item D from $\neg D$ to D in the subset $\mathcal{DB}_c$ of all records of the original dataset which completely support the rule $\neg D, B \to \neg C$ and have minimum impact on other rules to increase the denominator of Expression (5) while keeping the numerator and $conf(B \to C)$ fixed.

2. **Discriminatory attributes in the dataset.** For each redlining rule in this case, Inequality (4) should be enforced. Note that in this case $conf(r_{b2} : D, B \to A)$ and $conf(r_{b1} : A, B \to D)$ might not be constant. So it is clear that the values of both inequality sides are dependent; hence, a transformation is required that increases the left-hand side of the inequality without any impact on the right-hand side. A possible solution for increasing

$$conf(B \to C) = \frac{supp(B, C)}{supp(B)} \tag{6}$$

in Inequality (4) to the target value is to perturb item C from $\neg C$ to C in the subset $\mathcal{DB}_c$ of all records of the original dataset which completely support the rule $\neg A, B, \neg D \to \neg C$ and have minimum impact on other rules; this increases the numerator of Expression (6) while keeping the denominator and $conf(r_{b1} : A, B \to D)$, $conf(r_{b2} : D, B \to A)$, and $conf(r : D, B \to C)$ fixed.

In Algorithms 1 and 2, records in $\mathcal{DB}_c$ should be changed until the transformation requirement is met for each redlining rule. Among the records of $\mathcal{DB}_c$, one should change those with lowest impact on the other (non-redlining) rules. Hence, for each record $db_c \in \mathcal{DB}_c$, the number of rules whose premise is supported by db_c is taken as the impact of db_c, that is $impact(db_c)$; the rationale is that changing db_c impacts on the confidence of those rules. Then the records db_c with minimum $impact(db_c)$ are selected for change, with the aim of scoring well in terms of the four utility measures proposed in the next section.

Background Information. In order to implement the proposed data transformation method for indirect discrimination prevention, we simulate the availability of a large set of background rules under the assumption that the dataset contains the discriminatory items. Let BK_s be a database of background rules be defined as

$$\mathcal{BK} = \{r_{b2} : X(D, B) \to A | A \text{ discriminatory itemset and } supp(X \to A) \geq ms\}$$

In fact, $\mathcal{BK}$ is the set of classification rules $X \to A$ with a given minimum support ms and A a discriminatory itemset. Although rules of the form r_{b1} :

Algorithm 1.

Inputs: $\mathcal{DB}$, $\mathcal{FR}$, $\mathcal{RR}$, α, DI_s
Output: $\mathcal{DB}'$: the transformed dataset
for each $r : X(D, B) \rightarrow C \in \mathcal{RR}$ **do**
  $\gamma = conf(r)$
  for each $r' : (A \subseteq DI_s), (B \subseteq X) \rightarrow C$ **do**
    $\beta_2 = conf(r_{b2} : X \rightarrow A)$
    $\Delta_1 = supp(r_{b2} : X \rightarrow A)$
    $\delta = conf(B \rightarrow C)$
    $\Delta_2 = Supp(B \rightarrow A)$
    $\beta_1 = \frac{\Delta_1}{\Delta_2}$ $//conf(r_{b1} : A, B \rightarrow D)$
    Find $\mathcal{DB}_c$: all records in $\mathcal{DB}$ that completely support $\neg D, B \rightarrow \neg C$
    for each $db_c \in \mathcal{DB}_c$ **do**
      Compute $impact(db_c) = |\{r_a \in \mathcal{FR} | db_c \text{ supports the premise of } r_a\}|$
    end for
    Sort $\mathcal{DB}_c$ by ascending impact
    while $\gamma \geq \frac{\alpha \cdot \delta \cdot \beta_2}{\beta_1} - (\beta_2 + 1)$ **do**
      Select first record db_c in $\mathcal{DB}_c$
      Modify D item of db_c from $\neg D$ to D in $\mathcal{DB}$
      Recompute $\gamma = conf(r : X \rightarrow C)$
    end while
  end for
end for
Output: $\mathcal{DB}' = \mathcal{DB}$

$A, B \rightarrow D$ are not included in $\mathcal{BK}$, $conf(r_{b1} : A, B \rightarrow D)$ could be obtained as $supp(r_{b2} : D, B \rightarrow A)/supp(B \rightarrow A)$.

From each redlining rule $(r : X(D, B) \rightarrow C)$ in combination with background knowledge, more than one α-discriminatory rule $r' : A, B \rightarrow C$ might be generated because of two reasons: 1) existence of different sub-itemsets $D, B \subseteq X$ such that X can be written as D, B and 2) existence of more than one item in the set of predetermined discriminatory items (DI_s). Hence, given a redlining rule (r), proper data transformation should be conducted for all α-discriminatory rules $r' : (A \subseteq DI_s), (B \subseteq X) \rightarrow C$ ensuing from r.

4 Experimental Evaluation

This section presents an experimental evaluation of our solution for indirect discrimination prevention. First, we present the utility measures that we propose to evaluate our solution. Finally, we report the experimental results.

4.1 Utility Measures

Two aspects are relevant to evaluate the performance of our indirect discrimination prevention method, namely the success of the method in removing all evidence of indirect discrimination from the original dataset (degree of discrimination prevention) and the impact of the method on data quality (degree of

Algorithm 2.

Inputs: $\mathcal{DB}$, $\mathcal{FR}$, $\mathcal{RR}$, α, DI_s
Output: $\mathcal{DB}'$: the transformed dataset
for each $r : X(D, B) \rightarrow C \in \mathcal{RR}$ **do**
  $\gamma = conf(r)$
  for each $r' : (A \subseteq DI_s), (B \subseteq X) \rightarrow C$ **do**
    $\beta_2 = conf(r_{b2} : X \rightarrow A)$
    $\Delta_1 = supp(r_{b2} : X \rightarrow A)$
    $\delta = conf(B \rightarrow C)$
    $\Delta_2 = Supp(B \rightarrow A)$
    $\beta_1 = \frac{\Delta_1}{\Delta_2}$ $//conf(r_{b1} : A, B \rightarrow D)$
    Find $\mathcal{DB}_c$: all records in $\mathcal{DB}$ that completely support $\neg A, B, \neg D \rightarrow \neg C$
    for each $db_c \in \mathcal{DB}_c$ **do**
      Compute $impact(db_c) = |\{r_a \in \mathcal{FR} | db_c \text{ supports the premise of } r_a\}|$
    end for
    Sort $\mathcal{DB}_c$ by ascending impact
    while $\delta \leq \frac{\beta_1(\beta_2+\gamma-1)}{\beta_2 \cdot \alpha}$ **do**
      Select first record db_c in $\mathcal{DB}_c$
      Modify C item of db_c from $\neg C$ to C in $\mathcal{DB}$
      Recompute $\delta = conf(B \rightarrow C)$
    end while
  end for
end for
Output: $\mathcal{DB}' = \mathcal{DB}$

information loss). A discrimination prevention method should provide a good trade-off between both aspects above. We propose the following measures for evaluating our solution:

- *Discrimination Prevention Degree* (DPD). This measure quantifies the percentage of redlining rules that are no longer redlining in the transformed dataset. It is defined as

$$DPD = \frac{|\mathcal{RR}| - |\mathcal{RR}'|}{|\mathcal{RR}|}$$

  where $\mathcal{RR}$ is the database of redlining rules extracted from $\mathcal{DB}$, $\mathcal{RR}'$ is the database of redlining rules extracted from the transformed dataset $\mathcal{DB}'$, and $|\cdot|$ is the cardinality operator.
- *Discrimination Protection Preservation* (DPP). This measure quantifies the percentage of the non-redlining rules in the original dataset that remain non-redlining in the transformed dataset. It is defined as

$$DPP = \frac{|\mathcal{NR} \bigcap \mathcal{NR}'|}{|\mathcal{NR}|}$$

  where $\mathcal{NR}$ is the database of non-redlining rules extracted from the original dataset $\mathcal{DB}$, and $\mathcal{NR}'$ is the database of non-redlining rules extracted from the transformed dataset $\mathcal{DB}'$.

- *Misses Cost* (MC). This measure quantifies the percentage of rules among those extractable from the original dataset that cannot be extracted from the transformed dataset (side-effect of the transformation process). It is defined as

$$MC = \frac{|\mathcal{FR}| - |\mathcal{FR} \bigcap \mathcal{FR}'|}{|\mathcal{FR}|}$$

where $\mathcal{FR}'$ is the database of frequent classification rules extracted from the transformed dataset $\mathcal{DB}'$.

- *Ghost Cost* (GC). This measure quantifies the percentage of the rules among those extractable from the transformed dataset that could not be extracted from the original dataset (side-effect of the transformation process). It is defined as

$$GC = \frac{|\mathcal{FR}'| - |\mathcal{FR} \bigcap \mathcal{FR}'|}{|\mathcal{FR}'|}$$

where $\mathcal{FR}'$ is the database of frequent classification rules extracted from the transformed dataset $\mathcal{DB}'$.

The DPD and DPP measures are used to evaluate the success of the proposed method in indirect discrimination prevention; ideally they should be 100%. The MC and GC measures are used for evaluating the degree of information loss (impact on data quality); ideally they should be 0% (MC and GC may not be 0% as a side-effect of the transformation process). MC and GC were previously proposed and used as information loss measures for knowledge hiding in PPDM [8].

4.2 Results

We use the German Credit Dataset [12] in our experiments, since it is a well-known and frequently used dataset in the context of anti-discrimination. In this dataset, we consider the following set of predetermined discriminatory items (DI_s): female and not single as *personal status*, unemployed or unskilled non resident as *job*, the attributes marking the individual as *foreign worker* and *old-aged*.

In this section, we present the experimental evaluation of the proposed method. For the first phase we have used Apriori [9]. The algorithms and the utility measures corresponding to the second and third phases of the proposed solution, respectively, were implemented using Microsoft Visual Studio 2008 with C# programming language. The tests were performed on an 2.27 GHz Intel® Core^{TM}i3 machine, equipped with 4 GB of RAM, and running under Windows 7 Professional.

In order to evaluate our proposed solution we need to simulate the background knowledge rules. Hence, we assume that the original dataset $\mathcal{DB}$ contains discriminatory attributes and implement Algorithm 2. The values of utility measures for minimum support 0.5% and minimum confidence 10% are presented in Table 1. In this experiment, the number of frequent classification rules extracted

Table 1. Utility measures for minimum support 0.5% and minimum confidence 10%

	N. Redlining Rules	N. α-Disc. Rules	MC	GC	DPD	DPP	Execution time (sec)
α= 0.6	1	2	0	0.21	100	100	11
α= 0.5	2	5	0.34	0.49	100	100	27
α= 0.4	3	7	0.52	0.47	100	99.95	49
α= 0.3	11	28	1.62	1.97	90.90	99.81	125

from $\mathcal{DB}$ is 7690 and the number of background knowledge rules is 7416. As shown, the results are reported for different values of $\alpha \subset [0.3, 0.6]$. We selected the upper bound (0.6) because, with respect to our predetermined discriminatory items, redlining rules could be extracted from $\mathcal{DB}$. We restrict the lower bound to limit the number of redlining rules extracted form $\mathcal{DB}$. Other than utility measures, the number of redlining rules and the number of α-discriminatory rules that could be generated from those redlining rules are also reported for different values of α.

As shown in Table 1, the values of DDP and DPD demonstrate that the proposed solution achieves a high degree of indirect discrimination prevention in different cases (*i.e.* different values of α). In addition, the values of MC and GC demonstrate that the proposed solution incurs little information loss, especially when α is not too small. By decreasing the value of α, the number of redlining rules is increased, which causes more data transformation to be done, thereby increasing MC and GC. As presented in Table 1, the execution time of the algorithm increases linearly with the number of redlining rules and α-discriminatory rules.

5 Conclusions

To the best of our knowledge, we have presented the first method for preventing indirect discrimination in data mining due to biased training datasets. Our contribution in this paper concentrates on producing training data which are free or nearly free from indirect discrimination while preserving their usefulness to data mining algorithms. In order to prevent indirect discrimination in a dataset, a first step consists in discovering whether there exists indirect discrimination. If any discrimination is found, the dataset is modified until discrimination is brought below a certain threshold or is entirely eliminated. In the future, we want to present a unified discrimination prevention approach based on the discrimination hiding idea that encompasses both direct and indirect discrimination.

Disclaimer and Acknowledgments. The authors are with the UNESCO Chair in Data Privacy, but they are solely responsible for the views expressed in this paper, which do not necessarily reflect the position of UNESCO nor commit that organization. This work was partly supported by the Spanish Government through projects TSI2007-65406-C03-01 "E-AEGIS", CONSOLIDER INGENIO

2010 CSD2007-00004 "ARES" and TSI-020100-2009-720 "everification", by the Government of Catalonia under grant 2009 SGR 01135, and by the European Commission under FP7 project "DwB". The second author is partly supported as an ICREA Acadèmia Researcher by the Government of Catalonia.

References

1. Pedreschi, D., Ruggieri, S., Turini, F.: Discrimination-aware data mining. In: Proc. of the 14th ACM International Conference on Knowledge Discovery and Data Mining (KDD 2008), pp. 560–568. ACM, New York (2008)
2. Kamiran, F., Calders, T.: Classification without discrimination. In: Proc. of the 2nd IEEE International Conference on Computer, Control and Communication (IC4 2009). IEEE, Los Alamitos (2009)
3. Ruggieri, S., Pedreschi, D., Turini, F.: Data mining for discrimination discovery. ACM Transactions on Knowledge Discovery from Data 4(2) Article 9 (2010)
4. Pedreschi, D., Ruggieri, S., Turini, F.: Measuring discrimination in socially-sensitive decision records. In: Proc. of the 9th SIAM Data Mining Conference (SDM 2009), pp. 581–592. SIAM, Philadelphia (2009)
5. Kamiran, F., Calders, T.: Classification with No Discrimination by Preferential Sampling. In: Proc. of the 19th Machine Learning Conference of Belgium and, The Netherlands (2010)
6. Calders, T., Verwer, S.: Three naive Bayes approaches for discrimination-free classification. Data Mining and Knowledge Discovery 21(2), 277–292 (2010)
7. Pedreschi, D., Ruggieri, S., Turini, F.: Integrating induction and deduction for finding evidence of discrimination. In: Proc. of the 12th ACM International Conference on Artificial Intelligence and Law (ICAIL 2009), pp. 157–166. ACM, New York (2009)
8. Verykios, V., Gkoulalas-Divanis, A.: A survey of association rule hiding methods for privacy. In: Aggarwal, C.C., Yu, P.S. (eds.) Privacy- Preserving Data Mining: Models and Algorithms. Springer, Heidelberg (2008)
9. Agrawal, R., Srikant, R.: Fast algorithms for mining association rules in large databases. In: Proc. of the 20th International Conference on Very Large Data Bases, pp. 487–499. VLDB (1994)
10. Hajian, S., Domingo-Ferrer, J., Martínez-Ballesté, A.: Discrimination prevention in data mining for intrustion and crime detection. In: Proc. of the IEEE Symposium on Computational Intelligence in Cyber Security (CICS 2011), pp. 47–54. IEEE, Los Alamitos (2011)
11. Hajian, S., Domingo-Ferrer, J., Martínez-Ballesté, A.: Rule generalization and protection for discrimination prevention in data mining (submitted)
12. Newman, D.J., Hettich, S., Blake, C.L., Merz, C.J.: UCI repository of machine learning databases (1998), `http://archive.ics.uci.edu/ml`

A Comparison of Two Different Types of Online Social Network from a Data Privacy Perspective

David F. Nettleton[1,2], Diego Sáez-Trumper[2], and Vicenç Torra[1]

[1] Artificial Intelligence Research Institute, IIIA
Spanish National Research Council, CSIC
Campus Universitat Autònoma de Barcelona 08193 Bellaterra, Catalonia, Spain
{Dnettleton,vtorra}@iiia.csic.es
[2] Pompeu Fabra University, c/Tanger 122-140
08018 Barcelona, Spain
diego.saez01@upf.edu

Abstract. We consider two distinct types of online social network, the first made up of a log of writes to wall by users in Facebook, and the second consisting of a corpus of emails sent and received in a corporate environment (Enron). We calculate the statistics which describe the topologies of each network represented as a graph. Then we calculate the information loss and risk of disclosure for different percentages of perturbation for each dataset, where perturbation is achieved by randomly adding links to the nodes. We find that the general tendency of information loss is similar, although Facebook is affected to a greater extent. For risk of disclosure, both datasets also follow a similar trend, except for the average path length statistic. We find that the differences are due to the different distributions of the derived factors, and also the type of perturbation used and its parameterization. These results can be useful for choosing and tuning anonymization methods for different graph datasets.

Keywords: Social network, data privacy, descriptive statistics, risk of disclosure, information loss.

1 Introduction

Data Privacy in Social Network logs is now an important issue, given that millions of users worldwide are generating high volume data logs of their online social network activity and relations. This data offers a great analysis opportunity to data miners, but on the other hand, it may represent a threat to an individual's data privacy if it falls into the wrong hands. However, if we can sufficiently protect the data by anonymization techniques, then we can publish the social network log data for commercial and academic use.

In the current work we statistically compare and anonymize two real datasets represented as a graph, from a data privacy perspective: the Enron emails dataset [1] and the Facebook New Orleans dataset [2]. We calculate descriptive statistics for the graphs: degree, clustering coefficient and average path length. Then we anonymize/

V. Torra et al. (Eds.): MDAI 2011, LNAI 6820, pp. 223–234, 2011.

perturb the datasets by randomly adding links to the nodes and calculate the information loss and risk of disclosure for different degrees of perturbation.

The structure of the paper is as follows: in Section 2 we present the state of the art and related work; in Section 3 we define the basic data and derived factors used to describe the graphs; in Section 4 we present the statistics calculated for both graphs and make comments and comparisons; in Section 5 we calculate the information loss and risk of disclosure for both datasets, for different degrees of perturbation; finally, in Section 6 we summarize the present work.

2 State of the Art and Related Work

Privacy in on-line social networks is a relatively new area of research which however has a solid base in classic graph theory and data privacy concepts.

We will consider the state of the art from two main perspectives: the statistical analysis of online social networks, and data privacy analysis of online social networks. In terms of data privacy in general, we can cite Sweeney's paper on k-anonymity[3], and more recently [4], in which key definitions are given for information loss and risk of disclosure.

In the field of the statistical analysis of online social networks, some key authors are: Kumar[5], Ahn[6], Klienberg[7,8], Mislove[9], Shetty[1] and Viswanath[2]. In [1], Shetty et al. present some concepts related to 'graph entropy' and the identification of 'important' or 'interesting nodes. The study is specifically applied to the Enron email dataset. The basic idea is to measure the effect of removing a node from a graph, as the difference between the 'entropy' of the graph before and after removing the given node. In [9], Mislove defines some of the key metrics which characterize a social network. Viswanath in [2] performs a statistical analysis of the New Orleans Facebook dataset (the dataset we use in the present work), using the degree, clustering coefficient and average path length statistics to evaluate social network evolution over time. Klienberg[7,8] considers data mining of online social networks, defining different possible topologies within OSNs and making considerations about the computational cost of data processing.

In the field of data privacy analysis applied to online social networks, we can cite Hay[10], Zhou[11], Wondracek[12] and Liu[13]. Hay[10] presents a simple graph anonymization based on random addition and deletion of edges. The attack method attempts re-identification using two types of queries, vertex refinement and sub-graph knowledge. The risk measure is considered as the percentage of nodes whose equivalent candidate set falls into one of a given set of buckets (1 node, 2-4 nodes, 5-10 nodes, ...). The information loss measure calculates some common graph metrics (clustering coefficient, path length distribution, degree distribution, ...) in the graph before and after anonymization. The information loss is considered from the point of view of an analyst who consults these statistical properties. Zhou[11] presents a more sophisticated anonymization algorithm which firstly generalizes vertex labels and secondly adds edges. One of the precepts of the approach is to create local topologies which are isomorphic with other local topologies, achieved by adding edges to them. Wondracek[12] presents a different approach, in that the attacker uses a malicious website to obtain information about users of an on-line social network. Finally, Liu, in

[13], presents a defense method which is k-anonymous, that is, it produces k-degree anonymized degree sequences.

3 Definition of Basic Data and Derived Factors

In this Section we present the datasets used and their data format. We also define the derived statistical factors which we later use to calculate the information loss and risk of disclosure for the graphs.

The Enron email dataset[1] consists of a collection of 150 folders corresponding to the emails to and from senior management and others at Enron, collected over a period between 1998 to 2002. The total number of emails sent/received between users is approx. 1.5 million. We filtered the records so as to only include users with mutual links for which at least one email was sent and received along the link. This gave us a subset of 10630 users, which we used for all the analysis in the current work. Each email sender/recipient represents a node in the graph and the activity is represented by the number of emails sent-received along the edges which connect the users. We consider the email corpus as an extension of the idea of an "online social network", useful for comparison purposes with the Facebook data.

The Facebook New Orleans dataset was generated by Viswanath et al[2] by crawling the Facebook New Orleans regional network, and consists of approx. 63,000 users, 1.5 million links between users, and 800,000 logged interactions over a two year period. We filtered the records so as to only include users with mutual links for which at least one write to wall was sent in each direction. This gave us a subset of 31720 users, which we used for all the analysis in the current work. In contrast to the Enron dataset, for which a link between users is established when an email is sent/received between them, in the case of the Facebook users, a link is established by the explicit solicitation and acceptance of friendship. Also, in the Facebook dataset, 'writes to wall' is the activity indicator.

Basic Data - Facebook: the available data consists of one file which contains writes to wall between users and their corresponding timestamps. The format of the write to wall data is {user-id 1, user-id 2, timestamp}, where the user ids are anonymous numbers between 1 and 63000. For example, {1, 2, 3-4-2010} would signify that user 1 wrote on user 2's wall on the 3rd of April, 2010. All links are reciprocal, therefore, in the dataset there will be a corresponding record: {2, 1, ...}. This is assured by only including users who reciprocally wrote on each others' walls, at least once.

Basic Data - Enron: the available data consists of separate sender and recipient files which we merged into one file and used as input to create the graph. We anonymized the emails to sequential integers. In the original files, the 'to' and 'cc' type recipients are not distinguished, following Shetty's [15] approach. This gives us a unique file with two columns of anonymized id's, the first id is that of the sender and the second is that of the recipient. In order to construct the graph and the edges, we select unique id's between sender and recipient.

Note that we consider both graphs (Facebook and Enron) as undirected in the current study, that is the degree (total number of links to a node) is considered as the in-degree (number of incoming links) + the out-degree (number of outgoing links).

Derived Factors: in order to calculate the statistics, we have implemented the algorithms which process the graphs in Java. In the case of the 'apl' (average path length) statistic, we have used Dijkstra's algorithm[14]. The following basic statistics have been calculated to describe the graph:

(i) **Degree:** number of immediate neighbors which a node has.

(ii) **Clustering Coefficient:** is an indicator of how many of the "friends" of a user, are friends of each other.

$$CC = \frac{Number\_of\_mutual\_friends\_of\_user\_i}{Total\_number\_of\_friends\_of\_user\_i} \tag{1}$$

Example: if user 1 has 30 friends, and of those 30 friends, 7 have links between them, independently of the link with user 1, then the CC for this "group" will be 7 / 30 = 0.233. For the New Orleans Facebook dataset an average CC value of 0.0257 was reported., and for Enron, 0.15.

(iii) **Average path length:** For each node x this is the average of the sum of the shortest number of hops required to reach every other node y in the graph:

$$APL(x) = \frac{\sum_{i=1}^{n}(shortest\_path\_length\_from\_node\_x\_to\_node\_y_i)}{n} \tag{2}$$

4 Descriptive Statistics for Derived Factors

In this Section we present the descriptive statistics for the Facebook and Enron datasets, and compare the two.

Firstly we will comment the Enron and Facebook correlation statistics for 'degree', 'cc' (clustering coefficient) and 'apl' (average path length). For Enron, the highest correlation was between "degree" and "apl" (-0.49), some correlation between "degree" and "cc" (-0.12), and a negligible correlation between "cc" and "apl" (-0.001). With reference to the Facebook correlation statistics, the highest correlation was between "degree" and "apl" (-0.14), followed by the correlation between "degree" and "cc" (0.12), and a negligible correlation between "cc" and "apl" (-0.037).

In Table 1 (Enron) we observe a high standard deviation of degree with respect to the average value (2 times more than its average value), whereas "cc" shows a lesser deviation and "apl" shows a significantly smaller relative deviation (7.3 times less than its average value). In Table 1 (Facebook) we observe a high standard deviation of "cc" with respect to its average value (more than twice), whereas "degree" shows a deviation slightly greater than its average value and "apl" shows a significantly smaller relative deviation (3.37 times less than its average value).

In terms of the distributions, the degree displays a typical "power law" distribution for both datasets, with just a few nodes having a very high degree. The distribution of the clustering coefficient for Facebook and Enron have different characteristics: for Facebook, In the first two quartiles and half the third quartile, all the nodes have a "cc" equal to zero, which means that none of the neighbors are interconnected between each other. The distribution of the *average path length* for both datasets shows a characteristic 'S' pattern, but in the case of Facebook the left hand ascent is displaced to the right, which implies there are more nodes with a small average path length.

Table 1. Averages and standard deviations of statistical factors for Enron and Facebook datasets

		degree	**cc**	**apl**
Enron	average	31.035	0.1556	3.1516
	standard dev.	63.381	0.1121	0.4173
Facebook	average	5.0815	0.0257	6.001
	standard dev.	6.4705	0.0528	1.7782

5 Data Privacy: Information Loss and Risk of Disclosure - Enron vs. Facebook

In this Section we present the results of Information Loss and Risk of Disclosure for the Enron and Facebook datasets, and compare the two.

5.1 Information Loss

The objective of this test is to introduce a given percentage of random perturbation into the graph data and observe the change in the graph statistics. We interpret information loss as the deviation from the original data which a data analyst (end user of the data) would perceive. We measure the information loss by calculating the correlations between the three key descriptive variables for the original graph (degree, clustering coefficient and average path length) and then for the perturbed graph. The difference will then be the information loss. That is, if $C_d o$, $C_d p$, $C_{cc} o$, $C_{cc} p$, $C_{apl} o$ and $C_{apl} p$ are the correlations of the degree, clustering coefficient and average path length, for the original graph and the perturbed graph, respectively, then:

$$Inf.Loss = \frac{|(|C_d o| - |C_d p|)| + |(|C_{cc} o| - |C_{cc} p|)| + |(|C_{apl} o| - |C_{apl} p|)|}{3} \tag{3}$$

The correlation values are already normalized between -1 and 1, and we take the absolute value to obtain a number between 0 and 1. The difference between the correlation values is a typical statistic used in the data privacy literature. The perturbation method we have used, that of adding links to nodes, selected randomly in the graph, is also a common graph perturbation method used in the literature of graph privacy[10,11]. We add one link to each randomly selected node. Thus a perturbation of 25% means that we added one link between 25% of the nodes in the graph. Each node can only be selected once in any trial. We note that each trial (for each % perturbation) was repeated randomly three times as an experimental procedure to validate the results, and the average was taken.

Primary and Secondary (collateral) perturbation. Given the interrelated nature of graph data, if we modify a given (primary) node, other (secondary) nodes may also be affected. Our perturbation measure refers only to the number of primary nodes modified. However it is worthwhile to comment the aspect of secondary node modification, how it may affect the results, and how we could measure it. In this context, we propose that the way in which the results are affected depends on the way we define "risk of disclosure", which in our case is in terms of statistical properties such as degree, clustering coefficient and average path length, with a "hit" margin of 1%.

Given that, in our current work, the perturbation operator is "add link", then the only statistical value which will be directly modified (and which cannot be modified indirectly), is the *degree*. On the other hand, the *clustering coefficient*, in some cases, could change as a secondary effect (of joining two neighbors together, for example). Finally, the *average path length* is the statistic which would be most likely to change, if we add links to the graph. However, in general, from empirical observation of the data values before and after perturbation for the same nodes, the values only register relatively small alterations.

In conclusion, we propose that it would be reasonable to consider that the risk of disclosure (the percentage of hits), within the defined attacker "hit" margin of 1%, is equivalent to one minus the percentage of nodes affected both primarily and secondarily with a margin greater than 1%. That is:

$$DR = 1 - A_{PS} \tag{4}$$

For example, in Table 2 we show the relation between Risk of Disclosure and the total percentage of nodes affected, for the Enron dataset with 50% perturbation.

Table 2. Relation between Risk of Disclosure and Nodes affected for the Enron dataset and 50% perturbation

Attacker query	%Hits (Risk of Disc.)	%Nodes whose values are affected more than 1% (primary and secondary)
Degree	**0.54**	**0.46***
Degree, cc	**0.49**	**0.51**
Degree, cc, apl	**0.48**	**0.52**

We observe that the percentage of affected nodes is only 46% (with margin > 1%) for 50% perturbation. This is possible because there exist a percentage of nodes with more than 100 links, thus if we add just one link to one of these nodes, the change will be less than (or equal to) 1%, and thus with this criteria will not count as having being perturbed.

Enron. In Fig. 1a we see the information loss for different percentages of perturbation on the Enron graph. On the *y-axis* a value of 0.01 represents an information loss of 1%, and on the *x-axis* a value of 0.1 represents a grade of perturbation of 10%.

We observe a fairly linear relation between the two, with a slightly steeper gradient between 25% and 100% perturbation. We note that the maximum information loss is only 7.7% at 100% perturbation. This is correct given our definition of information loss and perturbation: adding just one link to a node in a graph when the average degree is 31.03 (see Table 1) will not have a great influence on the graph overall. This allows a comparison with the results of Facebook for which the average degree is much lower at 5.08 (Table 1) and therefore we would expect that adding one link to a node will have a significantly greater effect on the graph statistics.

In Table 3 (Enron) we see the results of the tests of perturbation versus information loss on the Enron graph dataset, which are also plotted in Fig. 1a. A clear increasing and linear trend for information loss is evident in relation with increasing perturbation values.

Facebook. In Fig. 1b we see the information loss for different percentages of perturbation on the Facebook graph. On the *y-axis* a value of 0,01 represents an information loss of 1%, and on the *x-axis* a value of 0.1 represents a grade of perturbation of 10%. Similarly to the Enron dataset (Fig. 1a), we observe a fairly linear relation between the two, with a slightly steeper gradient between 25% and 100% perturbation. However, in contrast to the Enron results, the information loss is significantly greater, ranking from 2.5%, for 10% perturbation, to 27.4% for 100%

Table 3. Results of tests of perturbation *versus* information loss on the Enron and Facebook graph datasets

	Perturbation				
	10%	**25%**	**50%**	**75%**	**100%**
Enron	0.00702	0.02056	0.04009	0.06010	0.07742
Facebook	0.02519	0.06387	0.14156	0.21435	0.27491

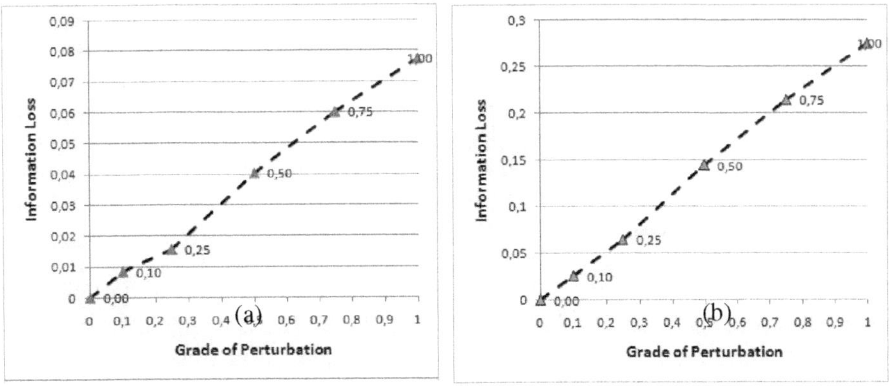

Fig. 1. Information Loss *versus* Grade of Perturbation: (a) Enron, (b) Facebook. The marker labels indicate the grade of perturbation.

perturbation. This is primarily due to the greater impact of adding one link to each node, given the different statistical characteristics of Enron with respect to Facebook, especially the smaller average degree of the nodes in Facebook (ratio of degree in Facebook Vs degree in Enron is 6.1 to 1).

In Table 3 (Facebook) we see the results of the tests of perturbation versus information loss on the Facebook graph dataset, which are also plotted in Fig. 1b. Again, a clear increasing and linear trend for information loss (2%, 6%, 14%, 21%, 27%) is evident in relation with increasing perturbation values (10%, 25%, 50%, 75% and 100%).

5.2 Risk of Disclosure

The risk of disclosure is calculated by launching a query on the graph to find a given sub-graph topology (node and its immediate neighborhood) in the complete graph, with a % margin. A check is made to determine if the target node is in the subset S returned, and how many nodes are in S (value equivalent to that given by k anonymity). We perceive the attacker as statistically knowledgeable and whose objective is to identify specific nodes and their immediate neighbors, in a simply anonymized graph.

Consider that if we do not consider the 'apl' statistic, then there are many low risk users, whose 'cc' is equal to zero and/or whose 'degree' is equal to one. The 'apl' statistic is much more expensive and difficult to obtain, because it needs access to the whole graph dataset, thus we have considered the risk with and without the 'apl' statistic. Thus, we have three different measures for the risk of disclosure, defined by three queries:

- Q_1 which searches for a given node based on 'degree'
- Q_2 which searches for a given node based on 'degree' and 'cc'
- Q3 which searches for a given node based on 'degree', 'cc' and 'apl'.

For Q_1 we only consider users with degree > 1, and for Q_2 and Q_3 we only consider users with degree > 1 and cc > 0.0. All queries are allowed a 1% margin of error.

The *risk of disclosure* for a given node *Ng* in the original graph is calculated by multiplying the % of correct hits on the perturbed dataset for node *Ng*, by the % of nodes which are returned by the query within a given margin with respect to node *Ng*. We apply a margin of 1% in all cases. That is, if the degree of node *Np* in the perturbed dataset is within 1% of the degree of node *Ng* in the original dataset, then it is returned by the query. The same margin of 1% applies to the 'cc' and 'apl' values. Finally, a 'hit' is considered when the unique id of a node *Np* returned by the query has the same unique id as the node *Ng* in the original graph.

Facebook. With reference to Table 4, we see the results of the three query types and grades of perturbation, on the risk of disclosure. In Table 4 we see that for the degree query, the risk of disclosure reduces from 90.11% risk for 10% perturbation, to 0.009% risk for 100% perturbation, a significant reduction, for a simple query based only on degree. For progressively more complex queries, we observe a faster reduction in risk. In the case of 'degree, cc' the risk reduces from 84% to 0.007%, for 10% to 100% perturbation. In the case of the 'degree, cc, apl' query the reduction of risk occurs earlier: from 71% to 0.00026% for 10% to 50% perturbation. This is because the 'apl' (average path length) statistic of a node is very sensitive to change when one link is added to the node. The 'apl' value is also much more statistically diverse than the 'degree' and 'cc' values.

In Fig. 2a (Facebook) we see a sharp drop for the risk of the 'degree,cc, apl' query, for increasing percentages of perturbation, whereas the other two queries, 'degree' and 'degree, cc' show a more gradual drop, from 90% and 85% risk respectively, for 10% perturbation, to 50% and 35% risk respectively, for 50% perturbation.

Enron. The results shown in Table 5 have the same format and calculation method as we have described previously for the Facebook data of Table 4. We see the results of the three query types and grades of perturbation, for the risk of disclosure.

In Table 5 (Enron) we see that for the degree query, the risk of disclosure reduces from 94.34% risk for 10% perturbation, to 11.74% risk for 100% perturbation, with a similar decreasing tendency as for the Facebook data (Table 4), but leaving a greater residual risk. For the query 'degree, cc' the risk reduces from 83% to 1.6%, for 10% to 100% perturbation, again with a similar decreasing tendency as for the Facebook data, but leaving a greater residual risk. However, the risk reduction of the query 'degree, cc, apl' behaves in a different way to the Facebook query. The reduction of risk is much less pronounced: from 81% to 28% for 10% to 50% perturbation. This is due to two factors: (i) the sensitivity of the 'apl' value and (ii) the difference in the 'apl' values for Facebook and Enron (see Table 1): the average 'apl' for the Enron dataset is much smaller than that of Facebook (3.1516 and 6.001, respectively), and the other statistics related to 'apl' are also different if we compare the datasets.

In Fig. 2b (Enron) we see a sharper drop for the risk of the 'degree, cc' and 'degree,cc, apl' queries (relative to the 'degree' query), for increasing percentages of perturbation, from 82% and 81% risk for 10% perturbation, to 1.6% and 0.7% risk, respectively, for 100% perturbation. Both queries follow a very similar line. On the other hand, the 'degree' query shows a more gradual drop for increasing perturbation. These tendencies are similar to the results for the Facebook dataset, as seen in Figure 2a, with the exception of the query including 'apl', which shows a much more gradual descent, as we have already discussed with reference to Table 4.

Information Loss vs Risk of Disclosure. With reference to Fig. 3, we see a plot of the results of Sections 5.1 and 5.2, for Information Loss and Risk of Disclosure for the Facebook (Fig. 3a) and Enron (Fig. 3b) datasets, respectively. Note that for information loss we have just one value for each degree of perturbation (see Sec. 5.1),

Table 4. Results of tests of perturbation *versus* risk of disclosure on the Facebook graph dataset

		Perturbation				
		10%	25%	50%	75%	100%
Risk of Dis-closure	degree	0.9011	0.7495	0.5014	0.2495	0.00009
	degree, cc	0.84834	0.6076	0.3415	0.1407	0.00007
	degree, cc, apl	0.71068	0.07333	0.00026	0.0000	0.00000

Table 5. Results of tests of perturbation *versus* risk of disclosure on the Enron graph dataset

		Perturbation				
		10%	25%	50%	75%	100%
Risk of Dis-closure	degree	0.9434	0.8570	0.6454	0.3943	0.11747
	degree, cc	0.82960	0.60519	0.31081	0.11363	0.01613
	degree, cc, apl	0.81819	0.58871	0.28832	0.09804	0.00798

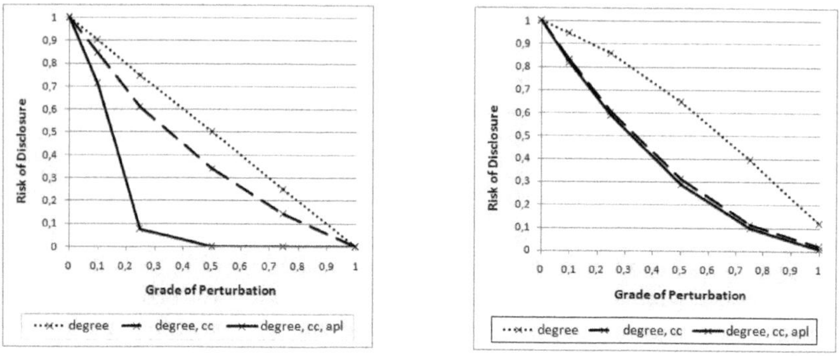

(a) (b)

Fig. 2. Risk of disclosure *versus* Grade of Perturbation: (a) Facebook, (b) Enron

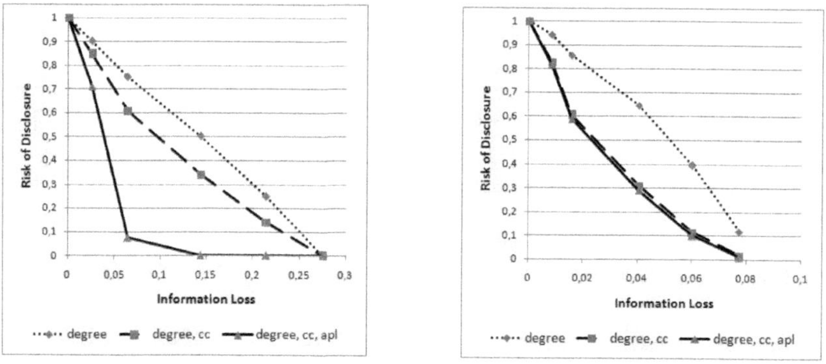

(a) (b)

Fig. 3. Information Loss *versus* Risk of Disclosure: (a) Facebook, (b) Enron

whereas for risk of disclosure we have three values (one for each query, see Sec. 5.2). We observe that the information loss of the Facebook dataset (Fig. 3a) ranges from 2.5% to 27.4% for a risk of disclosure which drops from 71% to 0.0%, in the case of the query 'degree, cc, apl'. On the other hand, the Enron dataset (Fig. 3b) has an information loss which rises from 0.7% to 7.7%, for a risk of disclosure which drops from 81.8% to 0.7%, in the case of the query 'degree, cc, apl'. Thus, we see that

Facebook has a greater reduction in risk of disclosure than Enron, especially for Q3 (degree, cc, apl). However, Facebook achieves this at a cost of four times the information loss, with respect to Enron. In summary, we can say that the information loss is relatively low for Enron (max. of 7.7%), whereas the Facebook result, with a maximum information loss of 27.4%, leaves room for improvement.

6 Conclusions

In this paper we have represented the Facebook and Enron user data and activity as a graph, which has allowed us to derive descriptive factors based on graph theory. We have introduced different percentages of perturbation into the data, by randomly adding links to the nodes. Then we have analyzed the information loss and risk of disclosure of the graphs from a data privacy point of view.

Lessons learnt: *firstly*, the perturbation method should be calibrated for each dataset. In our case, the perturbation method was 'add one link to node', and we could calibrate by varying the number of links added, based on the average degree value, for example; *second*, the risk of disclosure has to take into account the number of hits achieved within the subset of nodes returned by a query, rather than just the number of nodes returned (we note that this is distinct from k-anonymity); *thirdly*, it is important to filter the data, due to the presence of many nodes with just one link or with cc=0.0, in the graph. We filter these nodes because they are not interesting for a potential attacker because of their lack of interrelations (poor topology) and because they cannot be distinguished without the 'apl' (average path length) value, which is much more expensive and difficult to obtain. Also many values of 'degree' equal to one and 'cc' equal to zero would distort the graph statistics.

Future work: It would be interesting to try different perturbation methods on the graph, such as 'node aggregation' and compare this with 'add link'. For 'node aggregation' we could then consider 'k-anonymity' as a risk disclosure measure. Also it would useful to contrast the results for more online social network datasets, such as Twitter and a synthetic small-world graph.

Acknowledgements. This research is partially supported by the Spanish MEC (projects ARES CONSOLIDER INGENIO 2010 CSD2007-00004 -- eAEGIS TSI2007-65406-C03-02 -- and HIPERGRAPH TIN2009-14560-C03-01).

References

1. Shetty, J., Adibi, J.: Discovering Important Nodes through Graph Entropy - The Case of Enron Email Database. In: KDD 2005, Chicago, Illinois (2005)
2. Viswanath, B., Mislove, A., Cha, M., Gummadi, K.P.: On the Evolution of User Interaction in Facebook. In: Proc. 2nd ACM Workshop on Online Social Networks (WOSN), Barcelona, Spain, August 17 (2009), `http://socialnetworks.mpi-sws.org/`
3. Sweeney, L.: k-anonymity: a model for protecting privacy. International Journal of Uncertainty Fuzziness and Knowledge-Based Systems (IJUFKS) 10(5), 557–570 (2002)

4. Domingo-Ferrer, J., Rebollo-Monedero: Measuring Risk and Utility of Anonymized Data Using Information Theory. In: Int. Workshop on Privacy and Anonymity in the Information Society, PAIS 2009 (2009)
5. Kumar, R., Novak, J., Tomkins: Structure and evolution of online social networks. In: Proc. 12th ACM SIGKDD Int. Conf. on Knowledge Discovery and Data Mining. ACM, New York (2007)
6. Ahn, Y., Han, S., Kwak, H., Moon, S., Jeong: Analysis of topological characteristics of huge online social networking services. In: Proc. 16th Int. Conf. WWW 2007, USA (2007)
7. Kleinberg, J.: Challenges in Mining Social Network Data. In: Proc. of the 13th ACM SIGKDD Int. Conference on Knowledge Discovery and Data Mining (KDD 2007), pp. 4–5 (2007)
8. Kleinberg, J., Backstrom, L., Dwork, C., Liben-Nowell, D.: Algorithmic Perspectives on Large-Scale Social Network Data. In: Data-Intensive Computing Symposium, (March 26, 2008 - Hosted by Yahoo! and the CCC), http://research.yahoo.com/files/7KleinbergSocialNetwork.pdf
9. Mislove, A., Marcon, M., Gummadi, K.P., Druschel, P., Bhattacharjee, B.: Measurement and analysis of online social networks. In: Proceedings of the 7th ACM SIGCOMM Conference on Internet Measurement, San Diego, California, USA (2007)
10. Hay, M., Miklau, G., Jensen, D., Weis, P., Srivastava, S.: Anonymizing Social Networks. SCIENCE Technical Report 07-19, vol. 245, pp. 107–3, Computer Science Department, University of Massachusetts Amherst (2007)
11. Zhou, B., Pei, J.: Preserving Privacy in Social Networks against Neighborhood Attacks. In: IEEE 24th International Conference on Data Engineering (ICDE), pp. 506–515 (2008)
12. Wondracek, G., Holz, T., Kirda, E., Kruegel, C.: A Practical Attack to De-Anonymize Social Network Users. In: Proc. IEEE Symp. on Security and Privacy, pp. 223–238 (2010)
13. Liu, K., Terzi, E.: Towards Identity Anonymization on Graphs. In: SIGMOD 2008 (2008)
14. Dijkstra, E.W.: A note on two problems in connexion with graphs. Numer. Math. 1, 269–271 (1959)

On the Declassification of Confidential Documents

Daniel Abril[1], Guillermo Navarro-Arribas[2], and Vicenç Torra[1]

[1] Institut d'Investigació en Intel·ligència Artificial (IIIA), Consejo Superior de Investigaciones Científicas (CSIC)

[2] Dep. Enginyeria de la Informació i de les Comunicacions (DEIC), Universitat Autònoma de Barcelona (UAB)

Abstract. We introduce the anonymization of unstructured documents to settle the base of automatic declassification of confidential documents. Departing from known ideas and methods of data privacy, we introduce the main issues of unstructured document anonymization and propose the use of named entity recognition techniques from natural language processing and information extraction to identify the entities of the document that need to be protected.

Keywords: Privacy, Declassification, Anonymization, Named Entity Recognition.

1 Introduction

Declassification of text documents is a key issue for governments and organizations. Documents in plain text with sensitive information are kept by companies and official organizations. However, on the one hand, some documents might be required by third parties for a particular use, and, on the other, digital information can be copied easily. Therefore, it is foreseeable to expect unauthorized copies. In fact, the international non-profit organization, WikiLeaks, has been publishing thousands of classified information (about military and diplomatic issues) of many countries of the world.

The importance of this problem has attracted the attention of some international agencies. For example, the DARPA, the Defense Advanced Research Projects Agency of the United States Department of Defense, solicited for *new technologies to support declassification* of confidential documents [6].

The maturity of these technologies would permit partial or complete declassification of documents. In this way, documents could be transferred to third parties without any confidentiality problem, or with the only information really required by the third party. Aiming to make the possibility of sensitive information leakage minimal.

These technologies will also help the capability of departments to identify still-sensitive information and to make declassified information available to the public.

We introduce the anonymization of unstructured documents departing from more traditional data privacy approaches in statistical disclosure control and privacy preserving data mining. Moreover we propose and analyze the use of named entity recognition, used mainly in natural language processing and information extraction, to identify the parts of the document, which need to be protected.

V. Torra et al. (Eds.): MDAI 2011, LNAI 6820, pp. 235–246, 2011.

In Section 2 we introduce our motivations describing basic concepts of data privacy and named entity recognition. In Section 3 and 4 we introduce and discuss our proposal. Section 5 describes our experiments, and finally, Section 6 concludes the paper.

2 Motivations and Preliminary Notions

The work presented in this paper serves as a starting point which aims to provide tools for the anonymization of unstructured documents. This is a field that to our knowledge has not been investigated yet and will surely provide interesting research result. In this section we introduce some concepts of data privacy and named entity recognition, which are the base of our proposal.

We introduce the anonymization of unstructured documents departing from the well known foundations of the statistical disclosure control (SDC) [26] and privacy preserving data mining (PPDM) [2]. Currently, there are a great number of protection methods used in data privacy, both from SDC and PPDM. See [22] for a classification of such methods.

In practice there are some differences between SDC and PPDM. The former, was originated from statistical offices to be able to publish statistical data from census and questionnaires for researchers or policy makers. On the other hand, PPDM, was introduced by the data mining community, where the ability to mine anonymous data is very valuable to companies and researchers. Nevertheless both disciplines share a lot of similarities and most methods are actually used i both cases.

SDC is mainly concerned with the protection of *microdata* files. Files with several attributes (columns) for a set of individuals (rows). PPDM also uses similar microdata files, although sometimes the attributes come from other structured data sources such as computer logs [18].

In data privacy, attributes are commonly classified regarding their sensitiveness from a privacy point of view. *Identifiers* are those attributes which can unequivocally identify an individual such as a social security number. On the other hand *quasi-identifiers* [5] are attributes (or sets of attributes) that, in combination with external information can be used to re-identify individuals. Some authors also refer to *confidential* or private attributes, which are those that provide the sensitive information about the correspondent.

Most protection methods are concerned with quasi-identifier attributes. Identifiers are normally deleted or encrypted, and the main objective of protection methods is to introduce enough perturbation in quasi-identifiers to make it unfeasible to re-identify the correspondent from them, while preserving confidential attributes.

2.1 Overview of Popular Protection Methods

We overview here three common protection methods used in SDC and PPDM. This is not an exhaustive list, but an example of some popular methods. We assume to be working with a set of records, with V_i attributes (or variables) for each one.

Microaggregation. Microaggregation provides privacy by means of clustering the data into small clusters and then replacing the original data by the centroids of the corresponding clusters.

Privacy is achieved because all clusters have at least a predefined number of elements, and therefore, there are at least k records with the same value. Note that all the records in the cluster replace a value by the value in the centroid of the cluster. The constant k is a parameter of the method that controls the level of privacy. The larger the k, the more privacy we have in the protected data.

Microaggregation was originally [7] defined for numerical attributes, but later extended to other domains. E.g., to categorical data in [23] (see also [8]), and in constrained domains in [24].

Rank Swapping. In Rank Swapping, the values of a variable V_i are ranked in ascending order; then each ranked value of V_i is swapped with another ranked value randomly chosen within a restricted range (e.g., the rank of two swapped values cannot differ by more than p percent of the total number of records). The method was first described for numerical variables in [16].

Additive Noise. Additive Noise adds Gaussian noise to the original data to get the masked data [3]. For example, if the standard deviation of the original variable is σ, noise can be generated using a $N(0, p\sigma)$ distribution, where p is the parameter of the method determining the protection degree.

2.2 Named Entity Recognition

The term *named entity* (NE) is normally used to refer to an entity for which one or many rigid designators stand for the referent [17]. In an unstructured text typical named entity are proper names, locations, or organizations. The task of identifying named entities is called Named Entity Recognition and Classification, NERC, or simply NER. NERC is an important task of information extraction and was initiated in the MUC (Message Understanding Conference) conferences [9].

The original classification of NE called *enamex* considers three specializations of NE: "persons", "locations", and "organizations". Later works have extended this classes either by fine-grained types, e.i locations can be divided in subtypes such as city, state, and country, or by introducing new types. Currently, there are NE type hierarchies with about 200 categories. Nevertheless most current generic NERC systems focus the recognition to the initial enamex types.

In order to identify NEs, NERC systems rely on different features associated to the text at different levels, normally: word, list, and document. Common word-level features are: case, punctuation, digit, character, morphology, part-of-speech, ... List features, also called gazetteer, lexicon, and dictionary features, take into account information beyond the single word, for example entities from a general dictionary, lists of stop words, lists of common abbreviations, list of known organizations, celebrities, politicians, etc., or lists of entity cues such as person title, or typical words in organitzation names (Associates, inc., corp., ...). Finally, document features are defined over document content and structure, and normally are based in large collections of documents (corpora). Example document features are: multiple occurrences (presence of the same entity in the context, disambiguation of uppercase and lowercase occurrences, anaphora, etc.), corpus frequency, or meta information obtained in semi-structured documents (HTML, or XML tags and hierarchies).

NERC systems are normally classified into supervised, semi-supervised, and unsupervised learning. These systems aim to generate classification rules for distinctive features to recognize entities. Supervised learning NERC systems use techniques such a Hidden Markov Models, Decision Trees, Support Vector Machines, or Conditional Random Fields. More recent semi-supervised systems, an unsupervised systems (normally based on clustering) have also been designed. A detailed survey can be found in [17]. More generic information extraction systems that include NERC as a sub-task are also interesting [4].

> [*PER* Prof. Smith] asked [*ORG* Imagine Inc.] to start the project in [*LOC* New York], where the [*ORG* NY University] could provide a laboratory near the [*LOC* Washington Square Park].

Fig. 1. Example of NERC output

An example of the output of a NERC system recognizing person (PER), organization (ORG), and location (LOC) NEs is shown in Figure 1.

3 Anonymization of Unstructured Documents

In order to settle the base for the anonymization of unstructured documents, we introduce some definitions and concepts. We will focus on completely unstructured documents, assuming that the documents do not provide any kind of meta-information on both their content or their structure.

First of all, we consider the anonymization of an unstructured document as *the modification or perturbation of the document in order to preserve the privacy of the correspondents associated with documents*. Correspondents can be organizations or individuals, which are directly or indirectly mentioned or referenced in the document.

We consider the process of unstructured documents anonymization as composed of two main stages:

1. Private entity recognition: is the process that identifies the entities in the document, which can be protected in order to anonymize the document.
2. Private entity protection: concrete protection method applied to the previously detected entities.

In an analogy to data privacy methods, the first stage will be equivalent to determine the attributes of the data to be protected, something that in SDC and PPDM is normally not considered since one departs directly from the given attributes. The second stage is the application of a concrete protection method as the ones described in Section 2.1.

3.1 Private Entity Recognition

We introduce the concept of *private entity* (PE) as:

Definition 1. *A* private entity *in an unstructured document is an* entity *which reveals information about the correspondents directly or indirectly associated with the document.*

We have identified PEs to be very coincident with NEs (see Section 2.2). This makes the PE recognition stage potentially solvable by using a NERC system. Although not all named entities in a document can be considered private, and not all private entities will correspond to a normally recognized named entity, we assume so in favor of generalization, that is, $PE \approx NE$ (see experiments in Section 5).

Given the private entities of a document, we distinguish between two different types of entities from a privacy perspective: identifier entities, and quasi-identifier entities.

Definition 2. *An* identifier private entity *within an unstructured document is a PE, which by itself can unequivocally designate a correspondent.*

Definition 3. *A* quasi-identifier private entity *within an unstructured document is a PE, which in combination with contextual information and external knowledge can be used to re-identify a correspondent.*

It is important to note that just as the entity recognition process requires some semantic interpretation of the words or noun phrases appearing in the document, so does the distinction between identifier and quasi-identifier PEs.

Similarly to traditional data privacy, the distinction between identifier and quasi-identifier entities will determine how are they treated in the protection stage. Identifier PEs will simply be encrypted, deleted, substituted by their named entity type such as "person" or "organization" or by a meaningless string. On the other hand quasi-identifier PEs will be subjected to a concrete protection method as will be detailed in Section 3.3.

3.2 Some Consideration on Entity Recognition

As an example of quasi-identifier PEs, consider the proper name *Michael* as appearing in Figure 2. The name by itself does not reveal the identity of the correspondent. On the other hand by considering the contextual information provided in whole sentence (or document) and some prior knowledge one can conclude that the first Michael refers to Michael Vick the quarterback of the Philadelphia Eagles football team, while the second one refers to Michael Jackson the pop celebrity. In both cases the PE *Michael* will be considered a quasi-identifier.

Also, the named entities *Fredrik Reinfeldt*, and *Mr. Assange* in Figure 3, can be easily linked to the prime minister of Sweden, and the spokesperson and editor-in-chief of Wikileaks, but this identification requires contextual or prior knowledge, which will not necessarily be assumed.

In general, the distinction between identifier and quasi-identifier PEs will be very dependent on the concrete application and context where the documents are semantically

> Philadelphia Eagles quarterback **[PER Michael]** will sign his franchise tender Wednesday, [...]

> The doctor charged in **[PER Michael's]** death is due in court as a judge considers delaying the physician's upcoming trial. [...]

Fig. 2. Example of quasi-identifier PEs

> [PER Fredrik Reinfeldt] rejected claims that [LOC Sweden]s attempt to extradite [PER Mr Assange] from the [LOC UK] was politically motivated and hit back against attacks on the [ORG Swedish justice] system made during an extradition hearing in [LOC London] this week.

Fig. 3. Example of easily identifiable quasi-identifiers NEs

interpreted. In some cases a 9 digit number will be interpreted as a social security number and thus considered an identifier, and in some other contexts it will be just a number without identifying properties. Note that this is somehow equivalent to the same distinction made in SDC. There, an attribute "social security number" will be considered as an identifier, but not all attributes with 9 digit values will be interpreted the same way (for instance consider an attribute describing consumed liters of water per year, expressed with a 9 digit number).

3.3 Entity Protection

Once the PEs of the document are identified, they have to be protected. As mentioned earlier, the PEs considered identifiers are directly deleted or encrypted and the protection method is focused towards the quasi-identifier PEs.

One can see quasi-identifier PEs in the document as values for named entity types attributes. That is, we can consider that we have an attribute "person" with a set of values in a given document and so on. This view is obviously very simplistic since contextual information will surely influence the protection degree of a given named entity, but we consider it to be a good starting point. Moreover, it provides the ability to conduct a protection focused in a concrete type or category of named entities.

Definition 4. *We define* NE-type protection *as the protection mechanism which provides some degree of anonymization or protection to named entities of a given entity type.*

For example, we can describe methods for *PER-protection* which will operate on PEs of type "person", or *LOC-protection* operating on "location" PEs.

Following the ideas of the protection methods broadly described in Section 2.1, we introduce some proposals for PE protection.

Named entity generalization. PEs can be generalized to achieve some degree of privacy while preserving some of their semantic meaning in the document. A clear example can be found if we consider LOC-protection methods, where the location can be generalized to a broader named entity.

The suspect approached the van in **[LOC Old Nichol St.]**, and police, who were in **[LOC Arnold Circus]**, intervened. The suspect ran away and was later identified in an police control near **[LOC Gipsy Lane in Luton]** and chased by police until he was captured in **[LOC Castle St., Luton]**.

The suspect approached the van in **[LOC London]**, and police, who were in **[LOC London]**, intervened. The suspect ran away and was later identified in an police control near **[LOC Luton]** and chased by police until he was captured in **[LOC Luton]**.

Fig. 4. Example of LOC-protection by generalization

Figure 4 shows a simplistic example of LOC-protection where all locations have been generalized to city level, removing fine-grained information of the concrete streets.

Other examples are the generalization of proper names to names (removing the surname), or the generalization of organizations based on some given hierarchy. In the later case, for example, entities "IBM" and "Google" could be generalized to "Computer-based company", while "Médecins Sans Frontières" could be generalized to "NGO".

Similar techniques are widely described for SDC and PPM. They normally depart from Value Generalization Hierarchies (VGH), which are used to generalize textual attributes [19,11,13], or even set valued data [10,12]. This approaches normally depart from already established generalization hierarchies, but even if we do not have the hierarchy defined it can be determined based on generic semantic properties and ontologies [15,1].

Entity swapping. Another possible protection is to follow an strategy inspired by Rank Swapping (see Section 2.1). In this case we propose to swap PEs between documents of the same set, or within the same document depending on the concrete case.

In order for two PEs to be swappable they need to be relatively similar. First of all PEs are swapped only with other PEs of the same type, and secondly we can use a distance metric to rank them in order to perform the swapping.

For example given the set of PEs of type α, $\{\alpha_1, \ldots, \alpha_n\}$, where α is one of the identified types of PEs, in this case $\{PER, ORG, LOC\}$. First, we rank all values of PEs of type α so $PE_\alpha = (\alpha_{\sigma(0)}, \alpha_{\sigma(0)}, \ldots, \alpha_{\sigma(n)})$. The ordering can be predefined if the PEs form, for example, a complete or partial order [25], or it can be computed from a distance function.

In the later case, given a distance function d_α on the PEs of type α, we chose a initial PE $\alpha_{\sigma(0)}$, and use it as the starting point, then the other PEs of type α are ranked given their relative distance to $\alpha_{\sigma(0)}$. That is, given the initial vector $\alpha_{\sigma(0)}$ the ordering $\alpha_{\sigma(0)}, \alpha_{\sigma(1)}, \alpha_{\sigma(2)}, \ldots$ will be determined so:

$$d_\alpha(\alpha_{\sigma(0)}, \alpha_{\sigma(1)}) \leq d_\alpha(\alpha_{\sigma(0)}, \alpha_{\sigma(2)}) \leq d_\alpha(\alpha_{\sigma(0)}, \alpha_{\sigma(3)}) \leq \ldots$$

Once the vectors are ordered, we swap them. Given a term $\alpha_{\sigma(i)}$, it is randomly and uniformly swapped with another unswapped $\alpha_{\sigma(l)}$, given that $i < l \leq i + p$, where p is the parameter of the swapping method.

Entity noise addition. In analogy to the Additive Noise technique (see Section 2.1) we can actually introduce some *semantic noise* in the PEs to provide some degree of anonymity. That is, instead of swapping similar PEs as described in Section 3.3, we can substitute a PE by another similar PE, which is not present in the document.

The suspect approached the van in **[LOC Old Nichol St.]**, and police, who were in **[LOC Arnold Circus]**, intervened. The suspect ran away and was later identified in an police control near **[LOC Gipsy Lane in Luton]** and chased by police until he was captured in **[LOC Castle St., Luton]**.
The suspect approached the van in **[LOC Elmore St.]**, and police, who were in **[LOC De Beauvoir Square]**, intervened. The suspect ran away and was later identified in an police control near **[LOC Marsh Rd., Luton]** and chased by police until he was captured in **[LOC Alma St., Luton]**.

Fig. 5. Example of LOC-protection by noise addition

For example, in Figure 5 both the street Old Nichol and Arnold Circus from London are respectively randomly substituted by another street and square from London. And similarly, for the streets from Luton. If higher protection is required we could perform the random substitution at higher levels, for example with streets from other cities of the UK, Europe, and so on.

4 Some Desired Properties of Anonymized Documents

When anonymizing unstructured documents it is important to previously determine what exactly we want to anonymize. As previously stated the main purpose is to anonymize to some extend the information contained in unstructured documents about concrete persons, organizations, or locations, but maintaining the main semantics and information of the document. In other words, by reading a protected document the reader *should no be able to identify entities for which he/she did not have previous knowledge*.

For instance, in the protected example of Figures 4 the reader gets to know that the police has set off in pursuit of a suspect in London, who was later captured in Luton. The concrete locations at street level are anonymized. A reader previously knowing about a pursuit in Old Nichol St. in London will easily link the document with Old Nichol St. but he/she will not gain new knowledge about street-level locations.

Another possible controversial anonymization is shown in Figure 6, where entities have been anonymized by generalizing persons to their known occupation, organizations to their business market, and locations to continent level. The *utility* of the protected document, that is, the information intended to be revealed is that "a *fashion clothing company* has dismissed a *chief designer* due to an anti-Semitic incident that happened in a bar in *Europe*". But the concrete company, name of the designer, and location of the incident are anonymized.

The company **[ORG Fashion Clothing]** said Tuesday that it would dismiss **[PER chief designer]**, for his anti-Semitic outbursts at a **[LOC Europe]** bar[...]

Type of PE	Original PE	Anonymized PE
ORG	Christian Dior S.A.	fashion clothing company
PER	John Galliano	chief designer
LOC	Paris	Europe

Fig. 6. Example of protected document

Some people will argue that this text can easily be linked to the Christian Dior S.A. company, his chief designer John Galliano, and the incident was located in bar La Perle, in the Marais district of Paris. But it is important to note that this linkage requires previous knowledge about the concrete anonymized entities (and the incident). We argue that: *no new knowledge should be gained from the protected sentence beyond the intended information (utility).*

For example if the reader knows that Christian Dior has fired his designer John Galliano, he/she will gain knowledge that the cause was due to an anti-Semitic incident in a bar. But this information is actually part of the *utility* of the document, thus information deliberately intended to be revealed. On the contrary if the reader only knows that recently Dior has fired his chief designer, he/she will not gain knowledge about the name of the designer or the city of the incident from the sentence itself. Note that this is a very simplistic example and external knowledge that can undo the anonymization process is easily accessible by the reader, which will normally not be the case.

It is also important to note that in some cases it will be very difficult to estimate the actual full utility of the document, and that contextual information together with external knowledge could potentially leak non intended information. Special care has to be taken into account.

5 Experiments

In this section we describe some preliminary experiments about our work. These are focused to determine if named entity recognition and classification (see Section 2.2) is actually a good approach to determine the PEs of a document.

We have used the CoNLL02 dataset [21], which includes a set of documents from a news agency in Spain. The documents are annotated with 4 named entity types: PER, ORG, LOC, and MISC. For our first experiments we have used the first 10 documents. Due to the difficulty of determining the private entities of a document we have relied in human detection. That is, humans were asked to manually annotate the document with PEs, without previous knowledge of the named entities described in the dataset. Results are the average of two different human recognitions.

Table 1 shows the words from the document that where detected by a human to be PEs as compared to what the CoNLL02 dataset annotates as NEs (and thus the words to be identified as NEs by NERC systems). We can see that they are mostly coincident

Table 1. Words in PEs vs. NEs. The original dataset has 3092 words

	No. of words	Percentage of words from the total
Named entities	492	15.91%
Private entities	501	16.20%

with just 9 words identified as PEs, which were not annotated as NEs, which is a 0.29% from the total.

We have also evaluated the potential classification of NEs as a classification for PE. Table 2 shows the comparison of the correspondence of PEs and NEs per type. That is, the PEs recognized as revealing private information from persons by the human inspection as compared to the NEs annotated as PER, and so on. For each case we have depicted the *precision*, *recall*, *accuracy*, and the *balanced F-measure* (F_1) [14].

Table 2. Comparison of the classification of NEs types and PEs types

	Precision	Recall	Accuracy	F_1
PER	0.982	1.00	0.999	0.991
ORG	0.802	0.983	0.985	0.883
LOG	0.795	0.941	0.987	0.862

As we can see, preliminary results point to NERC systems as a good base to private entity recognition and classification. This is specially inspiring given the broad range of NERC systems available, making it possible to take advantage of the research already done in NERC systems.

6 Conclusions

In this paper we have introduced the anonymization of unstructured documents. We have departed from traditional data privacy approaches and proposed the use of named entity recognition and classification systems (NERC) to identify private entities in the documents. Once identified, these entities can then be protected. We also provide some empirical evaluation about the convenience on using NERC to identify such private entities.

We plan to further develop the anonymization of unstructured documents to settle the bases for production systems with the ability to assist in the declassification of confidential documents, and if possible to do it automatically.

Acknowledgments

Partial support by the Spanish MICINN (projects TSI2007-65406-C03-02, TIN2010-15764, ARES- CONSOLIDER INGENIO 2010 CSD2007-00004) is acknowledged.

References

1. Abril, D., Navarro-Arribas, G., Torra, V.: Towards Semantic Microaggregation of Categorical Data for Confidential Documents. In: Torra, V., Narukawa, Y., Daumas, M. (eds.) MDAI 2010. LNCS, vol. 6408, pp. 266–276. Springer, Heidelberg (2010)
2. Aggarwal, C.C., Yu, P.S. (eds.): Privacy-Preserving Data Mining. Springer, Heidelberg (2007)
3. Brand, R.: Microdata Protection through Noise Addition. In: Domingo-Ferrer, J. (ed.) Inference Control in Statistical Databases. LNCS, vol. 2316, pp. 97–116. Springer, Heidelberg (2002)
4. Chang, C., Kayed, M., Girgis, M.R., Shaalan, K.F.: A Survey of Web Information Extraction Systems. IEEE Trans. on Knowl. and Data Eng. 18(10), 1411–1428 (2006)
5. Dalenius, T.: Finding a needle in a haystack - or identifying anonymous census record. Journal of Official Statistics 2(3), 329–336 (1986)
6. DARPA, New technologies to support declassification. Request for Information (RFI) Defense Advanced Research Projects Agency. Solicitation Number: DARPA-SN-10-73 (2010)
7. Defays, D., Nanopoulos, P.: Panels of enterprises and confidentiality: The small aggregates method. In: Proc. of the 1992 Symposium on Design and Analysis of Longitudinal Surveys, Statistics, Canada, pp. 195–204 (1993)
8. Domingo-Ferrer, J., Torra, V.: Ordinal, Continuous and Heterogeneous k-Anonymity Through Microaggregation. Data Mining and Knowledge Discovery 11(2), 195–212 (2005)
9. Grishman, R., Sundheim, B.: Message Understanding Conference - 6: A Brief History. In: Proc. International Conference on Computational Linguistics (1996)
10. He, Y., Naughton, J.: Anonymization of Set-Valued Data via Top-Down. In: VLDB 2009: Proceedings of the Thirtieth International Conference on Very Large Data Bases. VLDB Endowment, Lyon (2009)
11. Iyengar, V.S.: Transforming data to satisfy privacy constraints. In: Proceedings of the Eighth ACM SIGKDD International Conference on Knowledge Discovery and Data Mining (KDD 2002), pp. 279–288 (2002)
12. LeFevre, K., DeWitt, D.J., Ramakrishnan, R.: Mondrian Multidimensional K-Anonymity. In: Proceedings of the 22nd International Conference on Data Engineering, p. 25. IEEE Computer Society, Los Alamitos (2006)
13. Li, T., Li, N.: Towards optimal k-anonymization. Data Knowledge Engineering 65(1), 22–39 (2008)
14. Manning, C.D., Raghavan, P., Schtze, H.: Introduction to Information Retrieval. Cambridge University Press, Cambridge (2008)
15. Martínez, S., Sánchez, D., Valls, A.: Ontology-Based Anonymization of Categorical Values. In: Torra, V., Narukawa, Y., Daumas, M. (eds.) MDAI 2010. LNCS (LNAI), vol. 6408, pp. 243–254. Springer, Heidelberg (2010)
16. Moore, R.: Controlled Data Swapping Techniques for Masking Public Use Microdata Sets, U. S. Bureau of the Census (unpublished manuscript) (1996)
17. Nadeau, D., Satoshi, S.: A survey of named entity recognition and classification. Lingvisticae Investigationes 30(1), 2–26 (2007)
18. Navarro-Arribas, G., Torra, V.: Privacy-preserving data-mining through microaggregation for web-based e-commerce. Internet Research 20(3), 366–384 (2010)
19. Samarati, P., Sweeney, L.: Protecting Privacy when Disclosing Information: k-Anonymity and Its Enforcement through Generalization and Suppression, Technical Report SRI-CSL-98-04, SRI Computer Science Laboratory (1998)
20. Sekine, S., Nobata, C.: Definition, Dictionaries and Tagger for Extended Named Entity Hierarchy. In: Proc. Conference on Language Resources and Evaluation (2004)

21. Tjong Kim Sang, E.: Introduction to the CoNLL-2002 shared task: language-independent named entity recognition. In: Proc. Conference on Natural Language Learning (2002)
22. Torra, V.: Privacy in Data Mining. In: Maimon, O., Rokach, L. (eds.) Data Mining and Knowledge Discovery Handbook, 2nd edn. (2010) (invited chapter)
23. Torra, V.: Microaggregation for Categorical Variables: A Median Based Approach. In: Domingo-Ferrer, J., Torra, V. (eds.) PSD 2004. LNCS, vol. 3050, pp. 162–174. Springer, Heidelberg (2004)
24. Torra, V.: Constrained microaggregation: Adding constraints for data editing. Transactions on Data Privacy 1(2), 86–104 (2008)
25. Torra, V.: Rank swapping for partial orders and continuous variables. In: International Conference on Availability, Reliability and Security, pp. 888–893 (2009)
26. Willenborg, L., de Waal, T.: Elements of Statistical Disclosure Control. Lecture Notes in Statistics, vol. 155. Springer, Heidelberg (2001)

Uncovering Community Structure in Social Networks by Clique Correlation

Xu Liu[1], Chenping Hou[1], Qiang Luo[2], and Dongyun Yi[1,⋆]

[1] Department of Mathematics and Systems Science, College of Science, National University of Defense Technology, China
{liusure2001,hcpnudt}@gmail.com, dongyunyi@nudt.edu.cn
[2] Department of Management, College of Information Systems and Management, National University of Defense Technology, China
luoqiang@nudt.edu.cn

Abstract. Community is tightly-connected group of agents in social networks and the discovery of such subgraphs has aroused considerable research interest in the past few years. Typically, a quantity function called modularity is used to guide the division of the network. By representing the network as a bipartite graph between its vertices and cliques, we show that community structure can be uncovered by the correlation coefficients derived from the bipartite graph through a suitable optimization procedure. We also show that the modularity can be seen as a special case of the quantity function built from the covariance of the vertices. Due the the heteroscedaticity, the modularity suffers a resolution limit problem. And the quantity function based on correlation proposed here exhibits higher resolution power. Experiments show that the proposed method can achieve promising results on synthesized and real world networks. It outperforms several state-of-the-art algorithms.

Keywords: social network, community structure, resolution limit, correlation analysis.

1 Introduction

Social networks are a paradigm of the complexity of human interactions [1,2], which can be represented in terms of a set of social agents related in pairs between them by a set of peer-to-peer relationships. This structure can thus be abstracted as a complex network [3,4] $G = (V, E)$, where $V = \{V_1, V_2, \cdots, V_n\}$ is the set of vertices and $E = \{(V_i, V_j)|V_i, V_j \in V\}$ is the set of edges. The vertices represent social agents and the edges stand for their mutual relations or interactions. The advantage of this abstraction is that any social organization can be represented as a mathematical object, on which we can implement various mathematical tools.

One of the most manifest feature of social networks is community structure [5,6,7], the gathering of vertices into groups such that there is a higher

⋆ Corresponding author.

V. Torra et al. (Eds.): MDAI 2011, LNAI 6820, pp. 247–258, 2011.

density of edges within groups than between them [8]. The density refers to the ratio of actual edges between the nodes of the community to the maximum possibility. It is common that people can be divide into groups along lines of interests, occupation, age, and so on. Besides, the phenomenon of assortativity [11,4] certainly suggests that this is the case.

This paper proposes a novel approach to uncovering community structure in social networks, which takes a new view that vertices in the same group tend to have strong correlation with each other. By building a bipartite graph from network nodes and the corresponding cliques, we introduce a measurement of correlation between pairs of nodes. The more the cliques shared by two nodes, the higher the correlation between them. Thereby, the community structure can be uncovered by maximizing the total correlation inside a community. Experiment on synthesized network and real social network shows the effectiveness of our method. Besides, our method is free from resolution-limit effect.

2 Uncovering Community Structures Based on Modularity

There are many definitions of communities [1,2] and many other definitions have been introduced by computer scientists and physicists [10,9]. The community discovery task can be viewed as a unsupervised learning and the partition can be measured by modularity introduced by Newman and Girvan [12]. The community structure is revealed by the comparison between the actual density of edges in a subgraph and the density of a random graph with the same degree sequence. Suppose we split the vertices set in to c disjoint clusters set $C = \{C_1, \cdots, C_c\}$ which satisfies $C_i \cap C_j = \phi, \forall i \neq j, 1 \leq i \leq c, 1 \leq j \leq c$ and $\bigcup_{k=1}^{c} C_k = V$. The modularity of C is [12]

$$Q(C) = \frac{1}{2m} \sum_{k=1}^{c} \sum_{i,j \in C_k} \left(A_{ij} - \frac{k_i k_j}{2m} \right) \tag{1}$$

Modularity has been employed as quality function in many algorithms: the CNM algorithm [8], the method based on edge betweenness centrality [12], the label propagating method [13], the method by leading eigenvector of the community matrix [14] and the spin glass method [15], to name a few. For a comprehensive review please see [7] and references therein.

Unfortunately all the modularity optimization approach has a resolution limit [16] that may prevent it from detecting clusters which are comparatively small with respect to the graph as a whole. Even when they are well defined communities like cliques [7,17]. Clique is a fully connected subgraph,thus has the most high density of edges.

By building a bipartite graph from network, nodes and the cliques that belong the community structure can be uncovered by the correlation based on this bipartite graph. A merit of this approach is that it dose not suffer from the

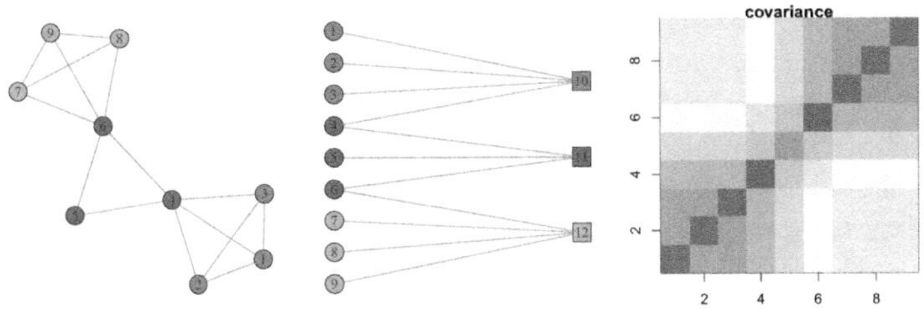

(a) (b) (c)

Fig. 1. Example of building vertex clique bipartite graph. (a)The original network. (b)The resulting bipartite graph(the 2-clique set is omitted for demonstration purpose). (c)The covariance matrix of the vertex-clique bipartite graph. See text for more details.

resolution limit problem. Here we focus on the detection of community structure of undirected and unweighted networks without overlapping.

The method proposed here can be also improved to cope with both directed and weight networks. By combining with fuzzy clustering algorithm such as c-means it is also capable of overlapping community detection. Since the main focus of this paper is the identification of non-disjoint community structure of simple networks,we leave these generalizations to future works.

3 Measuring the Correlation

The agents in the same community share many common features such as hobby, occupation,etc, which lead to a strong coherence between thema tendency known as homophily[18]. The dense connections between nodes in the same community reflect the strength of coherence and can be deduced from the topological structure of the social network. This approach is possible and there are many successful cases [19,20]. Here we propose a measurement based on cliques correlations to identify the community structure in social networks by focusing on the network cliques.

3.1 Building the Vertex-Clique Bipartite Graph

Unlike the clique expanding algorithm [6,21], which derives the overlapping community structure from cliques adjacency or inclusion relationship directly, here we adopt an indirect approach by building a bipartite graph from the original network then detecting community structure based on the numerical analysis of this new graph.

There is a common consensus that vertices within the same community bearing denser connecting than vertices between communities. The clique is the most densely connected subgraphs of given number of vertices thus should be partitioned in the same group. Thus we gather cliques in networks together and build a bipartite graph from it. The vertex set of the bipartite graph is made up of two nodes set,i.e. partition A and partition B, with partition consists of nodes in the original network and partition B consists of cliques. Denote these two sets as V_A and V_B, then we have $V_A = V$. Next step is to build set V_B.

A formal definition of a k-clique $Cl_k = (V_1, \cdots, V_k)$ of graph G is a subset of vertices with edges connecting each pair of them, and the set of all the k-clique is defined as $CL_k = \{(V_1, \cdots, V_k) \mid A_{ij} = 1, \forall 1 < i, j < k, i \neq j\}$. All the cliques of the network are collected together and denoted as $CL = \{CL_k \mid k = 1, 2, \cdots\}$. For a large and dense network CL may grows too huge if there is no upper bound for the size of cliques,costing tremendous computation to accomplish the gathering task [21]. The community uncovering algorithm always focus on certain subset of CL,avoiding unnecessary computations. Typically we should vary k from 2 to 4 then we have $V_B = CL_2 \bigcup CL_3 \bigcup CL_4$.

The edges connecting V_A and V_B is nature to define as belonging relationships. For node $a \in V_A$ and clique $b \in V_B$ if a is covered by b then add edge between a and b. Later we will allocate weights to these edges.

Summaries these up the vertices set of the vertex-clique graph is $V_A \cup V_B$. Assume that there are n vertices in partition A and m cliques in partition B,i.e.$|V_A| = n, |V_B| = m$. The adjacency matrix of the vertex-clique bipartite graph is $Y_{(m+n)\times(m+n)}$. Obviously we have

$$A = \begin{pmatrix} 0 & X \\ X^T & 0 \end{pmatrix} \tag{2}$$

since there is no edges within the same partition. Thus the $m \times n$ dimension matrix X is sufficient enough to depict the topological structure of the vertex-clique graph.

Fig.1. is simple example of the above progress. Fig.1(a) is a 9-vertex network with three cliques in $CL_3 \bigcup CL_4$ and Fig.1(b) is the resulting bipartite graph with $V_B = CL_3 \bigcup CL_4$. We omit the clique set CL_2 for demonstration purpose.

3.2 Weighting the Edges

As in the vertex-clique bipartite graph nodes in the same clique connect to the same clique node in partition B. Compared with small cliques, nodes connecting to big cliques should have higher possibility to belong to the same community. Therefore, edges in the vertex-clique bipartite graph should be weighted according to the size of the clique they connect. Let

$$A_{ij} = \begin{cases} f(|C_j|) & \text{if } V_i \in C_j, \\ 0 & \text{otherwise.} \end{cases} \tag{3}$$

where $C_j \in V_B$ is a clique with size $|C_j|$ and f is a weighting function assigning weights for different cliques.

As a prior common sense nodes in a common bigger clique are more akin to be members of the same community than nodes in a common smaller clique. So the function f should be designed to allocate a bit more weight on cliques with larger size, but a too drastic slop will destroy the relative importance of smaller cliques, since the number of latter is much more than the bigger one typically. In the experiment section we adopt a gently monotone increasing function $f(k) = \alpha^k$ with $\alpha = 1.1$.

3.3 Covariance and Correlation Coefficients

Nodes connecting to the same clique tend to have bigger correlation since they carry the weight on the same location. Numerical analysis on matrix X can be carried out to reveal the correlation between the nodes of G.

Denote x_i as the i-th column of matrix X and $cov(x_i, x_j)$ as the covariance between x_i and x_j,i.e.

$$cov(x_i, x_j) = \frac{1}{m}\sum_{e=1}^{m}(x_{ei} - \bar{x}_i)(x_{ej} - \bar{x}_j) \tag{4}$$

where $\bar{x}_i = \frac{1}{m}\sum_{e=1}^{m} x_{ei}$ is the average of the i-th column. When $i = j$, we get the variance of x_i: $var(x_i) = cov(x_i, x_i)$

Fig.1(c) is the covariance matrix for the bipartite graph of Fig.1(b). Dark(Red on line) color shows the most positive covariance and white shows the most negative covariance. From the image we can see that vertices in the same clique show much bigger covariances than vertices in different cliques.

Another measure of similarity of vertices is the correlation coefficients between x_i and x_j, i.e.

$$cor(x_i, x_j) = \frac{cov(x_i, x_j)}{\sqrt{var(x_i)var(x_j)}} \tag{5}$$

Unlike the covariance the correlation coefficients is re-scaled by the variance of each vector thus have a lower bound -1 and upper bound $+1$ which makes it more desirable for community uncovering. We will compare it with covariance in experiment section in more details.

Therefore we define a quantity function for the partition $C = \{C_1, \cdots, C_k\}$ as follows:

$$V(C) = \sum_{k=1}^{c}\sum_{i,j\in C_k, i\neq k} cov(x_i, x_j) \tag{6}$$

Similarly define $R(C)$ as

$$R(C) = \sum_{k=1}^{c}\sum_{i,j\in C_k, i\neq k} cor(x_i, x_j) \tag{7}$$

The community structure can be uncovered by maximizing $V(C)$ or $R(C)$ through any handy optimizing procedure such as k-means clustering. However, later we will show that $R(C)$ gives a better performance than to $V(C)$ in community uncovering.

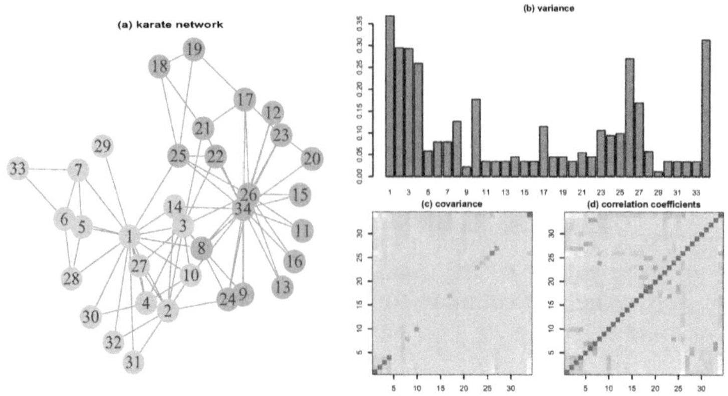

Fig. 2. The karate network. (a)The karate network is split into two communities by the proposed method. (b)The variance of vertices according the vertex-clique bipartite graph. (c) The covariance matrix. (d)The correlation matrix. The result shows that the covariances(c) are short of discriminative ability due to the fluctuations in variances(b), and the correlation coefficients(d) is more powerful. See text for more details.

3.4 Special Case: Modularity

Here we will show that the modularity defined by Eq.1 equal to a special case of covariance introduced in the previous subsection. If we put more strick restrictions on the partition B of the vertex-clique graph, i.e. just 2-clique are allowed and take the orientation into consideration i.e. each edge is split into two opposite orientated edges therefor there are $2m$ edges in total, then build the vertex-clique bipartite graph from them. Equivalently define matrix Y and Z as follows:

$$y_{ie} = \begin{cases} 1 & \text{if } \exists V_k \in V \quad s.t. \quad \overrightarrow{V_k V_i} = E_e, \\ 0 & \text{otherwise.} \end{cases} \tag{8}$$

$$z_{ie} = \begin{cases} 1 & \text{if } \exists V_k \in V \quad s.t. \quad \overrightarrow{V_i V_k} = E_e, \\ 0 & \text{otherwise.} \end{cases} \tag{9}$$

It is easy to check that $cov(y_i, z_j) = \frac{1}{2m}\left(A_{ij} - \frac{k_i k_j}{2m}\right)$. Define $q = -\sum_{k=1}^{n} \frac{k_i^2}{4m^2}$ then we can rewrite the modularity as follows:

$$Q(C) = q + \sum_{k=1}^{c} \sum_{i,j \in C_k, i \neq j} cov(y_i, z_j) \tag{10}$$

Since the value of q only depend on G, thus any optimizing procedure actually optimize $\sum_{k=1}^{c} \sum_{i,j \in C_k, i \neq j} cov(y_i, z_j)$. We conclude that the modularity can be reformulated as a special case of vertex-clique covariance.

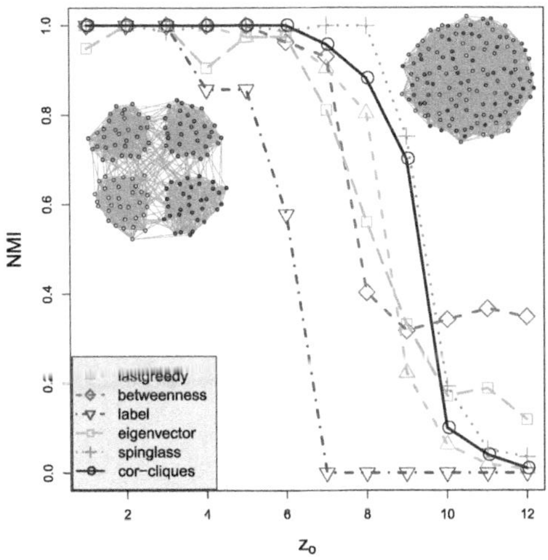

Fig. 3. The NMI of different algorithms on benchmark networks. The more similar the clusters uncovered and the original ones are, the larger the NMI is. The proposed method **cor-cliques** out performs most of the comparison algorithms with a narrow gap after the best performed. The embeding graphs are the benchmark network with $z_o = 1$ and $z_o = 12$ respectively.See text for details.

4 Uncovering Communities Structure by Eigenvectors

The existing state-of-the-art community detecting method based on matrix is proposed by Newman [14] which use the leading eigenvector of the so called modularity matrix to split the nodes into different groups.

The modularity matrix is defined as $B = A - \frac{dd^T}{2m}$,where the column vector $d = (k_1, \cdots, k_n)^T$ and k_i is degree of node V_i. The community uncovering approach is equal to maximizing the modularity $Q(C)$.Here we will give an analogy eigenvector clustering framework based on covariance and correlation coefficients defined by Eq.4 and Eq.5 respectively.

Denote the covariance matrix as $\Sigma = (cov(i,j))_{n\times n}$ and correlation coefficients matrix as $R = (cor(x_i, x_j))_{n\times n}$. Both of them are real and symmetric matrix, thus can be decomposed as product of orthogonality and diagonal matrices. Let W be a real and symmetric matrix,which stand for matrix Σ or R,then we have$W = U\Lambda U^T$, where $\Lambda = diag\{\lambda_1, \cdots, \lambda_n\}$ and $\lambda_1 \geq \cdots \geq \lambda_n$,$U = (u_1|\cdots|U_n)$ and $u_i^T u_i = 1, u_i^T u_j = 0, \forall i \neq j$. The matrix W can be approximated by a few leading column vectors of U,i.e. $W \approx \sum_{k=1}^{c} \lambda_k u_k u_k^T$. The parameter c is picked to maximize the eigengap

$$c = \arg\max_i (\lambda_{i-1} - \lambda_i) \tag{11}$$

Thus the major variation of the covariances of vertices are encoded in $U_c = (u_1|\cdots|u_c)$. The optimal number of communities is set to c, and then we us

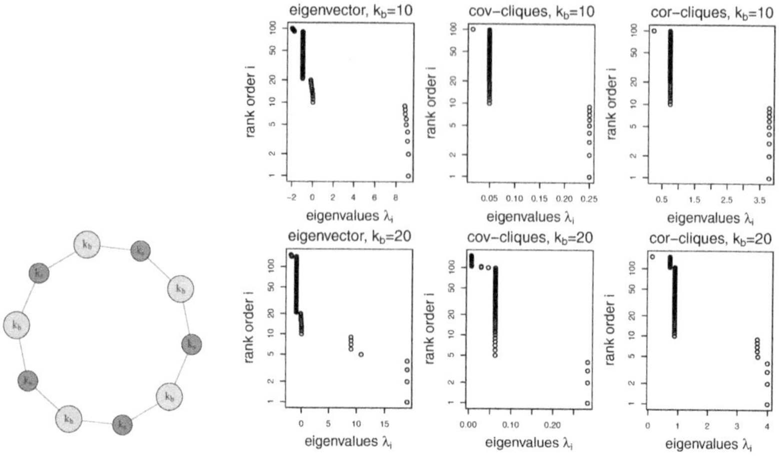

(a) (b)

Fig. 4. (a)Clique-circle network.The network consist of 10 cliques with two different size:k_s and k_b. (b)Eigenvalues of the clique-circle network obtained from different matrices. The largest eigengap indicates the optimal number of communities obtained by corresponding method and the idea location lies between the 9-th and the 10-th eigenvalues which recovery the number of communities correctly. When $k_b = k_s = 10$ all methods get the right number of communities successfully,but when $k_b = 20, k_s = 10$ only the *cor-cliques* method proposed here works well and both the other method fails. See text for details.

k-means clustering on the rows of U_c to identification the community structure of the underlying social network.

The problem of generating all the cliques,or maximal independent sets, of a given network is fundamental in graph theory.The complexity of finding all the cliques is $O(nmk)$[22],here n,m and k are the numbers of nodes, edges and the maximal size of cliques we want to find. In the experiment section we set $k = 4$,resulting in a complexity of $O(4nm)$ for cliques finding.The correlation analysis can be done within $O(n^2)$ time.Thus the total complexity is $O(n^2)$.

5 Experiment

In this section we will carry out experiment to examine the merit of the method proposed in the previous section. Both synthesis and real social networks are tested upon proposed method and a series of famous state-of-the-art methods. These methods are(short names are bracketed):

1. Divisive algorithm based on edge betweenness [12](**betweenness**)
2. Greedy algorithm based on modularity with advanced data structure [8](also known as CNM algorithm,**fastgreedy**)
3. Label propagation algorithm which runs in linear time [13](**label**)

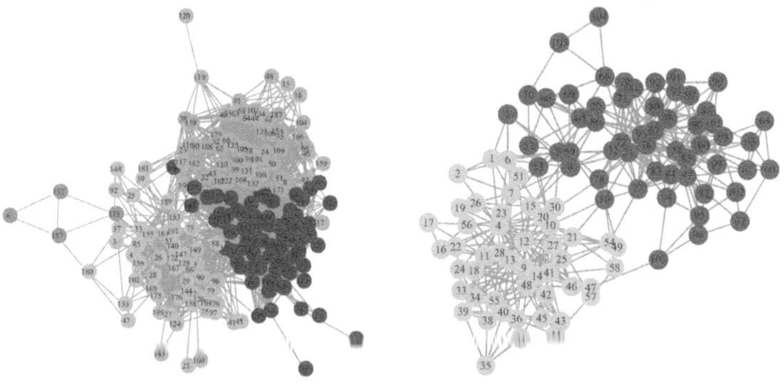

(a) (b)

Fig. 5. Community structure of the jazz network(a) and the polbooks network(b) uncovered by *cor-cliques* method.See text for details

4. Spectral clustering method based on modularity matrix [14](**eigenvector**)
5. Approach minimizing the energy of spinglass attached on the network [15] (**spinglass**)

5.1 Covariance vs. Correlation Coefficients

Here we will show the merit of correlation coefficients upon covariance with example on the well known karate club network [23] with 34 nodes and 78 edges, which has been extensively used as a benchmark for different algorithms aiming at discovering communities in social networks.

Result of the variances,covariances and correlation coefficients of karate network are displayed in Fig.2. From Fig.2(b) we can see that the variances of the vertices shows heteroscedasticity. $V_1, V_2, V_3, V_4, V_{26}, V_{34}$ have noticeable bigger variance than the rest vertices which lead to poor discriminative power of the covariances as plotted in Fig.2(c). The correlation coefficients show more discriminative power, benefiting from the rescaling operation described by Eq.5,which can be seen in Fig.2(d).

This observation is further verified by the modularity result: modularity obtained by covariance is $Q_{cov} = 0.2475345$ and the modularity obtained by correlation is $Q_{cor} = 0.3714661$. The community structure uncovered by clique correlation is displayed in Fig.2(a), distinguishing by different colors.

Due to the shortcomings of clique covariances, in the follow-up experiments we only report the result on clique correlation and denote it as *cor-cliques*.

5.2 Experiment on Computer Generate Networks

Here we generate two kinds of networks. The first benchmark network is a 4-group network proposed by Newman [5]. There are 128 nodes splitting in four equal groups and edges are generated randomly. Every node has an average of

z_o links with nodes outside its group and $16 - z_o$ links with the other nodes inside. When $z_o = 8$ the community structure is quite obscure since each node has equal number of links inside and outside its group.

We use normalized mutual information(NMI) [24] to evaluate the differences between uncovered community structure C_e and the original community structure C_o.

$$NMI(C_o, C_e) = \frac{H(C_o) + H(C_e) - H(C_o, C_e)}{\sqrt{H(C_o)H(C_e)}} \tag{12}$$

Here $H(C)$ is the entropy of the set C. When C_o and C_e are identical $NMI = 1$, and when C_o and C_e are totally different $NMI(C_o, C_e) = 0$.

The experiment results are showed in Fig.3,from which we can see that the performances of all the method declined as z_o increased. When $z_o \leq 6$ except for the *label* algorithms behavior quite desirable. When $z_o = 7$ the *label* algorithm and *eigenvector* algorithm show a quick drop in performance. When $z_o = 8$ the *betweenness* falls even more sharply than *eigenvector* and the *cor-cliques,spinglass* and *fastgredy* are still quite good. When $z_o = 9$ only *cor-cliques* and *spinglass* are stay above 0.6,despite the situation that there are more edges between groups than within them. *cor-cliques* falls behind *spinglass* but the gap is quite tight. When $z_o = 12$ there is no community structure at all which can be seen from the embeding graph. The *betweenness* still keep a sound result due to the global nature of edge betweenness. The proposed method is slightly fall behind *spinglass* but the *spinglass* algorithm is very time consuming [7].

The second benchmark network is the clique-circle network [25] shows in Fig.4(a), which consists two kinds of cliques with size k_b and k_s. This experiment tests performance the following algorithms, i.e.*eigenvector*,*cor-cliques* and the *cov-cliques*, as the other algorithms do not use eigenvectors. Each one of the three algorithm uses the largest eigengap (see Eq.11) to determine optimal number of communities.

We generate two version of the networks with $k_b = k_s = 10$ and $k_b = 20, k_s = 10$.The results are depicted in Fig.4(b). When $k_b = k_s = 10$ all three method detect corrected number of communities. However when $k_b = 20, k_s = 10$ there is two difference size of cliques and the *eigenvector* and *cor-cliques* algorithm detection 5 communities,showing a resolution limit. However the algorithm proposed here get the right number of communities.

5.3 Real Social Networks

Now we apply the method based on clique correlation on real world networks. despite the karate network we have also analyzed the community structure of two more real networks. The first is the jazz musicians network [26],where two musicians are connected if they have played in the same band. The second is the politic book co-purchasing network,where two books are connected if they are bought by a same customer[27].

Fig.5 shows the result by *cor-clique* algorithm. The optimal number of communities of jazz network is 4 with $Q_{cor} = 0.44$ and the optimal number of

Table 1. Modularity and number of communities uncovered by different algorithms. Numbers within brackets shows the number of communities uncovered by different methods.

	cor-cliques	fastgreedy	betweenness	label	eigenvector	spinglass
karate	0.37(**2**)	0.38(3)	0.40(5)	0.37(3)	0.38(5)	**0.42**(4)
jazz	0.44(**4**)	0.44(4)	0.41(39)	0.28(3)	0.35(8)	**0.44**(5)
polbooks	0.45(**2**)	0.50(4)	0.52(5)	0.50(4)	0.40(9)	**0.53**(6)

communities of polbooks network is 4 with $Q_{cor} = 0.45$. All results are displayed in Table.1, from which we can see that *cor-cliques* achieves promising performance compared to other methods. For the karate network, several methods tend to find more communities and return a higher modularity value. But the fact is that the network can be split into only two groups in real world. They correspond to a modularity value 0.37, the same as our method[7,26]. The main philosophy behind this is that our method tend to not split nodes in the same clique. The results on the other two networks tell the same story. Thanks to the weighting allocation method in Eq.3, our method keeps nodes in the same clique in the same group. This feature can be seen from Fig. 5(b) clearly.

6 Conclusion

We have discussed the problem community uncovering in social networks. By building a vertex-clique bipartite graph out from the original network and carrying out correlation analysis on the adjacency matrix of this bipartite graph, the community structure of the original networks can be recovered effectively. Unlike other modularity optimizing procedure tend to split cliques into difference communities[17] the method proposed here shows the ability of keeping them in the same group.

The method can be further modified to deal with directed and weighted networks to. Also the clustering method we used can be replaced with a fuzzy clustering algorithm making the method proposed here capable of uncovering overlapping communities. The cliques finding is the most time consuming operation in our algorithm and to our best knowledge the most efficiency strategy cost $O(nmk)$ time[22]. Better clique enumeration algorithm will be one of our possible future work.

Acknowledgments. The authors would like to thank the anonymous referees for their helpful comments and suggestions. This work is supported by the National Natural Science Foundation of China (NO.60902089, 61005003).

References

1. Wasserman, S., Faust, K.: Social Network Analysis: Methods and Applications. Cambridge University Press, Cambridge (1994)
2. Scott, J.: Social Network Analysis: A Handbook. Sage Publications, London (2000)

3. Albert, R., Barabási, A.-L.: Statistical mechanics of complex networks. Rev. Mod. Phys. 74(1), 47–97 (2002)
4. Newman, M.E.J.: The structure and function of complex networks. SIAM Review 45(2), 167–256 (2003)
5. Girvan, M., Newman, M.E.J.: Community structure in social and biological networks. Proc. Natl. Acad. Sci. USA 99(2), 7821–7826 (2002)
6. Palla, G., Derenyi, I., Farkas, I., Vicsek, T.: Uncovering the overlapping community structure of complex networks in nature and society. Nature 435, 814–818 (2005)
7. Fortunato, S.: Community detection in graphs. Physics Reports 486(3-5), 75–174 (2010)
8. Clauset, A., Newman, M.E.J., Moore, C.: Finding community structure in very large networks. Phys. Rev. E70, 066111 (2004)
9. Radicchi, F., Castellano, C., Cecconi, F., Loreto, V., Parisi, D.: Defining and identifying communities in networks. Proc. Natl. Acad. Sci. USA 101, 2658–2663 (2004)
10. Flake, G.W., Lawrence, S.R., Giles, C.L., Coetzee, F.M.: Self-organization and identification of Web communities. IEEE Computer 35, 66–71 (2002)
11. Newman, M.E.J.: Assortative mixing in networks. Phys. Rev. Lett. 89, 208701 (2002)
12. Newman, M.E.J., Girvan, M.: Finding and evaluating community structure in networks. Phys. Rev. E 69, 026113 (2004)
13. Raghavan, U.N., Albert, R., Kumara, S.: Near linear time algorithm to detect community structures in large-scale networks. Phys. Rev. E 76, 036106 (2007)
14. Newman, M.E.J.: Finding community structure in networks using the eigenvectors of matrices. Phys. Rev. E 74, 036104 (2006)
15. Reichardt, J., Bornholdt, S.: Statistical mechanics of community detection. Phys. Rev. E 74, 016110 (2006)
16. Fortunato, S., Barthélemy, M.: Resolution limit in community detection. Proc. Natl. Acad. Sci. USA 104(1), 36–41 (2007)
17. Reid, F., McDaid, A., Hurley, N.: Community finding:partitioning considered harmful. In: NIPS Workshop on Network Across Disciplines (2010)
18. McPherson, M., Smith-Lovin, L., Cook, J.M.: Birds of a feather: Homophily in social networks. Annual Review Of Sociology 27, 415–444 (2001)
19. Ravasz, E., Somera, A.L., Mongru, D.A., Oltvai, Z.N., Barabási, A.-L.: Hierarchical organization of modularity in metabolic networks. Science 297, 1553–2002 (2002)
20. Leicht, E.A., Holme, P., Newman, M.E.J.: Vertex similarity in networks. Phys. Rev. E 73, 026120 (2006)
21. Lee, C., Reid, F., McDaid, A., Hurley, N.: Detecting Highly Overlapping Community Structure by Greedy Clique Expansion. In: SNAKDD, vol. 10 (2010)
22. Tsukiyama, S., Ide, M., Ariyoshi, H., Shirawaka, I.: A new algorithm for generating all the maximal independent sets. SIAM J. Computing 6, 505–517 (1977)
23. Zachary, W.W.: An information flow model for conflict and fission in small groups. J. Anthropol. Res. 33(4), 452–473 (1977)
24. Danon, L., Duch, J., Diaz-Guilera, A., Arenas, A.: Comparing community structure identification. J. Stat. Mech., 09008 (2005)
25. Shen, H.W., Cheng, X.Q.: Spectral methods for the detection of network community structure:a comparative analysis. J. Stat. Mech., 10020 (2010)
26. Gleiser, P., Danon, L.: Community structure in jazz. Adv. Complex Syst. 6, 565–573 (2003)
27. Krebs, V.: `http://www.orgnet.com/` (unpublished)

Author Index

GPSR Compliance

The European Union's (EU) General Product Safety Regulation (GPSR) is a set of rules that requires consumer products to be safe and our obligations to ensure this.

If you have any concerns about our products, you can contact us on ProductSafety@springernature.com

In case Publisher is established outside the EU, the EU authorized representative is:

Springer Nature Customer Service Center GmbH
Europaplatz 3
69115 Heidelberg, Germany

Batch number: 10844619

Printed by Printforce, the Netherlands